Reliability Modelling with Information Measures

N. Unnikrishnan Nair, S.M. Sunoj and G. Rajesh

Department of Statistics
Cochin University of Science and Technology, Kerala, India

CRC Press
Taylor & Francis Group
Boca Raton London New York

CRC Press is an imprint of the
Taylor & Francis Group, an **informa** business

A SCIENCE PUBLISHERS BOOK

First edition published 2022
by CRC Press
6000 Broken Sound Parkway NW, Suite 300, Boca Raton, FL 33487-2742

and by CRC Press
4 Park Square, Milton Park, Abingdon, Oxon, OX14 4RN

CRC Press is an imprint of Taylor & Francis Group, LLC

Library of Congress Cataloging-in-Publication Data (applied for)

ISBN: 978-1-032-31413-6 (hbk)
ISBN: 978-1-032-31417-4 (pbk)
ISBN: 978-1-003-30963-5 (ebk)

DOI: 10.1201/9781003309635

Typeset in Times New Roman
by Radiant Productions

Preface

Ever since the introduction of Shannon entropy in the late nineteen forties, the scientific community witnessed a spontaneous growth in its applications to different fields of human activity. Apart from the areas in the classical sciences the impact of the notion of entropy attained universal acceptance through the substantial reformations it has brought about in topics like machine learning, pattern recognition, data mining, cryptography, neurobiology, medical physics, etc. The kind of flexibility imparted by entropy in representing uncertainty based on a large number of evidence patterns continues to make it an exciting prospect in more scientific ventures. Similar is the case with reliability analysis as an independent discipline developed after the Second World War that provides necessary tools to ensure the uninterrupted functioning of ordinary devices to complex systems that makes life easier. The concepts in reliability originally designed for lifetimes of devices was soon extended to cover duration variables that meant the time to occurrence of a specified event with the result that the notions in reliability found applications in economics, medicine, demography, physics, risk analysis, social sciences, etc. It has also contributed to new ideas in other subjects like survival analysis, information theory, statistics and mathematics besides expanding its own domain to reliability modelling, reliability engineering, reliability science and reliability management. The present monograph is an exposition of the interaction between these two subjects.

Even though the application of information theoretic concepts in reliability analysis has a history of only less than twenty years, the impact it has generated among researchers in allied areas has been phenomenal, and promises to be of immense potential in future. The material in this area is scattered over different journals in mathematics, statistics, reliability and information sciences. The need for a common platform where the prominent developments can be accessed is keenly felt and this desire has been culminated in the present volume. We have made earnest efforts to include the contributions from all sources in the book. If any of the important results have been left out by oversight, we sincerely apologize to the author concerned and to the readers.

The book is organized in to eight chapters. In Chapter 1, we present the role of information theory in reliability analysis as well as a review of the important results and definitions in reliability theory. This makes the book self-contained and easy to follow in its subsequent chapters. The Shannon's entropy is introduced in Chapter 2. After a brief introduction to the properties of differential entropy a detailed review of residual life is made. This is followed in Chapter 3 where the entropy of past life or reversed residual life is discussed. With the success of Shannon measure in unearthing uncertainty in several problems; many generalizations to it were suggested in literature. Chapter 4 gives a detailed account of such generalized entropies. A second category of information measures are in the form of divergences used to compare the randomness associated with two distributions. The Kullback-Leibler divergence belongs to this category. In Chapter 5 we consider the different divergence measures and their characteristics. Chapter 6 is devoted to the study of inaccuracy, its forms in residual and past life. Finally, the last two Chapters 7 and 8 cover the recently introduced cumulative residual entropy and its generalizations. A special feature of the discussions in all chapters is the comprehensive treatment given to the quantile-based approach to various measures and the new results arising there from. The authors will be grateful to the readers for pointing out the mistakes that might have been committed by us and also for any suggestions for improvements of the contents.

We take this opportunity to thank our colleagues and students in the department for the encouragement given to us during the course of writing this book. We also thank Mr. Sanjay Varma for the excellent secretarial assistance.

N. Unnikrishnan Nair,
S.M. Sunoj,
G. Rajesh

Contents

CHAPTER 1

Preliminaries

1.1. Reliability analysis and information theory

Generally we consider a device to be reliable if it works at the expected level whenever called upon to do so. However development of a proper intuitive theory, requires a logical definition that can translate our scientific knowledge to one capable of characterizing and quantifying the reliability of a device. Thus the notion of reliability is often taken as the probability that a device or system will perform its intended function under conditions specified for its operation for a given time period. When it does not perform satisfactorily, we say that the device has failed. With a wide range of products ranging from equipments for daily use to sophisticated machineries, the period of their successful operation is a matter of considerable interest to all. Reliability analysis has made rapid strides in the last few decades to become a major discipline. Broadly speaking reliability theory is a body of concepts, methodologies and practices that lead to a failure-free operation of products and systems. The objective is accomplished through five major areas, reliability modelling, analysis, reliability engineering, reliability science and reliability management. Our primary concern in the present work is reliability modelling that deals with the identification of the mechanism that generates the failure times of devices. This requires various concepts definitions and methodologies that can distinguish the occurring failure patterns that can shed some light relating to the law that governs the lifetimes of devices. By lifetime we mean the time interval during which a device functions before it ultimately fails and its lifetime mathematical model that is sought. In this endeavour, many results from probability theory, statistics, stochastic processes and operations research are employed to identify an appropriate model that satisfactorily explains the behaviour of the observations collected on the lifetimes of the device when put into test or operation. The model is then used to draw inferences on the lifetime and conclude the properties of the random phenomenon that generated the observed values. Among different possible mathematical models available to represent lifetimes, one that is more preferred is the probability distribution of lifetimes, which can then be employed to evaluate the probabilities associated with various lifetimes and summary measures that are of relevance. Throughout our discussions we treat lifetime as a continuous random variable. Though we talk about random lifetimes and their distributions, the deliberations here will be equally valid for the general class of duration variables where the time taken for the occurrence of a specific event is of interest. Thus the concepts and methodologies in reliability theory with different nomenclatures or interpretations consistent with the context, have found applications in actuarial science, economics, medical science, epidemiology and demography.

Information theory is a branch of knowledge dedicated to the discovery and exploration of mathematical laws governing the behaviour of information as it is transferred, stored or retrieved. Technically measures of information are classified as Fisher type and its variants, entropy type and Bayesian off which the entropy type is mainly discussed as part of information theory. While dealing with matters related to transmission, processing extraction and utilization of information, this field of study has an intersection with mathematics, computer science, physics, neurobiology, information and electrical engineering. It has found application in several disciplines like statistical inference, language processing, cryptography, human vision, bioinfomatics, model selection, thermal physics, quantum computing, linguistics, detection of plagiarism, data compression, biochemistry, cybernetics, political science and psychology. There is a close association between adaptive systems, anticipatory systems, artificial intelligence, machine learning, coding theory and information theory. Triggered by the publication of the seminal paper 'A mathematical theory of communication' by Shannon in the Bell System Technical Journal in July and October 1948, the information theoretic approach through its axiomatic foundation, versatility and simplicity continues to attract researchers from diverse fields. A recent addition to the application of information theory is the field of reliability. Notions of uncertainty and information generally concerns probability distributions through which the predictability of an unknown random event can be ascertained. In the context of reliability of devices the events of interest are their lifetime and predictability. Ebrahimi and Soofi (2004) have identified three information theoretic lines of research that have evolved in reliability analysis. They are (a) development of information functions that are specifically suitable for reliability (b) development of various entropy-based diagnostics and tests of hypotheses useful for reliability model building and (c) information theoretic research of wider applications in reliability in developing measures that quantify the amount of information about the immediate future contained in the past. The basic measures of information are the entropy and its generalizations, measures of discrimination and inaccuracy which are capable of handling problems in diverse fields. Of these, entropy is concerned with measuring information in distributions associated with a random variable while divergence deals with information in common with two random variables. In reliability and survival analysis, the time a device has spent while in operation (called the age of the device) is of importance, since it is relevant in calculating the length of life and reliability. The age of the device may contribute to deterioration or improvement in its remaining life. To deal with such devices, the information measures have to be suitably modified and their properties studied. One has to take up the task of interpreting various notions in reliability in terms of these modified measures and ensure that the two approaches lead to consistent results.

Attempts to link information theoretic concepts to reliability in its present form has a recent history with the work of Muliere et al. (1993) when they studied the entropy of the residual life. However, the role of entropy in life tests were recognized much earlier when Barlow and Hsiung (1983) computed the expected information from a life test experiment involving an exponential distribution. Also, Awad and Alawneh (1987) discussed the loss of entropy when a distribution is truncated to $(0, t)$. Research subsequent to the works of Muliere et al. (1993) and Ebrahimi (1996) attracted many to investigate the applications of various aspects of information theory in reliability modelling and analysis. Major results in this connection and some special problems of future interest will be discussed in the following chapters. In order to appreciate these developments covered in the sequel, it is essential to recall some basic definitions, concepts and results from reliability analysis and allied fields. These will be addressed in the next few sections of the present chapter.

1.2. Basic reliability functions

A general discussion of the material in this section is available in Lai and Xie (2006) and Nair et al. (2013a).

1.2.1. Hazard rate. Let X be a continuous non-negative random variable representing the lifetime of a device or organism, with distribution function $F(x) = P(X \leq x)$, survival function $\bar{F}(x) = P(X > x)$ and probability density function $f(x)$. The function defined by

$$h(x) = \frac{f(x)}{\bar{F}(x)}, \quad \bar{F}(x) > 0. \tag{1.1}$$

is called the hazard rate of X or F. Often the hazard rate is also called failure rate in reliability, mortality rate and force of mortality in actuarial science.

Writing (1.1) as

$$h(x) = \lim_{\Delta \to 0} \frac{P(x < X \le x + \Delta | X > x)}{\Delta}.$$

$\Delta h(x)$ is approximately the conditional probability that a device which has survival time (age) x will fail in the next small interval of time Δ. This interpretation of $h(x)$ makes the concept quite meaningful in theory and practice. From (1.1), we can recover $\bar{F}(x)$ using

$$\bar{F}(x) = \exp \left[-\int_0^x h(t)dt \right]. \qquad (1.2)$$

Thus the hazard rate determines the life distribution uniquely so that both provide equivalent information about X. From physical considerations or from an empirical plot of $h(x)$, one can approximately find the functional form of $h(x)$ and hence determine the distribution using (1.2). When (1.2) is used to find $\bar{F}(x)$ from $h(x)$, one should note that the chosen form of $h(x)$ satisfies the conditions (a) $h(x) \ge 0$ (b) $\int_0^x h(t)dt$ is finite for some x and (c) $\int_0^\infty h(t)dt = \infty$.

A function related to the hazard rate is its cumulated version

$$\Lambda(x) = \int_0^x h(t)dt = -\log \bar{F}(x)$$

called the cumulated hazard rate. Since X is non-negative $\Lambda(0-) = 0$, $\Lambda(x)$ is increasing and $\Lambda(x) \to \infty$ as $x \to \infty$. The distributions and their hazard rates which are of interest in reliability are presented in Table 1.1.

1.2.2. Residual life. In reliability theory the remaining life of a device which has not failed until age t plays a significant role. The residual life is defined as $X_t = (X - t | X > t)$ and has survival function

$$\bar{F}_t(x) = P(X_t > x) = \frac{\bar{F}(x+t)}{\bar{F}(t)}, \quad x \ge 0 \qquad (1.3)$$

at all points $t \ge 0$ for which $\bar{F}(t) > 0$. Obviously the density of (1.3) is

$$f_t(x) = \frac{f(x+t)}{\bar{F}(t)}$$

and hence the hazard rate becomes

$$h_t(x) = \frac{f(x+t)}{\bar{F}(x+t)} = h(x+t). \qquad (1.4)$$

Table 1.1. Hazard rates of some distributions.

Distribution	$\bar{F}(x)$	$h(x)$
exponential	$e^{-\lambda x}$	λ
	$x > 0; \lambda > 0$	
Weibull	$\exp[-(\frac{x}{\sigma})^\lambda]$	$\lambda\sigma^{-\lambda}x^{\lambda-1}$
	$x > 0; \lambda, \sigma > 0$	
Pareto II	$\alpha^c(x+\alpha)^{-c}$	$c(x+\alpha)^{-1}$
	$x > 0; \alpha, c > 0$	
rescaled beta	$(1-\frac{x}{R})^c$	$c(R-x)^{-1}$
	$0 \le x \le R; c, R > 0$	
power	$1-(\frac{x}{\alpha})^\beta$	$\beta x^{\beta-1}(\alpha^\beta - x^\beta)^{-1}$
	$0 \le x \le \alpha; \alpha, \beta > 0$	
Burr	$(1+x^c)^{-k}$	$kcx^{c-1}(1+x^c)^{-1}$
	$x > 0; c, k > 0$	
generalized Pareto	$(1+\frac{ax}{b})^{-\frac{a+1}{a}}$	$\frac{a+1}{ax+b}$
	$x > 0, b > 0, a > -1$	
inverse Weibull	$1-\exp[-(\frac{\sigma}{x})^\lambda]$	$\frac{\lambda\sigma^\lambda x^{-\lambda-1}e^{-(\frac{\sigma}{x})^\lambda}}{1-e^{-(\frac{\sigma}{x})^\lambda}}$
	$x > 0; \sigma, \lambda > 0$	
generalized exponential	$1-[1-e^{-\frac{x}{\sigma}}]^\theta$	$\frac{\theta(1-e^{-\frac{x}{\sigma}})^{\theta-1}e^{-\frac{x}{\sigma}}}{\sigma(1-(1-e^{-\frac{x}{\sigma}})^\theta)}$
	$x > 0; \sigma, \theta > 0$	
log logistic	$(1+(\alpha x)^\beta)^{-1}$	$\beta\alpha^\beta x^{\beta-1}(1+(\alpha x)^\beta)$
	$x > 0, \alpha, \beta > 0$	
exponential geometric	$(1-p)e^{-\lambda x}(1-pe^{-\lambda x})^{-1}$	$\lambda(1-pe^{-\lambda x})^{-1}$
	$x > 0, \lambda > 0, 0 < p < 1$	
Greenwich	$(1+\frac{x^2}{b^2})^{-\frac{a}{2}}$	$\frac{ax}{b^2+x^2}$
	$x \ge 0; a, b > 0$	
log Weibull	$\exp[-(\log(1+\rho x))^k]$	$\frac{K\rho}{1+\rho x}(\log(1+\rho x))^{k-1}$
	$x > 0; \rho, k > 0$	
modified Weibull	$\exp[-\alpha\sigma(e^{(\frac{x}{\sigma})^\lambda}-1)]$	$\alpha\lambda(\frac{x}{\sigma})^{\lambda-1}e^{(\frac{x}{\sigma})^\lambda}$
	$x > 0, \alpha, \lambda, \sigma > 0$	
generalized Weibull	$[1-\lambda(\frac{x}{\beta})^\alpha]^{[\frac{1}{\lambda}]}$	$\alpha x^{\alpha-1}(\beta^\alpha - \lambda x^\alpha)^{-1}$
Gompertz	$\exp[\frac{-B(C^x-1)}{\log C}]$	BC^x
quadratic hazard	$\exp(-(ax+\frac{bx^2}{2}+\frac{cx^3}{3})$	$a+bx+cx^2$
	$x > 0$	
Hjorth	$\frac{\exp(-\alpha\frac{x^2}{2})}{(1+\beta x)^{\frac{\theta}{\beta}}}$	$\alpha x + \frac{\theta}{1+\beta x}$
	$x > 0, \alpha, \beta, \theta > 0$	
	$\alpha + \theta > 0$	
generalized power	$\exp[1-(1+(\frac{x}{\beta})^\alpha)^\theta]$	$\frac{\theta\alpha}{\beta^\alpha}(1+(\frac{x}{\beta})^\alpha)^{\alpha-x}$
	$x \ge 0, \alpha, \beta, \theta > 0$	
logistic exponential	$[1+(e^{\alpha x}-1)^k]^{-1}$	$\frac{\lambda k e^{\lambda x}(e^{\lambda x}-1)^{k-1}}{1+(e^{\lambda x}-1)^k}$

The survival function $\bar{F}_t(x)$ tells the probability that a device which has survived age t will survive for x additional units of time. Various characterstics of this distribution are of fundamental interest in lifetime data modelling. Prominent among them is the mean of (1.3). If $E(X) < \infty$, the mean residual life of X is defined as

$$m(t) = E(X - t | X > t) = \int_0^\infty \frac{\bar{F}(x+t)}{\bar{F}(t)} \, dx$$

$$= \frac{1}{\bar{F}(t)} \int_t^\infty \bar{F}(x) \, dx \tag{1.5}$$

and it gives the average lifetime remaining to a device which is of age t. From (1.5), the survival function of X can be expressed in terms of the mean residual life function as

$$\bar{F}(x) = \frac{\mu}{m(x)} \exp\left[-\int_0^x \frac{dt}{m(t)}\right], \tag{1.6}$$

where $\mu = E(X) = m(0)$. The hazard rate can be obtained from $m(x)$ through the formula

$$h(x) = \frac{1 + m'(x)}{m(x)}. \tag{1.7}$$

Since $h(x)$ is always non-negative, $m'(x) \geq -1$.

Very much like the hazard rate, the mean residual life function also determines $F(x)$ uniquely. Thus from a modelling point of view the functional form of $m(x)$ has an important role, apart from it being an index of the remaining life. The merits of the mean residual life over the hazard rate, various theoretical properties of $m(x)$ and its applications are given in Muth (1977) and Guess and Proschan (1988). The mean residual life function has a simple closed form only for a few life distributions, see Table 1.2.

Table 1.2. Mean residual life functions.

Distribution	$\bar{F}(x)$	$m(x)$
exponential	$e^{-\lambda x}$	λ^{-1}
power	$1 - (\frac{x}{\alpha})^\beta$	$\frac{(\beta+1)\alpha^\beta(1-x)+x^{\beta+1}-1}{(\beta+)(\alpha^\beta-x^\beta)}$
rescaled beta	$(1 - \frac{x}{R})^c$	$\frac{(R-x)}{(R+1)}$
Pareto II	$\alpha^c(x+\alpha)^{-c}$	$\frac{x+\alpha}{c-1}$
half logistic	$2(1 + e^{\frac{x}{\sigma}})^{-1}$	$\sigma(1 + e^{-\frac{x}{\sigma}})$
	$x > 0, \sigma > 0$	
Pareto I	$(\frac{x}{\sigma})^{-\alpha}, x > \sigma > 0, \alpha > 0$	$(\alpha-1)^{-1}x$
exponential geometric	$(1-p)e^{-\lambda x}[1 - pe^{-\lambda x}]^{-1}$	$-(\lambda p)^{-1}e^{\lambda x}(1 - pe^{-\lambda x}) \log(1 - pe^{-\lambda x})$
modified Gompertz	$\exp[-\frac{1}{\alpha}(e^{\alpha x} - \alpha^2 x - 1)]$	$\frac{e^{-\alpha x}}{\alpha}$
	$x, \alpha > 0$	

Even though the identity (1.6) holds for all life distributions there are much simpler relationships that characterize specified distributions. For example $h(x)m(x) = C$, a constant if and only if X has a generalized Pareto distribution. More details can be obtained from Nair and Sudheesh (2006). In certain problems, the functional form of the mean residual function can be directly found from the data by choosing appropriate forms. The form so chosen must satisfy the following necessary and sufficient conditions for it to be a mean residual life (Guess and Proschan (1988)).

(i) $0 \leq m(x) < \infty$ for all $x \geq 0$

(ii) $m(0) = \mu = E(X)$

(iii) $m(x)$ is right continuous

(iv) $m(x) + x$ is increasing

(v) If $\lim_{x \to x_0} m(x) = 0$ then $m(x) = 0$ for all x in $[x_0, \infty)$. If there is no x_0 for which the above limit is zero then $\int_0^\infty \frac{dx}{m(x)} = \infty$.

There are some other characteristics of residual life that are of interest in reliability. The most important among them is the variance residual life defined as

$$V(x) = E((X - x)^2 | X > x) - m^2(x)$$
$$= \frac{2}{\bar{F}(x)} \int_x^\infty \int_u^\infty \bar{F}(t)dtdu - m^2(x).$$

Also

$$\frac{dV(x)}{dx} = h(x)(V(x) - m^2(x)).$$

For any $0 < \alpha < 1$, the αth percentile of the residual life distribution, called the αth percentile residual life of X is also used in various contexts like censored data, observations from heavy tailed distributions and cases of outliers. Denoted by $p_\alpha(x)$, it is defined as

$$p_\alpha(x) = F_x^{-1}(\alpha)$$
$$= \inf\{x | 1 - \frac{\bar{F}(x+t)}{\bar{F}(x)} \geq \alpha\}$$
$$= F^{-1}\{1 - (1 - \alpha)\bar{F}(x)\} - x.$$

The special case when $\alpha = \frac{1}{2}$, gives the median residual life as an alternative to the mean residual life.

1.2.3. Reversed hazard rate. The reversed hazard rate of X is defined as

$$\lambda(x) = \frac{f(x)}{F(x)}. \tag{1.8}$$

It can be written as

$$\lambda(x) = \lim_{\Delta \to 0} \frac{P(x - \Delta \leq X \leq x | X \leq x)}{\Delta}$$

or

$$\Delta\lambda(x) = P(x - \Delta < X < x | X \leq x) + o(\Delta).$$

Thus $\Delta\lambda(x)$ is the probability that a device whose lifetime does not exceed x fails within a small time interval Δ prior to x. Since $\lambda(x) = \frac{d \log F(x)}{dx}$, we can write

$$F(x) = \exp\left[-\int_x^\infty \lambda(t)dt\right] \tag{1.9}$$

enabling determination of $F(x)$ when $\lambda(x)$ is known. The function $\bar{C}(x) = \log F(x) = \int_0^x \lambda(t)dt$ is called the cumulative reversed hazard rate function. Often $X \leq x$, is called reversed life or past life.

1.2.4. Inactivity time. The random variable $X_{(t)} = (t - X | X \leq t)$ is called the inactivity time or reveresd residual life. It represents the time elapsed since the failure of a device given that its lifetime is at most x. The distribution function of $X_{(t)}$ is

$$\begin{aligned} F_{X_{(t)}}(t) = P(X_{(t)} \leq t) &= P((x - X) | x \leq t | X \leq t) \\ &= \frac{F(x) - F(x - t)}{F(t)} \end{aligned}$$

with density

$$f_{X_{(t)}}(t) = \frac{f(x - t)}{F(t)}. \tag{1.10}$$

Like the residual life various characteristics of (1.10) have also been extensively studied. The mean of (1.10) known as mean inactivity time (reversed mean residual life) is

$$\begin{aligned} r(x) &= \int_0^x \frac{tf(x - t)}{F(x)}dt \\ &= \frac{1}{F(x)} \int_0^x F(t)dt. \end{aligned}$$

We have

$$\lambda(x) = \frac{1 - r'(x)}{r(x)} \tag{1.11}$$

and

$$F(x) = \exp\left[-\int_x^\infty \frac{1 - r'(t)}{r(t)}dt\right]. \tag{1.12}$$

An arbitrary function $r(x)$ can be a mean inactivity time function if and only if it satisfies

(i) $r(x) \geq 0$ for all $x > 0$ and $r(0) = 0$

(ii) $r'(x) < 1$

(iii) $\int_0^\infty \frac{1-r'(x)}{r(x)} dt = \infty$, and,

(iv) $\int_x^\infty \frac{1-r'(t)}{r(t)} < \infty$ for $x > 0$.

Some importance is also attributed to the variance inactivity time (reversed variance residual life) function

$$\nu(x) = E((x - X)^2 | X \leq x) - r^2(x)$$
$$= \frac{2}{F(x)} \int_0^x \int_0^u F(t) dt du - r^2(x).$$

We refer to Kundu and Nanda (2010) for a discussion of the properties of this function. The percentiles of the distribution (1.9) have been studied by Nair and Vineshkumar (2010a) (see also Nair and Vineshkumar (2010b)) which give the reversed percentile residual life or percentile inactivity time

$$q_\alpha(t) = F_x^{-1}(\alpha) = \inf[t | F(x - t) \leq (1 - \alpha) F(x)]$$
$$= x - F^{-1}((1 - \alpha) F(x)).$$

It is related to the reversed hazard rate by

$$q_\alpha'(x) = 1 - \frac{\lambda(x)}{\lambda(x - q_\alpha(x))}.$$

Some distributions and their reversed hazard rates are given in Table 1.3 for ready reference.

Table 1.3. Reversed hazard rates.

Distribution function	$\lambda(x)$
$(\frac{x}{\alpha})^\beta, 0 \leq x \leq \alpha$	βx^{-1}
$\exp[-\frac{\lambda}{x}], x > 0, \lambda > 0$	λx^{-2}
$(1 + \frac{1}{\alpha x})^{-c}, x > 0, \alpha, c > 0$	$\frac{c}{x(1+\alpha x)}$
$(1 - \frac{1}{Rx})^c, 0 \leq x \leq R$	$\frac{c}{x(Rx-1)}$
$\exp[\frac{-B(\frac{c}{x}-1)}{\log c}]$	$\frac{B}{x^2} c^{\frac{1}{x}}$
$(1 - e^{-\lambda x})^\theta$	$\frac{\theta \lambda}{e^{\lambda x}-1}$
$(1 + x^{-c})^{-k}, x > 0, k, c > 0$	$\frac{kc}{x(1+x^c)}$
$(1 - x^{-\beta})^\theta, x > 1$	$\frac{\beta \theta}{x(x^\beta-1)}$
$\exp[-\theta(x^{-\beta-1})]$	$\frac{\theta}{x^{\beta+1}}$

1.3. Quantile-based reliability functions

A probability distribution is usually specified by its distribution function. An alternative way of specification is through the quantile function defined as

$$Q(u) = \inf\{x | F(x) \geq u\}, \quad 0 \leq u \leq 1. \tag{1.13}$$

Although both quantile and distribution functions convey the same information about the distribution with different interpretations, the latter is mostly used in reliability analysis. However several unique properties of quantile functions make them more convenient for use in lifetime data analysis and moreover several results that are difficult to arrive at by using distribution functions can be obtained if quantile functions are employed. A detailed discussion of these aspects can be seen in Gilchrist (2000) and Nair et al. (2013a).

If $F(x)$ is continuous and strictly increasing $Q(u)$ is the unique value of x satisfying $F(x) = u$ and so by solving $F(x) = u$, $Q(u)$ is the value of x. For example, in the case of the exponential distribution

$$F(x) = 1 - \exp(\lambda x),$$

setting $F(x) = u$ and solving $Q(u) = x = -\frac{1}{\lambda}\log(1 - u)$. Table 1.4 contains some common distributions and their quantile functions. We also have $F(Q(u)) = u$. On differentiation,

$$f(Q(u))\frac{dQ(u)}{du} = 1.$$

The derivative of $Q(u)$ is denoted by $q(u)$ is called the quantile density function and satisfies the relationship

$$q(u)f(Q(u)) = 1 \tag{1.14}$$

with the density function of X taken at $Q(u)$.

Some important properties possessed by quantile functions are

(i) $Q(u)$ is left continuous and non-decreasing on $(0, 1)$ with $Q(F(x)) \leq x$ for all $-\infty < x < \infty$ for which $0 < F(x) < 1$.

(ii) Generally $F(Q(u)) \geq u$ with equality holding in the case mentioned above.

(iii) If U is a uniform random variable over $[0, 1]$, $X = Q(U)$ has distribution function $F(x)$.

(iv) If Q_1 and Q_2 are quantile functions, $Q_1 + Q_2$ is also a quantile function. Similarly if Q_1 and Q_2 are positive, their product $Q_1 Q_2$ is also a quantile function.

(v) If $T(u)$ is a non-decreasing function satisfying $T(0) = 0$ and $T(1) = 1$ then $Q(T(u))$ is a quantile function of some random variable with the same support as X.

Table 1.4. Quantile functions of life distributions.

Distribution	$\bar{F}(x)$	$Q(u)$
exponential	$e^{-\lambda x}$	$\lambda^{-1}(-\log(1-u))$
Weibull	$\exp[-(\frac{x}{\sigma})^{\lambda}]$	$\sigma(-\log(1-u))^{\frac{1}{\lambda}}$
Pareto II	$\alpha^c(x+\alpha)^{-c}$	$\alpha[(1-u)^{-\frac{1}{c}}-1]$
rescaled beta	$(1-\frac{x}{R})^c$	$R[1-(1-u)^{\frac{1}{c}}]$
half-logistic	$2[1+e^{\frac{x}{\sigma}}]^{-1}$	$\sigma\log\frac{1-u}{1+u}$
power	$1-(\frac{x}{\alpha})^{\beta}$	$\alpha u^{\frac{1}{\beta}}$
Burr II	$(1+x^c)^{-k}$	$[(1-u)^{\frac{1}{k}}-1]^{\frac{1}{c}}$
log logistic	$(1+(\alpha x)^{\beta})^{-1}$	$\alpha^{-1}(\frac{u}{1-u})^{\frac{1}{\beta}}$
exponential geometric	$(1-p)e^{-\lambda x}(1-pe^{-\lambda x})^{-1}$	$\frac{1}{\lambda}\log\frac{1-pu}{1-u}$
generalized exponential	$1-[1-e^{-\frac{x}{\sigma}}]^{\theta}$	$\sigma[-\log(1-u^{\frac{1}{\theta}})]$
inverse Weibull	$1-exp[-(\frac{\sigma}{x})^{\lambda}]$	$\sigma(-\log u)^{-\frac{1}{\lambda}}$
generalized Pareto	$(1+\frac{ax}{b})^{-\frac{a+1}{a}}$	$\frac{b}{c}[(1-u)^{-\frac{a}{a+1}}-1]$
exponential Power	$\exp[e^{-(\lambda x)^{\alpha}}-1]$	$\frac{1}{\lambda}[-\log(1+\log(1-u))]^{\frac{1}{\alpha}}$
log Weibull	$\exp[-(\log(1+\rho x))^k]$	$e^{-1}[\exp(-\log(1-u))^{\frac{1}{k}}-1]$

(vi) If $T(\cdot)$ is non-decreasing (non-increasing), $T(Q(u))$ $(T(Q(1-u)))$ is a quantile function.

These properties become handy when we discuss the properties of information measures of residual life. With the aid of the above definitions we can translate the reliability functions discussed in the previous section by means of quantile functions. The definitions and results in this connection are taken from Nair and Sankaran (2009).

1.3.1. Hazard quantile function. The hazard quantile function is obtained by setting $x = Q(u)$ is (1.1). Then

$$h_Q(u) = h(Q_X(u)) = \frac{f(Q(u))}{\bar{F}(Q(u))} \qquad (1.15)$$
$$= [(1-u)q(u)]^{-1}.$$

In this definition $H(u)$ is interpreted as the conditional probability of failure of a device in the next small interval of time given the survival of the device at the $100(1-u)\%$ point of the distribution of X. From (1.15),

$$q(u) = [(1-u)h_Q(u)]^{-1}$$

and so

$$Q(u) = \int_0^u \frac{dp}{(1-p)h_Q(p)}. \qquad (1.16)$$

Thus the distribution is uniquely determined by the hazard quantile function. Equation (1.16) enables determination of distributions corresponding to a given functional form of $h_Q(u)$. As an illustration Midhu et al. (2013) found that the hazard quantile function is of the linear form $h_Q(u) = a + bu$, $a > 0$ if and only if

$$Q(u) = \log \left(\frac{a + bu}{a(1 - u)} \right)^{\frac{1}{a+b}} \qquad (1.17)$$

and called it the linear hazard quantile distribution. It contains as special cases the exponential, half-logistic and the exponential-geometric distributions.

1.3.2. Mean residual quantile function. The equivalent form of mean residual life in the quantile setting is the mean residual quantile function obtained as

$$m_Q(u) = m(Q_X(u)) = \frac{1}{1 - u} \int_u^1 (1 - p)q(p)dp. \qquad (1.18)$$

We interpret $m_Q(u)$ as the average remaining life beyond the $100(1 - u)\%$ of the distribution. An alternative form of (1.17) is

$$m_Q(u) = (1 - u)^{-1} \int_u^1 [Q(p) - Q(u)]dp.$$

For the exponential distribution $m_Q(u) = \lambda^{-1}$, a constant.

EXAMPLE 1.1. *The generalized Pareto distribution specified by*

$$Q(u) = \frac{b}{a} \left[(1 - u)^{-\frac{a}{a+1}} - 1 \right],$$

$$q(u) = \frac{dQ(u)}{du} = \frac{b}{a + 1}(1 - u)^{-\frac{a}{a+1} - 1}$$

so that (1.18) *becomes*

$$m_Q(u) = \frac{1}{1 - u} \int_u^1 \frac{b}{a + 1}(1 - p)^{-\frac{a}{a+1}} dp = b(1 - u)^{-\frac{a}{a+1}}.$$

From (1.18) *and* (1.15) *we have the identity*

$$[h_Q(u)]^{-1} = m_Q(u) - (1 - u)m_Q'(u). \qquad (1.19)$$

Also the distribution is determined from $m_Q(u)$ through the formula

$$Q(u) = \mu + \int_0^u \frac{m_Q(p)}{1 - p} dp - m_Q(u)$$

$$= \int_0^u \frac{m_Q(p) - (1 - p)m_Q'(p)}{1 - p} dp.$$

Some simple forms of (1.19) can characterize distributions. As examples $h_Q(u)m_Q(u) = C$, a constant and $m_Q(u) = (1 - u)^{-1}[A + B \log H(u)]$

characterize respectively the exponential and linear hazard quantile distributions.

The function $m_Q(u)$ can also be derived from the quantile function of the residual life $X_t = (X - t | X > t)$. Setting $F(t) = u_0$, $F(x + t) = v$ and $F_t(x) = u$, from (1.3) we have

$$v = u_0 + (1 - u_0)u$$

and the quantile function of X_t is

$$Q_1(u) = Q(u_0 + (1 - u_0)u) - Q(u_0). \qquad (1.20)$$

The mean residual quantile function is now the mean of (1.20) or $\int_0^1 Q_1(u)du$. The residual variance quantile function is

$$V_Q(u) = \frac{1}{(1-u)^2} \int_u^1 Q^2(p)dp - \left[\frac{1}{1-u} \int_u^1 Q(p)dp \right]^2$$

$$= \frac{1}{1-u} \int_u^1 Q^2(p)dp - (m_Q(u) + Q(u))^2.$$

The interesting identities realized from the above are

$$m_Q^2(u) = V_Q(u) - (1-u)V_Q'(u)$$

and

$$V_Q(u) = (1-u)^{-1} \int_u^1 m_Q^2(p)dp.$$

Also we can write the αth percentile residual quantile function as,

$$Q_\alpha(u) = q_\alpha(Q(u))$$
$$= Q[1 - (1 - \alpha)(1 - u)] - Q(u).$$

1.3.3. Quantile functions in reversed time. Various reliability functions in reversed time can also be expressed in quantile form in much the same way as above. For example the reversed hazard quantile function is

$$\lambda_Q(u) = \lambda(Q(u)) = [uq(u)]^{-1}. \qquad (1.21)$$

It determines $Q(u)$ through the formula

$$Q(u) = \int_0^a [p\lambda_Q(p)]^{-1}dp. \qquad (1.22)$$

The mean inactivity quantile function, also called the reversed mean residual quantile function derives from (1.10) as

$$r_Q(u) = r(Q(u)) = u^{-1} \int_0^u pq(p)dp. \qquad (1.23)$$

Among these functions we have the identities given below as,

$$Q(u) = r_Q(u) + \int_0^u p^{-1} r_Q(p) dp$$

$$[\lambda_Q(u)]^{-1} = r_Q(u) + u r'_Q(u)$$

$$r_Q(u) = u^{-1} \int_0^u [\lambda_Q(p)]^{-1} dp.$$

Along with the mean, the variance of the inactivity time in quantile formulation is sometimes needed in data analysis problems. This is

$$v_Q(u) = u^{-1} \in_0^u Q^2(p) dp - (Q(u) - r_a(u))^2$$

and satisfies

$$r_a^2(u) = v_a(u) + u v'_Q(u)$$

and

$$v_Q(u) = u^{-1} \int_0^u r_a^2(p) dp.$$

1.4. Quantile function models

In equation (1.13) we have defined the quantile function as the inverse of the distribution function. However this is not the only a way in which a quantile function can be constructed. Properties (iv), (v) and (vi) in Section 3 enable us to obtain new quantile functions from the existing ones and infact this is a crucial advantage in model building. See Gilchrist (2000) for details. A left continuous non-decreasing function can also qualify for a quantile function. Various researchers have proposed several such functions in literature as models of random phenomena. Generally these distributions do not possess a closed form expression for their distribution function, but at the same time they have special cases that invert to standard distribution functions. An attractive property common to all such models is that they provide very good approximations to several standard distributions by varying the parameters values. In this section we briefly discuss some such models.

The basic model from which many others were derived was given by Tukey (1962) known as the Tukey lambda distribution with

$$Q(u) = \frac{1}{\lambda} \left[u^\lambda - (1-u)^\lambda \right], \ 0 \le u \le 1, \lambda \ne 0. \tag{1.24}$$

As $\lambda \to 0$, $Q(u) = \log(\frac{u}{1-u})$ which is the logistic distribution. For $\lambda = 1, 2$, (1.24) becomes uniform. This distribution being symmetric several asymmetric extensions were subsequently suggested like

$$Q(u) = a u^\lambda - (1-u)^\lambda$$

and
$$Q(u) = au^\lambda + b(1-u) + c.$$

Ramberg and Schmeiser (1974) suggested the generalized lambda distribution

$$Q(u) = \lambda_1 + \frac{1}{\lambda_2} \left(u^{\lambda_3} - (1-u)^{\lambda_4} \right) \qquad (1.25)$$

where λ_1 is the location parameter, λ_2 is the scale parameter and λ_3 and λ_4 are shape parameters controlling the two tails of the distribution. Since (1.25) defines a distribution for $-\infty < x < \infty$, for it to be a life distribution the constraint $Q(0) = \lambda_1 - \frac{1}{\lambda_2} \geq 0$ is essential. Since (1.25) is valid only for certain regions of the parameter space, Freimer et al. (1988) proposed the modified generalized lambda family

$$Q(u) = \lambda_1 + \frac{1}{\lambda_2} \left[\frac{u^{\lambda_3} - 1}{\lambda_3} - \frac{(1-u)^{\lambda_4} - 1}{\lambda_4} \right], \ \lambda_2 > 0. \qquad (1.26)$$

Another four parameter family due to van Staden and Loots (2009) is

$$Q(u) = \lambda_1 + \lambda_2 \left[(1-\lambda_3)\frac{u^{\lambda_4} - 1}{\lambda_4} - \lambda_3 \frac{(1-u)^{\lambda_4} - 1}{\lambda_4} \right], \ \lambda_2 > 0. \ (1.27)$$

Taking the product of the quantile function of the power and Pareto distributions we have the power-Pareto distribution (Gilchrist (2000); Hankin and Lee (2006)),

$$Q(u) = \frac{Cu^{\lambda_1}}{(1-u)^{\lambda_2}}, \qquad (1.28)$$

where $C > 0$, $\lambda_1, \lambda_2 > 0$ where one of the λ's may be taken as zero. The uniform, power, log logistic and Pareto distributions are special cases of (1.28). Another flexible model is the Govindarajulu distribution studied in detail by Nair et al. (2012b). It is given by

$$Q(u) = \theta + \sigma((\beta + 1)u^\beta - \beta u^{\beta+1}), \ \sigma, \beta > 0 \qquad (1.29)$$

and the support of the distribution is $(\theta, \theta + \sigma)$. Two other simple models are expressed as quantile density functions (Nair et al. (2012b))

$$q(u) = Ku^\alpha (1-u)^\beta \qquad (1.30)$$

and

$$q(u) = K(1-u)^{-A}(-\log(1-u))^{-M}. \qquad (1.31)$$

Distribution (1.30) contains the exponential, Pareto II, rescaled beta, log logistic and Govindarajulu distributions as special cases, while (1.31) subsumes the Weibull, exponential, Pareto II, rescaled beta models. Midhu et al. (2013) obtained the distribution that has linear mean residual quantile function in the form,

$$Q(u) = -(c + \mu)\log(1 - u - 2cu), \ \mu = E(X) > 0, -\mu < c < \mu \quad (1.32)$$

for which $q(u) = \frac{c+\mu}{1-u} - 2c$ and $m_Q(u) = cu + \mu$. Some other simple forms for reliability functions have also been discussed in the literature to generate new quantile functions. These are hazard quantile function $h_Q(u) = \frac{A+Bu}{C+Du}$ and quadratic mean residual quantile function with respective distributions (Sankaran et al. (2015); Sankaran and Dileep Kumar (2018))

$$Q(u) = \frac{(A - B)\log(1 + Au) - A(B + 1)\log(1 - u)}{A(A + 1)K} \quad (1.33)$$

and

$$Q(u) = -\alpha\log(1 - u) + (\beta - \alpha)u + \frac{\gamma - \beta}{2}u^2, \ \alpha, \beta, \gamma \geq 0.$$

Two other distributions of interest to reliability are

$$Q(u) = A(-\log(1 - u))^\alpha + \frac{(B + 1)k(1 - (1 - u)^B)}{B}$$

which has Weibull, Pareto II, rescaled beta and linear mean residual quantile functions as special cases and

$$Q(u) = \alpha\log\frac{1 - pu}{1 - u} + \beta\log\frac{1 + u}{1 - u}, \ 0 \leq p < 1, \alpha, \beta > 0,$$

the generalization of the linear hazard quantile function distribution, see Sankaran et al. (2016) and Sankaran and Dileep Kumar (2018) for details.

Some remarks about using quantile functions instead of distribution functions in analyzing data with descriptive measures is also essential. The conventional moments of the distribution are

$$\mu'_r = E(X^r) = \int_0^1 Q^r(p)dp$$

from which the mean and the variance can be calculated. The difficulties in using the moment-based sample counterparts of these quantities can be overcome if the following measures are employed. As a measure of locations, the median $M = Q(0.5)$, the inter quantile range $I = Q(0.75) - Q(0.25)$ for dispersion, the Galton's coefficient

$$S = \frac{Q(0.75) + Q(0.25) - 2Q(0.50)}{Q(0.75) - Q(0.25)}$$

for skewness and the Moor's measure

$$T = \frac{(a(0.875) - Q(0.625) + Q(0.375) - Q(0.125))}{I}$$

to evaluate Kurtosis. These measures are less influenced by outliers, more stable and much easily calculated than those based on moments.

A third alternative is the analysis through L-moments. They are expected values of linear functions of order statistics and have lower sampling variances and robust against outliers. The first four L-moments are

$$L_1 = \int_0^1 Q(p)dp = \mu$$

$$L_2 = \int_0^1 (2p-1)a(p)dp$$

$$L_3 = \int_0^1 (6p^2 - 6p + 1)Q(p)dp$$

$$L_4 = \int_0^1 (20p^3 - 30p^2 + 12p - 1)Q(p)dp.$$

All L-moments exist whenever $E(X)$ is finite, whereas existence of μ_r' requires $E(X^r)$ to be finite. Since $L_2 = \frac{1}{2}\Delta$, Δ being the mean difference is a measure of dispersion. The L-coefficient of skewness and kurtosis are $\tau_3 = \frac{L_3}{L_2}$ and $\tau_4 = \frac{L_4}{L_2}$. For a discussion of the properties of the L-measures and their merits over the conventional moments we refer to Hosking (1990, 1996).

1.5. Ageing and associated criteria

Recall that the age of a device is reckoned as the time during which a device is functioning satisfactorily. By the term ageing we mean the phenomenon by which the life remaining to the device is affected by its current age in some probabilistic sense. In this connection we have three classifications; positive ageing, negative ageing and no ageing. These three occur in situations when the residual lifetime is decreasing, increasing or remains the same when the age of the device is increasing. Various equipments in common use that deteriorate in efficiency and gradually fail due to wear and tear belong to the category of positive ageing devices since their remaining life span decreases with usage time. There are also situations when the life length increases with age, examples being the case of human beings whose longevity at birth increases for some time thereafter decreases and equipments that undergo efficient preventive maintenance strategies. Here the case is that the residual lifetime shows an increase with age increase. Lastly no ageing occurs when the device has the same residual life irrespective of age or when failures occur at random independent of their age. Among continuous distributions, the exponential distribution is the only one possessing no-ageing property.

A large part of reliability theory is concerned with ageing and various concepts that describe the patterns of ageing, compare life distributions and explain the data generating mechanism. Notions of ageing developed in this context give an indication of the manner in which ageing can be described, life distributions can be classified and distinguished and appropriate models can be chosen when observations are available. In the next section we present criteria for ageing based on the reliability functions already discussed in Section 1.1.

1.6. Concepts based on hazard rate

The behaviour and properties of the hazard rate form the foundations for defining ageing classes in this category.

1.6.1. Increasing (decreasing) hazard rates.

DEFINITION 1.1. *A lifetime X or its distribution F belongs to the increasing (decreasing) hazard rate, IHR (DHR) class if $h(x)$ is increasing (decreasing) for all x. This is equivalent to the residual life distribution*

$$\bar{F}_t(x) = \frac{\bar{F}(t+x)}{\bar{F}(t)}$$

is decreasing (increasing) in t for all $x > 0$. The IHR (DHR) class posseses some important properties

(a) *X is IHR if and only if $c(x) = -\log \bar{F}$ is convex or $\bar{F}$ is log concave.*

(b) *If $h(x)$ is differential then X is IHR (DHR) if $h'(x) > (<)0$.*

(c) *Since $u = F(x)$ increases whenever x does so, the monotonicity of $h(x)$ and $h_Q(u)$ are identical.*

(d) *X is IHR if and only if the determinant*

$$\begin{vmatrix} \bar{F}(x_1 - y_1) & \bar{F}(x_1 - y_2) \\ \bar{F}(x_2 - y_1) & \bar{F}(x_2 - y_2) \end{vmatrix} \geq 0$$

when $x_1 \leq x_2$, $y_1 \leq y_2$. This property referred to as '$\bar{F}$ is a polya frequency function of order 2'.

(e) *If X is IHR then X has finite moments of all orders. Further the residual life distribution F_t has an increasing hazard rate*

(f) *If $\bar{F}$ has a log convex density then it has a decreasing hazard rate. Among various distributions given in Table 1.1, the exponential distribution has a constant hazard rate, rescaled beta, power, generalized Pareto $a > 0$, modified Weibull for $\alpha > 1$ and Weibull for $\lambda > 1$ have IHR and log Weibull $0 \leq k \leq 1$, exponential-geometric, log logistic for $\beta \leq 1$, Burr XII for $c \leq 1$ etc., are DHR distributions.*

Every hazard rate function is not monotonic and they may belong to one of the other categories, bathtub (BT), upside-down bathtub (UBT), periodic, roller coaster or polynomial type. Of these both BT and UBT shaped hazard rates have attracted substantial attention in literature.

DEFINITION 1.2. *A random variable is said to have BT (UBT) hazard rate if for some $0 \leq a < b$, the hazard rate $h(x)$ is decreasing (increasing) in x, $0 \leq x \leq a$, is constant in $a \leq x \leq b$ and is increasing (decreasing in x for $x \geq b$. The flat points from $[a, b]$ in which $h(x)$ is constant are called the change points of $h(x)$. Most of the cases we consider are those with one change point. In cases where $h(x)$ is differentiable, in one change point, X is BT (UBT) if $h'(x) < (<)0$ in $(0, x_0)$ and $h'(x) = 0$ at $x = x_0$ and $h'(x) > (<)0$ in (x_0, b), $b \leq \infty$. Some natural situations where bathtub-shaped hazards appear in biological and mechanical systems are explained in Marshall and Olkin (2007).*

Among distributions listed in Table 1.1, the power distribution for $\beta > 1$, the Hjorth distribution for $0 < d < \theta\beta$, generalized Weibull for $\beta < 1$, $\lambda > 0$ and others have BT shaped hazard rates while Burr type XII $(c > 2)$, log logistic $(\beta > 1)$ and inverse Weibull have UBT hazard rates. A considerable amount of literature exists in constructing lifetime distributions with BT (UBT) hazard rates, for example see Lai and Xie (2006) and Nair et al. (2013a).

1.6.2. Montone hazard rate averages.

DEFINITION 1.3. *A random variable X with $F(0) = 0$ has increasing (decreasing) hazard rate average IHRA (DHRA) if $\frac{C(x)}{x}$ is increasing (decreasing) in $x > 0$. Notice that*

$$\frac{C(x)}{x} = \frac{1}{x} \int_0^x h(t)dt \qquad (1.34)$$

is the average of the hazard rate $h(x)$ in $[0, x]$.

Some properties of the IHRA/DHRA distributions are given below.

(1) If X is IHR (DHR) then it is IHRA (DHRA). This follows from the fact that the average of an increasing function is also increasing. But the converse need not be true. For example,

$$\bar{F}(x) = e^{-\lambda_1 x} + e^{-\lambda_2 x} - e^{-(\lambda_1 + \lambda_2)x}; \ \lambda_1, \lambda_2 > 0, \ x > 0$$

has

$$h(x) = \frac{\lambda_1 e^{\lambda_2 x} + \lambda_2 e^{\lambda_1 x} - 1}{e^{\lambda_2 x} + e^{\lambda_1 x} - 1}.$$

Using (1.34), X is IHRA, but $h(x)$ is UBT as can be verified graphically also.

(2) $(\bar{F}(x)))^{1/x}$ is decreasing (nonincreasing) if X is IHRA (DHRA) or equivalently

$$\bar{F}(kx) \geq (\leq)(\bar{F}(x))^k, \ 0 \leq k \leq 1 \text{ or } C(kx) \leq (\geq)kC(x)$$

(3) X is IHRA is equivalent to $h(x) \geq \frac{C(x)}{x}$ and $\bar{F}(x) \geq e^{-xh(x)}$

(4) An IHRA distribution has finite moments of order $r, 0 < r < \infty$

(5)

$$\mu'_r \leq (\geq)\Gamma(r+1)\mu^r, \quad 0 < r < 1, \ \mu = E(X)$$
$$\geq (\leq)\Gamma(r+1)\mu^r, \quad 1 < r < \infty$$

accordingly as X is IHRA (DHRA) and the coefficient of variation is $\leq (\geq) 1$.

(6) When X is IHRA

$$\bar{F}(x) = \begin{cases} 1, & x \leq \mu \\ e^{-wx}, & x > \mu \end{cases}$$

where $1 - w\mu = e^{-wx}$.

When $X_1, X_2 \ldots X_n$ are independent, identically distributed and IHRA (Ahmad and Mugadi (2004)),

$$E(\min(X_1, \ldots, X_n))^r \geq \frac{\mu'_{r+1}}{(r+1)}.$$

1.6.3. New better than used in hazard rate. Positive ageing can also be indicated if the hazard rate at the initial age is less than that of a used one. This brings in the concept of new better (worse) than used in hazard rate NBUHR (NWUHR) defined through the relationship $h(x) \geq (\leq)h(0)$ for all $x \geq 0$. In the quantile formulation this is equivalent to $h_Q(u) \geq (\leq) h_Q(0)$ for all $0 \leq u \leq 1$. Further, if

$$h(0) \leq (\geq)\frac{1}{x} \int_0^x h(t)dt.$$

X is said to be new worse (better) than used in hazard rate average, NWUHRA (NBUHRA). When the distribution is specified by the quantile function, X is NBUHRA (NWUHRA) if

$$Q(u) \leq -\frac{\log(1-u)}{h_Q(0)}$$

meaning the quantile function of X is less than that of the exponential distribution. It may be noticed that

$$IHR \Rightarrow NBUHR \Rightarrow NBUHRA$$

and a similar classification for the dual classes.

1.6.3.1. *Increasing hazard rate (2)*. Another type of increasing hazard rate can be considered if the concept of stochastic dominance is utilized. Deshpande et al. (1986) defined increasing (decreasing) hazard rate (2) based on the condition that

$$\int_0^x \frac{\bar{F}(t+s)}{\bar{F}(t)} dt \text{ is non-decreasing (increasing) in } s$$

for every fixed $x \geq 0$. This class of distributions is weaker than the IHR class so that IHR (DHR) $\Rightarrow$ IHR (2) (DHR (2)).

1.7. Concepts based on residual life

In this section we discuss briefly notions of aging based on mean, variance and percentile of residual life.

1.7.1. Decreasing mean residual life.

DEFINITION 1.4. *A lifetime X is said to have a decreasing (increasing) mean residual life DMRL (IMRL) if $m(x)$ is decreasing (increasing) for all $x \geq 0$.*

The above definition is equivalent to saying

$$\int_0^\infty \frac{\bar{F}(x+t)}{\bar{F}(t)} dx \text{ is decreasing (increasing) in } t \geq 0,$$

and if $m(x)$ is differentiable X is DMRL (IMRL) if $m'(x) \leq (\geq)0$. The DMRL class has the following properties.

(i) If X is IHR (DHR) then it is DMRL (IMRL). Also X is DMRL (IMRL) if $m_Q(u)$ is decreasing (increasing)
(ii) IHRA (DHRA) does not imply DMRL (IMRL)
(iii) When X is DMRL (IMRL)

$$(r+1)E(X_1 \min(X_1, X_2)^r] \geq (\leq)(r+2)p_2^{r+1}, \quad p_r = E(\min X_1, X_2)^r.$$

1.7.2. Net decreasing mean residual life.

DEFINITION 1.5. *We say that X has a net decreasing (increasing) mean residual life function, NDMRL (NIMRL) if $m(x) \leq (\geq)m(0)$ for all $x \geq 0$.*

Obviously

$$\text{DMRL (IMRL)} \Rightarrow \text{NDMRL (NIMRL)}.$$

1.7.3. Decreasing harmonic mean residual life.

DEFINITION 1.6. *The random variable X has a decreasing (increasing) mean residual life in harmonic average, DMRLHA (IMRLHA) if (Deshpande et al. (1986))*

$$\left[\frac{1}{x}\int_0^x \frac{dt}{m(t)}\right]^{-1} \quad \text{is decreasing (increasing) in } x.$$

We have DMRL $\Rightarrow$ DMRLHA.

1.7.4. Used Better than Aged (UBA).
When a unit is working with unknown age, to assess its future ageing behaviour the used better (worse) than aged UBA (UWA) concept is helpful.

DEFINITION 1.7. *When $E(X) < \infty$ and $0 < m(\infty) < \infty$, the UBA (UWA) class is specified by (Alzaid (1994))*

$$\bar{F}(x+t) \geq (\leq)\bar{F}(t)e^{-\frac{x}{m(\infty)}}, \ x, t \geq 0.$$

We have several useful properties for the UBA class

    (a) IHR $\Rightarrow$ DMRL $\Rightarrow$ UBA

    (b) The moments satisfy

$$\frac{\mu_{r+s+2}}{(r+s+2)!} \geq \frac{\mu_{r+1}(m(\infty))^{s+1}}{(r+1)!}$$

    (c) the moment generating function $G(t)$ satisfies

$$G(t) \leq 1 + \frac{\mu t}{1 - tm(\infty)}$$

    (d) When $E(X) < \infty$, all moments exist and are finite

A weaker concept than UBA (UWA) is used better (worse) than aged in expectation, UBAE (UWAE).

DEFINITION 1.8. *The random variable X said to be UBAE (UWAE) if $m(x) \geq (\leq)m(\infty), \ 1 < m(\infty) < 1.$*

1.7.4.1. *Decreasing Variance Residual Life (DVRL).* Recall that the variance residual life function of X is

$$v(x) = \frac{2}{\bar{F}(x)}\int_x^\infty \int_u^\infty \bar{F}(t)dtdu - m^2(x).$$

We say that X has decreasing (increasing) variance residual life DVRL (IVRL) if $v(x)$ is a decreasing (increasing) function of x for all $x > 0$. Notice that as $v(x)$ is decreasing the uncertainty in the residual life is decreasing and is also indicative of positive ageing. We have the implication

$$\text{DMRL (IMRL)} \Rightarrow \text{DVRL (IVRL)}$$

and if X is DVRL (IVRL)

$$E(X^2|X \geq x) \leq (\geq)E^2(X|X \geq x) + E[(X|X \geq x) - x]^2$$

and

$$X \text{ is DVRL (IVRL)} \Rightarrow X \text{ is UBAE (UWAE)}.$$

1.8. Classes based on survival function

1.8.1. New better (worse) than used. We say that X is new better (worse) than used NBU (NWU) if $\bar{F}(x+t) \leq (\geq)\bar{F}(x)\bar{F}(t)$ for all $x, t \geq 0$. In terms of the cumulative hazard rate function, this is the same as

$$C(x + t) \leq (\geq)C(x) + C(t) \tag{1.35}$$

We note that the definition of NBU can be rewritten as

$$\bar{F}(x) \geq \bar{F}_t(x), \quad \text{for all } x, t \geq 0,$$

which in turn implies that a new item has a greater life than an unfailed item of age t. In this sense X is ageing positively. If X is NBU then

   (i) $\bar{F}(\alpha t) \leq [\bar{F}(t)]^\alpha$, $\alpha = 1, 2, \ldots$
   (ii) it has finite moments of all positive orders
   (iii) $\int g(\alpha x)k(\bar{\alpha}x)f(x)dx \leq \int g(x)f(x)dx \int k(x)f(x)dx$ for all non-negative increasing functions g and k, $0 < \alpha < 1$ and $\bar{\alpha} = 1 - \alpha$
   (iv) $h(x) \geq h(0)$ or NBU $\Rightarrow$ NBUHR
   (v) $Q(u + v - uv) \leq Q(u) + Q(v)$
   (vi) IHRA $\Rightarrow$ NBU
   (vii)

$$\frac{\mu_{r+s+2}}{\Gamma(r+s+3)} \geq \frac{\mu_{r+1}}{\Gamma(r+2)} \frac{\mu_s}{\Gamma(s+2)}, \quad r, s \geq 0.$$

A concept associated with NBU is NBU(2).

DEFINITION 1.9. *A lifetime X is NBU (2) (NWU (2)) if*

$$\int_0^x \bar{F}(y)dy \geq (\leq) \int_0^x \frac{\bar{F}(t+y)}{\bar{F}(t)}dy.$$

Obviously NBU $\Rightarrow$ NBU (2).

There are three other variants of the NBU concept. They are the NBU-t_0 (Hollander et al. (1985)), NBU*-t_0 (Li and Li (1998)) and SNBU (Stochastically New Better than Used) introduced by Singh and Deshpande (1985).

1.8.2. New better than used in convex order. Cao and Wang (1991) proposed this new class of life distributions called New Better (Worse) than Used in Convex order, NBUC (NWUC). The survival function of this class satisfies

$$\int_x^\infty \bar{F}_y(t)dy \le (\ge) \int_x^\infty \bar{F}(t)dt.$$

Some interesting features posed by the NBUC concept are

    (i) NBU (NWU) $\Rightarrow$ NBUC (NWUC)

    (ii) $(r+2)!(s+1)!E(X^{r+s+3}) \le (\ge)(r+s+3)!E(X^{r+2})E(X^{s+1})$.

Some extensions of NBUC class are

    (a) the NBUCA (NWUCA) - new better (worse) than used in convex order average satisfying

$$\int_0^\infty \int_x^\infty \bar{F}(u+t)du\,dx \le (\ge)\bar{F}(t)\int_0^\infty \int_x^\infty \bar{F}(u)du\,dx$$

       for $t \ge 0$.

    (b) the NBU(2)-t_0 and NBUC-t_0 classes defined in Elabatal (2007), and

    (c) the NBUL (NWUL) concepts based on Laplace transforms

$$\int_0^\infty e^{-sx}\bar{F}(t+x)dx \le (\ge) \int_0^\infty e^{-sx}\bar{F}(x)dx$$

       studied by Yue and Cao (2001). They found that NBU $\Rightarrow$ NBU(2) $\Rightarrow$ NBUL and also that if X and Y have survival functions $\bar{F}$ and $e^{-\lambda x}$ and $W = \min(X, Y)$ then X is NBUL (NWUL) if and only if W is NBUE (NWUE).

1.8.3. New better than used in expectation.

DEFINITION 1.10. *$\mu = E(X) < \infty$, X is said to be new better (worse) than used in expectation if and only if $m(x) \ge (\le) m(0)$. Equivalently, X is NBUE (NWUE) if*

$$\mu \ge (\le) \int_0^\infty \frac{\bar{F}(x+t)}{\bar{F}(t)}dx = m(x)$$

for all $t > 0$.

    The positive ageing notion arises from the fact that a device of any age has a smaller mean residual life than a new one with the same life distribution. We have

    (a) NBU (NWU) $\Rightarrow$ NBUC (NWUC) $\Rightarrow$ NBUE (NWUE)

    (b) NBU (2) (NWU (2)) $\Rightarrow$ NBUE (NWUE)

    (c) DMRL (IMRL) $\Rightarrow$ NBUE (NWUE)

    (d) $\int_x^\infty \bar{F}(t)dt \le \mu e^{-\frac{x}{\mu}}$ when X is NBUE

(e) When X is NBUE,

$$\bar{F}(x) \geq 1 - \frac{x}{\mu}, \ x \leq \mu,$$

$$F(x) \geq \frac{\sigma^2 + \mu^2 - x^2}{\sigma^2 + (\mu + x)^2 - x^2}, \ x \leq \mu_2^{1/2}, \sigma^2 = V(X),$$

$$\frac{\mu_{r+1}}{\gamma(r+2)} \leq \frac{\mu_r}{\Gamma(r+1)}\mu.$$

1.8.4. Harmonically new better than used. The harmonically new better (worse) than used in expectation HNBUE (HNWUE) distributions defined by Rolski (1975) consists of distributions for which

$$\int_x^{\infty} \bar{F}(t)dt \leq (\geq)\mu \bar{e}^{\frac{x}{\mu}}, \ x \geq 0.$$

The concept says that the harmonic mean residual life of a unit of age x is not greater than the harmonic mean life of a new unit. The HNBUE distributions satisfy

    (i) NBUE $\Rightarrow$ HNBUE

    (ii) A necessary and sufficient condition that X is HNBUE (HNWUE) is that $E\phi(X) \leq E(\phi(Y))$, for all non-decreasing convex functions ϕ on $(0, G)$ with $\phi(0) = 0$ and Y is exponentially distributed with the same mean as X

    (iii) $\mu^{r+3} \geq \frac{\mu_{r+3}}{(r+3)!}$.

The HNBUE class can be further weakened if we consider stochastic dominance of order three by defining the HNBUE (3) (HNWUE (3)) classes as those which satisfy

$$\int_0^{\infty} \int_t^{\infty} \bar{F}(u)dudt \leq (\geq)\mu^2 e^{-\frac{x}{\mu}},$$

for all $x, t \geq 0$. Evidently HNBUE (HNWUE) $\Rightarrow$ HNBUE (HNWUE (3)). Two more generalizations of the HNBUE class are the $\mathcal{L}$ and $\mathcal{M}$ classes. For definitions, properties and applications of these classes we refer to Klefsjö (1983); Bhattacharjee and Sengupta (1996); Klar and Muller (2003).

1.9. Concepts in reversed time

The similarity between the definitions of the basic reliability functions and their counterparts gives scope for developing classes of life distributions by means of reversed hazard rates and mean inactivity time, among others. However, such notions do not have behaviour like the usual reliability functions. For example an increasing reversed hazard or a decreasing mean inactivity time does not exist on the positive real axis. Hence lifetime random variables with $(0, \infty)$ as support, cannot permit monotonically decreasing and increasing classes. Moreover, there is difficulty in interpreting such classes in terms of positive or negative ageing of devices. However, the classes have other properties that makes there study worthy in the context of reliability.

1.9.1. Decreasing reversed hazard rate.

DEFINITION 1.11. *A lifetime random variable is decreasing reversed hazard rate (DRHR) if $\lambda(x) = \frac{f(x)}{F(x)}$ is decreasing for all x.*

The following theorem summarizes the properties of DRHR distributions.

THEOREM 1.1. *The following statements are equivalent*
  (i) *X is DRHR*
  (ii) *$\frac{F(x+t)}{F(x)}$ is decreasing for all $x \geq 0$*
  (iii) *$F(x)$ is log concave*
  (iv) *$r_Q(u)$ is decreasing*

Further properties are (i) if X is DHR then X is DRHR (ii) if a sequence of DRHR distributions converges to a limiting distribution, that distribution is DRHR (iii) if F is concave, then $\log F$ is concave and if F has a density then X is DRHR.

1.9.2. Increasing mean inactivity time.

DEFINITION 1.12. *A life distribution has increasing mean inactivity time (IMIT) if $r(x)$ is increasing in x. We also have DRHR $\Rightarrow$ IMIT. The converse need not be true. However if $E(X|X < x)$ is concave then IMIT $\Rightarrow$ DRHR. A similar definition holds for variance inactivity time. We say that X has increasing variances inactivity time (IVIT) if $v(x)$ is decreasing in x. Notice that IMIT $\Rightarrow$ IVIT.*

1.10. Order statistics

The order statistics of a random sample of size n where, $X_1, X_2, \ldots, X_n$ are the sample values placed in ascending order and are denoted by $X_{1:n}, \ldots, X_{n:n}$. Then $X_{1:n} \leq X_{2:n} \ldots X_{n:n}$ and $X_{1:n} = \min_{1 \leq i \leq n} X_i$ and $X_{n:n} = \max_{1 \leq i \leq n} X_i$. In the reliability setting, n units are put on test and the quantities of interest are their failure times $X_1, \ldots, X_n$ which constitutes a random sample from a population of units with distribution function $F(x)$. Since $X_{r:n}, r = 1, 2, \ldots, n$ are random variables, there is an interest in their distributions.

THEOREM 1.2. *Let* $X_{1:n}, \ldots, X_{n:n}$ *denote the order statistics in a random sample from a continuous population with distribution function* $F(x)$ *and probability density function* $f_X(x)$. *Then the density function of* $X_{r:n}$ *is*

$$f_{X_{r:n}}(x) = \frac{n!}{(r-1)!(n-r)!} f_X(x)[F_X(x)]^{r-1}[1 - F(x)]^{n-r} \quad (1.36)$$

and the joint density function of $X_{(r)}$ *and* $X_{(s)}$ *is*

$$f_{X_{r:n}, X_{s:n}}(x, y) = \frac{n!}{(r-1)!(s-r-1)!(n-s)!} f_X(x) f_X(y)[F_X(x)]^{r-1}$$
$$[F_X(y) - F_X(x)]^{s-r-1}[1 - F_X(y)]^{n-r}. \quad (1.37)$$

EXAMPLE 1.2. *Assume* X *is uniform in* $(0, 1)$ *so that* $F_X(x) = x$, $f_X(x) = 1$. *Then*

$$f_{X_{r:n}}(x) = \frac{n!}{(r-1)!(n-r)!} x^{r-1}(1-x)^{n-r}, \ 0 < x < 1.$$

Thus $X_{r:n}$ *has a beta distribution with parameters* $(r, n-r+1)$.

There are several topics in reliability analysis in which order statistics appear quite naturally. One such instance is system reliability where we consider a system consisting of n components with lifetimes $X_1, \ldots, X_n$ that are independent and identically distributed. This system is said to have a series structure if the system functions only when all the components are functioning. Then the lifetime of the system is the smallest among $X_1, \ldots, X_n$ or $X_{1:n}$. From (1.36) the reliability of the system is

$$P(X_{1:n} > x) = 1 - (1 - F(x))^n = 1 - \bar{F}^n(x). \quad (1.38)$$

Another system is called parallel system which functions if and only if at least one of the components is such that such a system has lifetime $X_{n:n}$. Again (1.36)

$$F_{X_{n:n}}(x) = P(X_{n:n} \leq x) = F^n(x). \quad (1.39)$$

and the reliability function is

$$\bar{F}_{X_{n:n}}(x) = 1 - F_{X_{n:n}} = 1 - F^n(x). \quad (1.40)$$

The above two systems are embedded in a more general k-out-of-n system which functions if and only if at least k of the n components function. Obviously such a system's lifetime is $X_{n-k+1:n}$ and its distribution follows from (1.36). In life testing experiments, it is the case that the experiments need not wait till all units put to test have failed. We may observe a prefixed number of failures, say r units and the experiment is terminated as soon as the rth unit fails. This is called type II censoring and that data is the observations $(X_{1:n}, \ldots, X_{r:n})$. Another sampling strategy is to fix a time T and observe the failures upto time T, which is called type I censoring. The number of failures observed is obviously random. One can also fix the time of termination as $\min(T, X_{r:n})$, wherein the experiment is ended as soon as r failures occurs or at time T which ever occurs first.

In the quantile formulation of order statistics one looks at the distribution function of $X_{r:n}$,

$$F_{X_{r:n}}(x) = \sum_{k=r}^{n} \binom{n}{k} F_X^k(x)(1 - F_X(x))^{n-k}$$

and the relationship

$$\sum_{k=r}^{n} \binom{n}{k} p^k (1-p)^{n-k} = I_p(r, n-r+1)$$

where

$$I_x(m, n) = \frac{\int_0^x t^{m-1}(1-t)^{n-1}}{\int_0^1 t^{m-1}(1-t)^{n-1}},$$

the incomplete beta function ratio. Denoting the quantile function of $X_{r:n}$ by $Q_r(u_r)$, we find

$$Q_r(u_r) = Q_X[I_{u_r}^{-1}(r, n-r+1)] \qquad (1.41)$$

where $u_r = I_u(r, n-r+1)$, so that I^{-1} is the inverse of the beta function ratio mentioned above. In particular for the order statistics $X_{n:n}$ and $X_{1:n}$,

$$Q_n(u_n) = Q_X(u_n^{1/n})$$

and

$$Q_1(u_1) = Q_X(1 - (1 - u_1)^{1/n}).$$

Equation (1.41) is a special feature of the quantile function, where the quantile of the rth order statistic has an explicit functional form independent of $F(x)$ which is not the case when we use the distribution function for the purpose, see (1.36). Another immediate advantage of the representation (1.41) is that the moments of order statistics can be computed as

$$\mu_{r:n} = E(X_{r:n}) = \frac{n!}{(r-1)!(n-r)!} \int_0^1 u^{r-1}(1-u)^{n-r} Q(u) du. \qquad (1.42)$$

1.11. Log concave and convex distributions

The geometric concepts of convexity and concavity and their logarithmic versions play a significant role in information theory and reliability analysis. In this section we give a brief account of some important results in this connection. Two classical works that treat this topic are Rockafeller (1970) and Roberts and Varberg (1973).

DEFINITION 1.13. *A function $a(x)$ of a real variable x is said to be convex (concave) in a closed interval $[a, b]$ if for two points x_1, x_2 in this interval*

$$a(\alpha x_1 + (1 - \alpha)x_2) \leq (\geq)\alpha a(x_1) + (1 - \alpha)a(x_2), \; 0 \leq \alpha \leq 1 \quad (1.43)$$

A continuous function is convex (concave) in $[a, b]$ if its graph is always below (above) or on the line joining any two points in it. A linear function is both convex and concave. If the inequality in (1.35) is strict, that is $< (>)$, then $a(x)$ is strictly convex (concave). Further if $a(x)$ is differentiable then it is convex (concave) if $a''(x) \geq (\leq)0$ or $a'(x)$ is increasing (decreasing) in $[a, b]$. Some properties of convex functions are

- (i) *If $a(x)$ id convex (concave) the $ca(x)$ is also convex (concave) for $c > 0$ and $-ca(x)$ is concave (convex) when $c > 0$.*
- (ii) *For convex (concave) functions $a_1(x), a_2(x), \ldots, a_n(x)$, $\sum c_i a_i(x)$ is also convex (concave) where $c_i \geq 0$ and atleast one $c_i > 0$.*
- (iii) *If $a(x)$ is convex (concave) and $g(x)$ is linear then $a(x) + g(x)$ is also convex (concave).*
- (iv) *If $a(x)$ is convex (concave) and $b(x)$ is increasing and convex (increasing and concave) then $b(a(x))$ is convex (concave) and if $a(r)$ is convex (concave) and $b(x)$ is decreasing and concave (decreasing and convex) then $b(a(x))$ is concave (convex).*
- (v) *If $a(x)$ is strictly increasing convex function in (a, b) then $Q^{-1}(x)$ is strictly increasing and concave.*

DEFINITION 1.14. *The function $a(x)$ is said to be log convex (log concave) on $[a, b]$ if $\log a(x)$ is convex (concave) on $[a, b]$. Log concavity of $a(x)$ is equivalent to (i) $\frac{a'(x)}{a(x)}$ is decreasing on $[a, b]$ or (ii) $(\log a(x))'' \leq 0$.*

THEOREM 1.3.   (i) *Let $a(x)$ be strictly monotone on $(c, d]$ and either $a(c) = 0$ or $a(d) = 0$. Then if $a'(x)$ is log concave then $a(x)$ is also log concave.*
- (ii) *If the density function $f(x)$ of X is log concave then $F(x)$ is also log concave. Further if $F(x)$ is log concave so is $G(x) = \int_a^x F(t)dt$. Thus if f is decreasing then F and G are both log concave.*
- (iii) *If f is log concave then $\bar{F}$ is also log concave and $H(x) = \int_x^b \bar{F}(t)dt$ is also log concave.*

The impact of (ii) and (iii) in reliability analysis is revealing

- (a) If f is log concave then the hazard rate $h(x)$ is increasing or X is IFR
- (b) The mean inactivity time $r(x)$ is increasing *i.e.*, X is IMIT if and only if $G(x)$ is log concave

(c) Either of the following conditions is sufficient for the mean residual life time $m(x)$ to be decreasing or X is DMRL (i) $f(x)$ is log concave (ii) $h(x)$ is increasing or X is IFR.

It is to be noted further that log concave densities are unimodal, that is they are non-decreasing upto some point x_0 and decreasing thereafter. If f is log concave it is seen that both F and $\bar{F}$ are log concave, but if f is log convex then $\bar{F}$ is log convex. In the case of log convex densities on $[0, \infty)$, the hazard rate is decreasing on $[0, \infty]$. Consequently the density is decreasing on $[0, \infty)$ and it cannot be decreasing on the whole real line since $\int_{-\infty}^{\infty} f(x)dx = 1$ and hence a density cannot be log concave on $(-\infty, \infty)$. The log concavity or otherwise for some distributions in common use are presented in Table 1.5 for easy reference.

Two other categories of functions needed in the sequel are star-shaped and super additive functions.

DEFINITION 1.15. *A real valued function $a(x)$ defined on $[0, \infty)$ is star shaped if $a(\alpha x) \leq \alpha a(x)$ for $0 \leq \alpha \leq 1$ and all $x \geq 0$.*

It is easier to determine whether a given function is star shaped by verifying whether it satisfies $a(0) \leq 0$ and $\frac{a(x)}{x}$ is increasing for all $x > 0$. Another equivalent condition is $a(0) \leq 0$ and $a(x)$ is convex.

DEFINITION 1.16. *A real valued function $a(x)$ is said to be super (sub) additive if*
$$a(x + y) \geq (\leq)a(x) + a(y).$$
The implication among these categories of functions is $a(0) \leq 0$ and $a(x)$ is convex $\Rightarrow a(x)$ is star shaped $\Rightarrow a(x)$ is super additive.

1.12. Weighted distributions

The origin of weighted distributions can be traced back to Fisher (1934), but it was popularized by Rao (1965). When observations are recorded they will not have the same distribution unless each observation has the same probability of being included in the set. Weighted distributions arise in this context. Let X be the original random variable with probability density (distribution, survival) function $f(x)$ $(F(x), \bar{F}(x))$ then the weighted distribution of X with weight function $w(x)$ has density

$$f_w(x) = \frac{w(x)f(x)}{Ew(X)}, \quad Ew(X) < \infty \tag{1.44}$$

Table 1.5. Log concavity of distributions.

Distribution	$f(x)$	$F(x)$	$\int_a^x F(t)dt$	$\bar{F}(x)$	$\int_x^b \bar{F}(t)dt$
half-normal	log concave	log concave	log concave	log concave	log concave
half-logistic	log concave	log concave	log concave	log concave	log convex
exponential $f(x)=\lambda e^{-\lambda x}$	log concave	log concave	log concave	log concave	log concave
Weibull $f(x)=cx^{c-1}e^{-x^c}$	log concave $c\geq1$	log concave	log concave	log concave $c\geq1$	log concave $c\geq1$
	log convex $c<1$			log convex $c<1$	? $c<1$
power $f(x)=\beta x^{\beta-1}$	log concave $\beta\geq1$	logconcave	logconcave	?	?
	log convex $\beta<1$				
gamma $f(x)=\frac{\theta e^{-\theta x}x^{m-1}}{\Gamma(m)}$	log concave $m\geq1$	log concave	log concave	log concave $m\geq1$	log concave $m\geq1$
	log convex $m<1$			log convex $m<1$	?
beta $f(x)=\frac{x^{a-1}(1-x)^{b-1}}{B(a,b)}$	log concave $a,b\geq1$	log concave	log concave	log concave	log concave
	log convex $a,b=\frac{1}{2}$	mixed	?	mixed	?
	mixed $a=2,b=\frac{1}{2}$	mixed	mixed	log convex	log convex
Pareto $f(x)=\beta x^{-\beta-1}$	log convex	log concave	log concave	log convex	log convex
log normal $f(x)=\frac{1}{x\sqrt{2\pi}}e^{-\frac{(\log x)^2}{2}}$	mixed	logconcave	log concave	mixed	mixed
uniform $f(x)=1$	log concave	log concave	log concave	log concave	log concave

1. ?– indicates neither log concave or log convex for specific values.

2. For the beta distribution f is log convex for $a<1$, $b<1$. For $a>1$, $b<1$, $a<1$, $b>1$ neither log convex or log concave.

and the random variable corresponding to f_w is denoted by X_w. The distribution and survival function of X_w are

$$F_w(x) = \frac{\int_{-\infty}^{x} w(x)f(x)dx}{Ew(X)} = \frac{E[w(X)|X \leq x]}{Ew(X)}F(x) \qquad (1.45)$$

and

$$\bar{F}_w(x) = \frac{\int_{x}^{\infty} w(x)f(x)}{Ew(X)}dx = \frac{E(w(X)|X > x)}{Ew(X)}\bar{F}(x). \qquad (1.46)$$

From (1.44) and (1.45), the hazard rate of X_w is

$$h_w(x) = \frac{w(x)}{E(w(X)|X > x)}h_X(x) \qquad (1.47)$$

and hence the mean residual lives of X and X_w satisfy

$$\frac{1 + m'_w(x)}{m_w(x)} = \frac{w(x)}{m(x)}\frac{(1 + m'(x)}{E(w(X)|X > x)}.$$

The measures in reversed time are the reversed hazard rate

$$\lambda_w(x) = \frac{w(x)}{E(w(X)|X \leq x)}\lambda(x) \qquad (1.48)$$

and the mean activity time

$$\frac{1 - r'_w(x)}{r_w(x)} = \frac{w(x)}{r(x)}\frac{(1 - r'(x))}{E(w(X)|X \leq x)}.$$

For details, see Gupta and Kirmani (1990).

Two important special cases of weighted distributions are the length-biased model and the equilibrium distribution.

DEFINITION 1.17. *A random variable X_L has a length-biased distribution corresponding to X, if it has a density function*

$$f_L(x) = \frac{xf(x)}{\mu}, \ \mu = E(X) < \infty.$$

Thus it is a subclass of (1.43) when $w(x) = x$. This arises in reliability through a sampling mechanism that selects units with a probability proportional to the lengths of lifetimes of the units put to test. When $w(x) = x^n$, in (1.43) we have length-biased distributions of order n.

DEFINITION 1.18. *The equilibrium distribution of a random variable X with $\mu = E(X) < \infty$, is defined as*

$$f_E(x) = \frac{\bar{F}(x)}{\mu}. \qquad (1.49)$$

The random variable that correspond to $f_E(x)$ will be denoted by X_E.

There are several interpretations to (1.49). It is a weighed distribution with weight $w(x) = \frac{1}{h(x)}$. Another way it appears is from renewal theory. Assume that n units are available for testing and we start with working with one of them at time zero, replace it with another upon failure and so on. If the failure times are X_i, $i = 1, 2, \ldots, n$ which are independent and identically distributed then the sequence $\langle S_n \rangle$, $S_n = X_1 + \cdots + X_n$ constitute a renewal process. Let $F(x)$ be the common distribution function of X_i with $F(0) = 0$. Upon denoting the age and residual life of the unit in use at time T by U_T and V_T, the asymptotic distributions of both U_T and V_T turn out to be (1.49). Since f_E is a proper distribution we can talk about its equilibrium distribution which is called the second order equilibrium model. Thus by iteration one can arrive at the nth order equilibrium distribution with survival function (Nair and Preeth (2009))

$$\bar{F}(x) = \mu_{n-1}^{-1} \int_x^\infty \bar{F}_{n-1}(t)dt, \qquad (1.50)$$

where $\mu_n = \int_0^\infty \bar{F}_n(t)dt$. See Nair and Preeth (2009) for further details of (1.50).

The survival function of (1.49) is

$$\bar{F}_E(x) = \frac{\int_x^\infty \bar{F}(t)dt}{\mu}$$

and hence the hazard rate of X_E is

$$h_E(x) = \frac{\bar{F}(x)}{\int_x^\infty \bar{F}(t)dt} = \frac{1}{m_X(x)}. \qquad (1.51)$$

Arising from the above

$$m_X(x) = \frac{m_E(x)}{1 + m'_E(x)},$$
$$h_X(x) = h_E(x) - \frac{h'_E(x)}{h_E(x)}.$$

When quantile functions are used

$$h_{E,Q}(u) = (m_Q(u))^{-1}.$$

We also have

(i) X_E is IHR (DHR) whenever X is IHR (DHR)
(ii) X is DMRL (IMRL) if and only if X_E is IHR (DHR)
(iii) X is DMRLHA $\Leftrightarrow$ X_E is IFRA
(iv) X is UBAE $\Leftrightarrow$ X_E is UBA
(v) If $\bar{F}(x)$ is strictly decreasing then X is DVRL (IVRL) if and only if X_E is DMRL (IMRL)

(vi) X_E is DVRL (IVRL) $\Rightarrow X_E$ in UBA $\Rightarrow X_E$ is UBAE

(vii) X_E in NBU $\Rightarrow X$ is NBUC

More generally if (1) X is IFR (DFR) and $w(x)$ is increasing and concave (decreasing and convex) then X_w is IFR (DFR) (2) X is DRHR and $w(x)$ is decreasing and log concave then X_w is DRHR (3) X is IMIT $\Rightarrow X_w$ is IMIT if $E(w(X)|X < x)$is decreasing and log convex (4) X is IMIT $\Rightarrow X_w$ is DRHR if $w(x)$is decreasing and logconcave and $r(x)$ is log convex.

1.13. Stochastic orders

There are many real situations in which we need to compare two life distributions by means of some reliability characteristics. Stochastic orders are important tools in this context. For a given property A, a stochastic order says that the distribution F_X of a random variable X has lesser or greater A than the distribution F_Y of another random variable Y. We express this as $X \leq_A Y$ ($X \geq_A Y$) or equivalently $F_X \leq_A F_Y$ ($F_X \geq_A F_Y$). In reliability theory one has to compare the reliabilities of devices used for the same purpose, but with different manufacturing processes that are governed by different lifetime distributions. If the entire lifetime is available it may be their mean lives chosen as the objects of comparison and if the comparison is sought when the devices were working for a specific time, it could well be their mean residual lives. In this section we discuss certain stochastic orders used in reliability, their properties and some applications.

1.13.1. Usual stochastic order. This is the basic stochastic order that compares the distribution functions of X and Y.

DEFINITION 1.19. *If X and Y are two random variables with distribution functions F_X and F_Y, then X is smaller than Y in the usual stochastic order denoted by $X \leq_{st} Y$ if $F_X(x) \geq F_Y(x)$ for all x.*

It may be noted that

(a) $X \leq_{st} (\geq_{st})Y \Leftrightarrow \bar{F}_X(x) \leq \bar{F}_Y(x)$ for all $x \Leftrightarrow Q_X(u) \leq Q_Y(u)$ for all u in $(0, 1)$

(b) If $X \leq_{st} (\geq)Y$ and g is an increasing (decreasing) function then $g(X) \leq_{st} (\geq_{st})g(Y)$

(c) X is IHR (DHR) if and only if $X_t \leq_{st} (\geq_{st})X'_t$, $t < t'$ where $X_t = X - t|X > t$ is the residual life

(d) X is NBU (NWU) if and only if $X \geq_{st} (\leq_{st})X_t$

(e) X is NBUE (NWUE) if and only if $X \geq_{st} (\leq_{st})X_E$

(f) X is DMRL (IMRL) if and only if $X_t \geq_{st} (\leq_{st})X_{E,t}$

1.13.1.1. *Hazard rate order.* The idea behind the hazard rate order is that when the hazard rate becomes larger the lifetime becomes stochastically smaller.

DEFINITION 1.20. *Let X and Y be absolutely continuous non-negative random variables. Then X is said to be smaller than Y in hazard rate order if $h_X(x) \geq h_Y(x)$ for all $x > 0$ and is denoted by $X \leq_{hr} Y$.*

In reliability analysis the hazard rate order has several implications. Some equivalent conditions for $X \leq_{hr} Y$ to hold are given below.

THEOREM 1.4. *Some necessary and sufficient conditions for $X \leq_{hr} Y$ are*

  (i) $u^{-1} F_Y(Q_X(1-u))$ *is decreasing in u*
  (ii) $\frac{\bar{F}_Y(x)}{\bar{F}_X(x)}$ *is increasing in x*
  (iii) $\bar{F}_X(x)\bar{F}_Y(y) \geq \bar{F}_X(y)\bar{F}_Y(x)$ *for all $x \leq y$*
  (iv) $\frac{\bar{F}_X(x+y)}{\bar{F}_X(x)} \leq \frac{\bar{F}_Y(x+y)}{\bar{F}_Y(x)}$ *for all $x, y \geq 0$*
  (v) $(X|X > x) \leq_{st} (Y|Y > x)$.

In some cases stochastic orders are useful in identifying whether X is IFR or DHR and has other ageing properties.

THEOREM 1.5. *The random variable X is IHR (DHR) if and only if one of the following conditions are satisfied*

  (i) $X_t \geq_{hr} (\leq_{hr}) X_s$ *for all $t \leq s$*
  (ii) $X \geq_{hr} (\leq_{hr}) X_t$ *for all $t \geq 0$*
  (iii) $X + t \leq_{hr} X + s, t \leq s$
  (iv) *If X_i, $i = 1, 2, , \ldots n$ are independent IHR variables $\sum_{i=1}^{N} X_i \leq_{hr} \sum_{i=1}^{M} X_i$, where M and N are discrete integer valued random variables.*

THEOREM 1.6. *If $E(X) < \infty$, then*

  (i) X *is DMRL (IMRL) $\Leftrightarrow X \geq_{hr} (\leq_{hr}) X_E$*
  (ii) $X_E \geq_{hr} X_{E,t} \Leftrightarrow X$ *is DMRL.*

There are some interesting results involving order statistics. If $X_1, \ldots, X_n$ and $Y_1, \ldots, Y_n$ are two sets of independent random variables and $X_i \leq_{hr} Y_j$, $i, j = 1, 2, \ldots, n$ then $X_{r:n} \leq_{hr} Y_{r:n}, r = 1, 2, \ldots n$.

1.13.2. Reversed hazard rate order.

DEFINITION 1.21. *If X and Y are two absolutely continuous random variables with reversed hazard rates λ_X and λ_Y, then X is smaller than Y in reversed hazard rate order denoted by $\lambda_X \leq_{rh} \lambda_Y$ if $\lambda_X(x) \leq \lambda_Y(x)$ for all x.*

Some equivalent conditions for the rth order $\leq_{rh}$ are

(1) $\frac{F_Y(x)}{F_X(x)}$ is increasing in x in the interval $(\min(l_x, l_y), \infty)$ where l_X and l_Y are the lower ends of the supports of X and Y respectively
(2) $F_X(x)F_Y(y) \geq F_Y(x)F_X(y)$ for all $x \leq y$
(3) $\frac{F_Y Q_X(u)}{u} \leq \frac{F_Y Q_X(v)}{v}$ for all $0 \leq u < v < 1$ and
(4) $(X|X \leq x) \leq_{st} (Y|Y \leq x)$.

Some other properties are

(1) X is DRHR if an only if either $(x - X|X \leq x) \leq_{st} (y - X|X \leq y)$ or $(x - X|X \leq x) \leq_{st} (y - X|X \leq y)$ for $0 < x < y$ or $X + t \leq_{st} X + y, t \leq y$
(2) if $\phi(x)$ is a continuous strictly decreasing function, then

$$X \leq_{rh} Y \Rightarrow \phi(X) \geq_{hr} \phi(Y)$$
$$X \leq_{hr} Y \Rightarrow \phi(X) \geq_{rh} \phi(Y)$$

(3) if $\phi(x)$ is increasing $X \leq_{rh} Y \Rightarrow \phi(X) \leq_{rh} \phi(Y)$.
(4) if Z is independent of X and Y and DRHR then $X \leq_{rh} Y \Rightarrow X + Z \leq_{rh} Y + Z$
(5) If $X_1, \ldots, X_n$ are independent and $X_i \leq_{rh} X_j$ for $j = 1, 2, \ldots, i - 1$ then

$$X_{r+1:n-1} \leq_{rh} X_{r:n}, \quad r = 2, 3, \ldots, n.$$

1.13.3. Likelihood ratio order. This is a stochastic order that compares the density functions.

DEFINITION 1.22. *If $\frac{f_Y(x)}{f_X(x)}$ increases over the union of supports of X and Y, we say that X is smaller than Y in likelihood ratio order, denoted by $X \leq_{lr} Y$.*

When X and Y have the same support, by differentiation we get

$$X \leq_{lr} Y \Leftrightarrow F_Y(Q_X(u)) \text{ is convex .}$$

The $\leq_{lr}$ order is related to the hazard rate and reversed hazard rate orders as follows.

THEOREM 1.7.   (i) *If $X \leq_{lr} Y$ then $X \leq_{hr} Y$ and $X \leq_{rh} Y$. Conversely, if $X \leq_{hr} Y$ $(X \leq_{rh} Y)$ and $\frac{h_Y(x)}{h_X(x)} \frac{\lambda_Y(x)}{\lambda_X(x)}$ is increasing in x, then $X \leq_{lr} Y$.*
(ii) *$X \leq_{hr} Y \Leftrightarrow X_E \leq_{lr} Y_E$.*
(iii) *X is IFR (DFR) if and only if $X \leq_{lr} (\geq_{lr})X_E$.*

Another interesting result concerns the order statistics. Let $X_1, X_2, \ldots, X_n$ be independent with the same support. If $X_1 \leq_{\mathrm{lr}} X_2 \leq_{\mathrm{lr}} \cdots \leq_{\mathrm{lr}} X_n$, then

$$X_{r-1;n} \leq_{\mathrm{lr}} X_{r:n}, \ 2 \leq r \leq n$$

$$X_{r-1:n-1} \leq_{\mathrm{lr}} X_{r:n}, \ 2 \leq r \leq n$$

and if $X_1 \geq_{\mathrm{lr}} X_2 \geq_{\mathrm{lr}} \cdots \geq_{\mathrm{lr}} X_n$, then

$$X_{r:n} \leq_{\mathrm{lr}} X_{r:n-1}, \ r = 1, 2, \ldots n - 1.$$

1.13.4. Mean residual life order.

DEFINITION 1.23. *X is said to be smaller than Y if mean residual life order, $X \leq_{mrl} Y$ if $m_X(x) \leq m_Y(x)$ for all x.*

Some properties of the order $\leq_{\mathrm{mlr}}$ are given below

a) $X \leq_{\mathrm{mrl}} Y \Leftrightarrow \frac{\int_x^\infty \bar{F}_Y(t)dt}{\int_x^\infty \bar{F}_X(t)dt}$ increases in it for values of x for which the denominator is positive

b) $X \leq_{\mathrm{mrl}} Y$ then $X \leq_{\mathrm{mrl}} Y$. Conversely if $X \leq_{\mathrm{mrl}} Y$ and $\frac{m_X(x)}{m_Y(x)}$ increases in x, then $X \leq_{\mathrm{hr}} Y$.

c) For every convex function ϕ, $X \leq_{\mathrm{mrl}} Y \Rightarrow \phi(X) \leq_{\mathrm{mrl}} \phi(Y)$

d) $X \leq_{\mathrm{mrl}} Y \Leftrightarrow X_E \leq_{\mathrm{hr}} Y_E$

e) If $X_1, X_2, \ldots, X_n$ are independent then (i) $X_i \leq_{\mathrm{mrl}} X_n$, $i = 1, 2, \ldots, n - 1$, (ii) $X_{n-1:n-1} \geq_{\mathrm{mrl}} X_{r:n}, r = 1, 2, \ldots, n - 1$

f) X is DMRL if either $X_t \geq_{\mathrm{mr}} X_t'$ for $t' \geq t$ or $X + t \leq_{\mathrm{mrl}} X + t'$

g) $1 - e^{-sX} \leq_{\mathrm{mrl}} 1 - e^{-sY}$ for all $s > 0 \Rightarrow X \leq_{\mathrm{hr}} Y$, provided X and Y are non-negative.

1.13.5. Mean inactivity time order.

DEFINITION 1.24. *The random variable X is said to be smaller than the random variable Y in mean inactivity time (reversed mean residual lifetime) order, $X \leq_{MIT} Y$ if $r_X(x) \geq r_Y(x)$ for all x.*

Notice that $X \leq_{\mathrm{rh}} Y \Rightarrow X \leq_{MIT} Y$ and if $\frac{r_X(x)}{r_Y(x)}$ is increasing in x, $X \leq_{MIT} Y \Rightarrow X \leq_{\mathrm{rh}} Y$. Li and Zuo (2007) have shown that X is IMIT if and only if $X \leq_{MIT} X + Y$, where Y is independent of X. Moreover if ϕ is an increasing concave function with $\phi(0) = 0$ then $X \leq_{MIT} Y \Rightarrow \phi(X) \leq_{MIT} \phi(Y)$. A characterization of thereversed hazard rate order is given in terms of the MIT order as (Ortega (2008))

$$X \leq_{\mathrm{rh}} Y \Leftrightarrow e^{sX} \leq_{MIT} e^{sY} \text{ for all } s > 0$$

and a reverse characterization

$$X \leq_{MIT} Y \Leftrightarrow \log X^{\frac{1}{s}} \leq_{\mathrm{rh}} \log Y^{\frac{1}{s}}.$$

Further for a strictly decreasing positive function

$$X \leq_{\mathrm{MIT}} Y \Rightarrow \phi(X) \geq_{\mathrm{mrl}} \phi(Y).$$

1.13.6. Transform orders. Assume that X and Y are non-negative random variables with interval support. Then we have three inter-related orders that describe the intensity of ageing.

DEFINITION 1.25. *We say that X is smaller than Y in (i) convex transform order, $X \leq_c Y$, if $F_Y^{-1}(F_X(x))$ is convex (ii) star order, $X \leq_* Y$ if $F_Y^{-1}(F(x))$ is star shaped and (iii) super additive order, $X \leq_{su} Y$ if $F_Y^{-1}(F(x))$ is super additive.*

$$X \leq_c Y \Rightarrow X \leq_* Y \Rightarrow X \leq_{su} Y.$$

These orders are used to characterize IFR, IFRA and NBU distributions

$$X \text{ is IFR } \Leftrightarrow X \leq_c Y, \ Y \text{ exponential}$$

$$X \text{ is IFRA } \Leftrightarrow X \leq_* Y, \ Y \text{ exponential}$$

and

$$X \text{ is NBU } \Leftrightarrow X \leq_{su} Y, \ Y \text{ exponential}.$$

X is also said to be more IFR (IFRA, NBU) if $X \leq_c Y$ ($X \leq_* Y$, $X \leq_{su} Y$) respectively and thus give criteria for understanding which one of the variables is ageing more positively.

1.13.7. Quantile function orders. Although the hazard rate $h(x)$ and the hazard quantile function $h_Q(u)$ have the same type of monotonicity, when distributions are compared in terms of $h(x)$ and $h_Q(u)$, they may not show the same pattern. This aspect calls for separate orderings for $h_Q(u)$ as done in Vinesh Kumar et al. (2015).

DEFINITION 1.26. *Let X and Y be two non negative random variables such that $h_{Q_X}(u) \geq h_{Q_Y}(u)$ for all u in $(0,1)$. Then X is smaller than Y in hazard quantile function order and is written as $X \leq_{HQ} Y$.*

The hazard quantile function is mathematically equivalent to the dispersive order defined as $X \leq_{\mathrm{disp}} Y$ given by

$$Q_X(v) - Q_X(u) \leq Q_Y(v) - Q_Y(u), \ 0 < u \leq v < 1.$$

We have

(1) $X \leq_c Y \Rightarrow X \leq_{HQ} Y$
(2) If X and Y have the same lower end of their supports $X \leq_{HQ} Y \Rightarrow X \leq_{st} Y$
(3) Let Z be a random variable with quantile function $Q_X(u) + (1 - \alpha)Q_Y(u), 0 \leq \alpha \leq 1$. Then $X \leq_{HQ} Z \leq_{HQ} Y$

(4) In general $\leq_{hr}$ and $\leq_{HQ}$ are not equivalent. If X or Y is IFR (DFR) then $H \leq_{HQ} Y \Rightarrow (\Leftarrow) X \leq_{hr} Y$

(5) Let $X_i \leq_{HQ} Y_i, i = 1, 2, \ldots, n$. Then $W \leq_{HQ} Z, Q_W = \sum Q_{X_i}(u)$, $Q_Z = \sum Q_{Y_i}(u)$ and $P \leq_{HQ} R, Q_p = \prod Q_{X_i}(u)$ and $Q_R = \prod Q_{Y_i}(u)$.

(6) If ϕ is non-negative and increasing convex function, $X \leq_{HQ} Y \Rightarrow T(X) \leq_{HR} T(Y)$.

DEFINITION 1.27. *We say that X is smaller than Y in mean residual quantile function order, $X \leq_{MQ} Y$, if $m_{Q_X}(u) \leq m_{Q_Y}(u)$ for all u in $(0, 1)$.*

Some properties of the $\leq_{MQ}$ order are

i) $X \leq_{MQ} Y \Leftrightarrow X \leq_{ew} Y$ where $\leq_{ew}$ is the excess wealth order defined by

$$\int_{Q_X(u)}^{\infty} \bar{F}(x)dx \leq \int_{Q_Y(u)}^{\infty} \bar{F}_Y(x)dx.$$

ii) $X \leq_{HQ} Y \Rightarrow X \leq_{MQ} Y$ and conversely if $\frac{m_{Q_X}(u)}{m_{Q_Y}(u)}$ is increasing then $X \leq_{MQ} Y \Rightarrow X \leq_{HQ} Y$.

iii) If $E(X) - E(Y) \leq M_X(u) - M_Y(u)$, then $X \leq_{MQ} Y \Rightarrow X \leq_{ST} Y$.

iv) For the random variable Z defined in (3) above, $X \leq_{MQ} Z \leq_{MQ} Y$.

DEFINITION 1.28. *For two non-negative random variables with reversed hazard quantile functions $\lambda_{Q_X}(u)$ and $\lambda_{Q_Y}(u)$, X is smaller than Y in reversed hazard quantile function order if $\lambda_{Q_X}(u) \leq \lambda_{Q_Y}(u)$ and this fact is denoted as $X \leq_{RHQ} Y$.*

Most of the properties of the $\leq_{RHQ}$ order hails from the important implication $X \leq_{RHQ} (\geq)y \Rightarrow X \leq_{HQ} (\leq_{HQ})Y$. We have mentioned some orderings in this section that are directly linked to reliability problems. There are many other stochastic orders that are either related to or have implications with what we have described. For such details we refer to Shaked and Shanthikumar (2007); Belzunce et al. (2005a).

CHAPTER 2

Residual Entropy

2.1. Introduction

The word 'entropy' appears to have originated from the words 'en' in English which means inside and 'tropē' in Greek with meaning transformation. It is believed that German physicist and mathematician Rudolph Clausius (1882–1888) while discussing the concept of a thermodynamic system, developed his ideas of lost energy and coined the term entropy which was later used to explain the second law of thermodynamics. However the foundations of entropy as it is in the present day information theory was laid down by the Swedish mathematician Harry Nyquist (1889–1976) in connection with his work on asserting the bandwidth required for transmission of information. A quantitative measure of information was proposed by Hartley (1928) in the form $H = \log s^n$. where s is the number of symbols available for selection and n is the number of selections. The definition, interpretation and the properties of entropy as a measure of uncertainty which formed the basis of information theory was formulated by Shannon (1948). Though Shannon used the nomenclature entropy for his measure at the instance of the celebrated mathematician John Van Neumann, it has no connection initially with the entropy concept in thermodynamics, except for the similarity in expression. It was only much later that Shannon's entropy's role in thermodynamics was identified. For a discussion on this topic we refer to Chapter 12 in Kapur (1989). Soon after the properties of entropy were spelt out, mathematicians and engineers were on the look out for postulates that characterize entropy. The postulates given in Shannon (1948) were not enough for such a characterization and as such several modifications were suggested. These include the works of Khinchin (1953); Faddeev (1956); Renyi (1959); Chaundy and McLeod (1960); Aczel and Daroczy (1963); Lee (1964); Pintacuda (1966); Havrda and Charvát (1967); Daroczy (1967); Borges (1967); Forte and Daroczy (1968); Daroczy (1969); Daroczy and Katai (1970); Forte and Ng (1973); Aczel et al. (1974). Almost during the same periods, several attempts were made to generalize the Shannon entropy through parametric forms and non-additive forms to give a broader framework for measuring uncertainty. Some important papers in this regard are Renyi (1961); Aczel and Daroczy (1963); Havrda and Charvát (1967); Kapur (1967, 1983); Belis and Guiasu (1968); Rathie (1970); Behara and Nath (1973); Behara and Chawla (1974); Sharma and Taneja (1975); Sharma and Mittal (1975); Ferreri (1980); Tsallis (1988); Varma (1966); Landsberg and Vedral (1998); Abe (1997); Taneja (2001); Kaniadakis (2002); Mathai and Haubold (2006); Taneja (2011a,b); Kittanch et al. (2016). The expressions of these entropies are given in Table 2.1. Research on the two streams mentioned above have resulted in several properties of Shannon entropy which in turn contributed to a wide variety of applications in various disciplines. Apart from communication theory from where the concept originated, the domain of application is scattered over statistical physics, nuclear reactions, statistics, biology, chemistry, psychology, economics and operation research. A detailed exposition of the influence of Shannon's work in the above areas with references can be found in Verdu (1998). In spite of the wide popularity the measures of uncertainty have received several objections with regard to its capability of accessing information beyond the analysis of information transfer or exchange. Various problems associated with this aspect are discussed in Schroeder (2004). In the next section we present the definition and some important properties of Shannon's entropy as background material for later deliberations of their usefulness in the context of reliability modelling.

2.2. Definition and properties of entropy

Let X be a discrete random variable taking values x_i such that

$$\Pr(X = x_i) = p_i, \ i = 1, 2, \ldots, n, \ p_i \geq 0, \ \sum_{i=1}^{n} p_i = 1. \qquad (2.1)$$

and $\mathcal{A}_n$ be the set of all distributions of the form (2.1). Then the Shannon's entropy of X or $P = (p_1, p_2, \ldots, p_n)$ is defined as

$$H_n(P) = -\sum_{i=1}^{n} p_i \log p_i \qquad (2.2)$$

for $P \in \mathcal{A}_n$ and for all $n = 1, 2, \ldots, n$. It is understood by convention that $0 \log 0 = 0$ and further the summation symbol $\sum$, unless otherwise mentioned is for $i = 1, 2, \ldots, n$. There are several conditions on H_n that characterize (2.2) as pointed out in the introduction and each characterization supplies a set of properties of $H_n(P)$. Some important properties of entropy are given below, (Taneja, 2001).

(1) $H_n(P) \geq 0$ with equality sign holding if and only if
$P = (0, 0, \ldots, 1, 0, \ldots, 0)_z \in \mathcal{A}_n$ (non-negativity)

(2) $H_n(P)$ is a continuous function of P (continuity)

(3) $H_n(p_1, p_2, \ldots, p_n) = H_n(p_{i_1}, p_{i_2}, \ldots, p_{i_n})$, where $i_1, \ldots, i_n$ are permutations of the integers $1, 2, \ldots, n$. (symmetry)

(4) $H_{n+1}(p_1, p_2, \ldots, p_n, 0) = H_n(p_1, p_2, \ldots, p_n)$ (expansible)

(5) $H_2(1, 0) = H_2(0, 1) = 0$ (decisive)

(6) $H_2(\frac{1}{2}, \frac{1}{2}) = 1$ (normality)

(7) $H_n(P) = \sum f(p_i)$, $f(p) = -p \log p$, $0 \leq p_i \leq 1$ (sum representation)

(8) $H_n(P) = H_{n-1}(p_1+p_2, p_3, \ldots, p_n) + (p_1+p_2)H_2\left(\frac{p_1}{p_1+p_2}, \frac{p_2}{p_1+p_2}\right)$,
$p_1 + p_2 > 0$, $n \geq 3$ (recursive)

(9) $H_{mn}(p_1q_1, \ldots, p_1q_m, p_2q_1, \ldots, p_2q_m, p_nq_1, \ldots, p_nq_m) = H_n(P) + H_m(Q)$ where $Q = (q_1, \ldots q_m) \in \mathcal{A}_m$. This means that the entropy of the combination of two independent experiments is the the sum of the entropies of the constituent experiments. (additive)

(10)
$$H_{mn}(p_{11}, \ldots, p_{1m}, \ldots, p_{n1}, \ldots, p_{nm}) = H_m(p_1, \ldots, p_m)$$
$$+ \sum_{j=1}^{m} p_j H_n\left(\frac{p_{1j}}{p_j}, \ldots, \frac{p_{nj}}{p_j}\right), p_{ij} \geq 0,$$

$p_j = \sum_{i=1}^{n} p_{i,j}$, $\sum_i \sum_{j=1}^{m} p_{ij} = 1$ (strongly additive)

(11) If $\psi(p) = H_2(p, (1-p))$, then
  (a) $\psi(p) = \psi(1-p)$,
  (b) $\psi(1) = \psi(0)$,
  (c) $\psi\left(\frac{1}{2}\right) = 1$,
  (d) $\psi(p) + (1-p)\psi\left(\frac{q}{1-p}\right) = \psi(q) + \psi\left(\frac{p}{1-q}\right)$, $p, q \in [0, 1)$, $p + q \leq 1$,
  (e) $\psi(p) \leq K$, for some K, $\psi(p)$ is non-decreasing in $(0, \frac{1}{2}]$,
  (f) $\lim_{q\to 0+} H_2(p_i, q) = h(p_0)$ for all $0 < p_0 < 1$, $\lim_{p\to 0} \psi(p) = 0$, and
  (g) $H_n(P) = \sum_{t=2}^{n}(p_1 + \cdots + p_t)\psi\left(\frac{p_t}{p_1+\cdots+p_t}\right)$ (binary entropic)

(12)
$$H_n(P) = H(p_1 + \cdots + p_r + p_{r+1} + \cdots + p_n)$$
$$+ \sum_{i=1}^{r} p_i\left(\frac{p_1}{\sum_{i=1}^{r} p_i}, \ldots, \frac{p_r}{p_1+\cdots+p_r}\right)$$
$$+ \sum_{i=r+1}^{n} p_i\left(\frac{p_{r+1}}{\sum_{i=r+1}^{n} p_i}, \ldots, \frac{p_n}{p_{r+1}+\cdots+p_n}\right) \quad \text{(grouping)}$$

(13)

$$H(p_1 \ldots p_{r_1}, p_{r_1+1} \ldots p_{r_2}, \ldots p_{r_{n-1}} + 1, \ldots p_{r_n})$$
$$= H(p_1 + \cdots + p_{r_1} + \cdots + p_{r_{n-1}} + 1 + \cdots + p_{r_n})$$
$$+ \sum_{i=1}^{n} (p_{r_{i-1}+1} + \cdots + p_{r_i}) H \left(\frac{p_{r_{i-1}} + 1}{\sum_{j=r_{i-1}+1}^{r_i} p_j} \cdots, \frac{p_{r_i}}{p_{r_{i-1}+1} + \cdots + p_{r_i}} \right).$$

(generalized grouping)

(14) $H_n(P) \leq H_n(\frac{1}{n}, \frac{1}{n}, \ldots, \frac{1}{n})$ (generalized grouping) with equality if and only if $p_i = \frac{1}{n}$.

(15) For the uniform distribution, $H_n(U) = H(\frac{1}{n}, \frac{1}{n}, \ldots, \frac{1}{n})$, $n \geq 2$ we have

(a) $H_2(\frac{n-n_1}{n}, \frac{n_1}{n}) = -\frac{n_1}{n}[H_{n_1}(U) - H_n(U)] - \frac{n-n_1}{n}[H_{n-n_1}(U) - H_n(U)]$, $1 \leq n_1 \leq n$,

(b) $H_n(U) \leq H_{n+1}(U)$; $nH_n(u) \leq (n+1)H_{n+1}(U)$ and $\lim_{n\to\infty}[H_{n+1}(U) - \frac{n+1}{n}H_n(U)] = 0$.

(16) If $P = (p_1, p_2, \ldots, p_n)$ and $Q = (q_1, q_2, \ldots, q_n) \in \mathcal{A}_n$

$$\sum p_i \log \frac{p_i}{q_i} \geq 0 \quad \text{(Shannon's inequality)}$$

(17) $H_n(P)$ is a concave function of P in $\mathcal{A}_n$

(i) $H_{nm}(a_{11}, \ldots, a_{1m}, \ldots, a_m, \ldots, a_{nm}) \leq H_m(\sum a_{i1}, \ldots, \sum a_{im})$
$+ H_n(\sum_{j=1}^{n} a_{ij}, \ldots, \sum_{j=1}^{m} a_{nj})$, (sub additive)
$P \in \mathcal{A}_n, Q \in \mathcal{A}_n$

(a) $H_{nm}(a_{11}, \ldots, a_{nm}) \leq H_{nm}(p_1 q_1, \ldots, p_1 q_m, \ldots, p_n q_1, \ldots, p_n q_m)$,
where $p_i = \sum_{i=1}^{m} a_{ij}$ and $q_j = \sum_{j=1}^{m} a_{ij}$. (independence)

(18) The function $g(p) = H_2(p, 1-p) = -p \log p - (1-p) \log(1-p)$, $0 \leq p \leq 1$, called Shannon's entropy function satisfies the functional equation

$$g(x) + (1-x)g\left(\frac{y}{1-x}\right) = g(y) + (1-y)g\left(\frac{x}{1-y}\right),$$

$0 \leq x, y < 1, 0 \leq x + y \leq 1, g(0) = g(1)$ and $g(\frac{1}{2})$.

(19) If

$$\sum_{i=1}^{n} |p_i - q_i| \leq \theta \leq \frac{1}{2}$$

then

$$|H_n(P) - H_n(Q)| \leq -\theta \log \frac{\theta}{n}.$$

REMARK 2.1. *All the above properties are not independent of one another. For example, (10) implies (9), (4) and (10) together imply (8) and more. See Taneja (2001) for details.*

2.3. Differential entropy

So far we have considered the entropy of a probability distribution in which the random variable X takes a finite number of values. In this section we discuss the cases when X assumes a countable set of values or when X is a continuous random variable. In general when P is a probability measure defined on a measurable space $(\Omega, \mathbb{B})$, the entropy of P is defined as

$$H(p) = -\int_{\Omega} p(x) \log p(x) d\lambda(x) \tag{2.3}$$

where λ is a σ-finite measure such that P is absolutely continuous with respect to λ and $p(x) = \frac{dP}{d\lambda}$ is the Radon-Nikodym derivative of P with respect to λ. Accordingly, when X has an absolutely continuous distribution function $F(x)$ with probability density function $f(x)$, the entropy is defined as

$$H_F = -\int_{-\infty}^{\infty} f(x) \log f(x) dx. \tag{2.4}$$

It may be observed that (2.4) satisfies many of the properties of $H_n(P)$ in (2.2), but there are some subtle differences too. While (2.2) measures the uncertainty due to the probabilistic nature of the concerned phenomenon in a well directed manner, the measure (2.4) is relative to the underlying coordinate system. The p_i's occurring in $H_n(p)$ lie in $[0, 1]$ where as the density occurring in the continuous case can exceed unity and as a result H_F can turn out to be negative. An example is the uniform distribution in $[0, \frac{1}{3}]$ in which $H_F = -3 \log 3$. Some other properties of H_F are

A. If $f(x)$ and $g(x)$ are probability density functions with support (a, b),

$$\int_a^b f(x) \log \left(\frac{f(x)}{g(x)} \right) dx \geq 0$$

with equality holding good if and only if $f(x) = g(x)$ almost every where in (a, b). Also

$$\int_a^b (c + df(x)) \log(c + df(x)) dx \geq \int_a^b (c + df(x)) \log(c + dg(x)) dx$$

for real constants c and d.

B. Let $Y = \phi(X)$. Then the entropies of X and Y satisfy

$$H_G = H_F - \int g(y) \log |J| dy,$$

where $G(\cdot)$ is the distribution of Y and $|J| = |\frac{dx}{d\phi(x)}|$, the Jacobian of the transformation. In particular, when $G = a + F$,

$$H_G = H_F$$

Table 2.1. Entropy of quantile functions.

Distribution	H_Q
Power-Pareto	$(\lambda_1 + \lambda_2 - 2) + (\lambda_1 - \lambda_2)^{-1}(\lambda_2 \log \lambda_2 - \lambda_1 \log \lambda_1)$ $- \log C$
Govindarajulu	$\beta - 2 - \log \sigma$
Nair et al.	$-A - \log K$
linear mean residual life	$1 + (2b)^{-1}[(a + b) \log(a + b) - (a - b) \log(a - b)], a > b$
lambda	$\log 2 - (\lambda_2 - 1) + B(\frac{\lambda_2 - 2}{\lambda_2 - 1}, \frac{1}{\lambda_2 - 1}) - B\frac{1}{2}(\frac{\lambda_2 - 2}{\lambda_2 - 1}, \frac{1}{\lambda_2 - 1}), \lambda_2 > 2$ $B_x(p, q) = \int_0^x t^{p-1}(1 - t)^{q-1}dt$
uniform-exponential	$\lambda^{-1}[(1 - \lambda)(\log(1 - \lambda) - 1) - (1 + \lambda)(\log(1 + \lambda) - 1) - 2]$

and when $G = aF$,

$$H_G = H_F + \log |a|.$$

The expressions for the entropy of continuous distributions can be seen in many publications, for example, Verdugo and Rathe (1978) or Kapur (1989).

When quantile functions are considered instead of distribution functions the transformation $x = Q(u)$ in (2.4) leads to the entropy

$$H_Q = \int_0^1 \log q(p)dp, \tag{2.5}$$

where $q(p) = \frac{dQ(p)}{dp}$ is the quantile density function of X, (see Section 1.3).

The expressions for H_Q and H_F are identical and therefore can be obtained from H_F in the references given above. In all cases $Q(u)$ is invertible into a closed form F. However there are cases where $Q(u)$ cannot have an analytically tractable expression for F and hence H_F is difficult to find, but H_Q has simple forms. In Table 2.4 we provide such examples. The form of the quantile functions are the same as in Table 1.4 and is therefore not given.

From the properties of quantile functions mentioned in Chapter 1, we have the following results.

(a) Since $Q_{aX+b}(u) = aQ_X(u) + b$, the entropy of $Y = aX + b, a > 0$ can be written in terms of the entropy of X as

$$H_Y = H_X - \log a$$

(b) If $Q(u) = Q_1(u) + Q_2(u)$, where Q_1 and Q_2 are quantile functions,

$$H_Q = -\int_0^1 \log(q_1(p) + q_2(p))dp.$$

For example, if $Q_1(u) = u$ and $Q_2(u) = -\frac{1}{\lambda}(\log(1-u))$ represent the uniform and exponential distributions respectively

$$H_Q = -\int_0^1 \log \frac{1+\lambda-\lambda p}{\lambda - \lambda p} dp$$

$$= \lambda^{-1}[(1-\lambda)(\log(1-\lambda) - 1) - (1+\lambda)(\log(1+\lambda) - 1) - 2].$$

(c) If $Q(u) = Q_1(u)Q_2(u)$, where Q_1 and Q_2 are positive quantile functions

$$H_Q = -\int_0^1 \log(Q_1(u)q_2(u) + Q_2(u)q_1(u))du.$$

An example in this case is the power-Pareto distribution with entropy given in Table 2.1.

(d) If $Q(u)$ is the quantile function of X_1 then the entropy of $\frac{1}{X}$ is

$$H_{1/X} = -\int_0^1 \log \frac{q(1-u)}{Q^2(1-u)} du$$

since $1/X$ has quantile function $[Q(1-u)]^{-1}$.

To illustrate the result, let X follow the power distribution with $Q(u) = u^\alpha$. Then $\frac{1}{X}$ has Pareto I distribution with quantile function $Q_X(u) = (1-u)^{-\alpha}$. Hence the entropy is

$$H_{\frac{1}{X}}(u) = \int_0^1 \frac{\alpha(1-u)^{\alpha-1}}{(1-u)^{2\alpha}} du$$
$$= (\alpha+1) - \log \alpha.$$

(e) Let $Y = \phi(X)$ be a non-decreasing function of X. Then $\phi(Q(u))$ is the quantile function of Y. Thus the entropy of Y is

$$H_Y = -\int_0^1 \log[\phi'(Q(u))q(u)]du.$$

Assume that X has loglogistic distribution with quantile function $Q(u) = (\frac{u}{1-u})^\beta$, $\beta > 0$ and take $\phi(X) = X^\alpha$, $\alpha > 0$. Then $\phi(Q(u)) = (\frac{u}{1-u})^{\alpha\beta}$. Thus

$$H_Y = -\int_0^1 \log[\alpha\beta u^{\alpha\beta-1}(1-u)^{-\alpha\beta-1}]du$$
$$= -2 - \log(\alpha\beta).$$

Properties (a) through (e) are specific to quantile functions that help in computing entropies as well as in proving theoretical results.

2.4. Residual entropy

2.4.1. Definition and properties. As mentioned earlier the concept of entropy of residual life and past life were introduced by Muliere et al. (1993) in their efforts to study the uncertainty associated with truncated random variables. Let X be a non-negative random variable representing the lifetime of a device with survival function $\bar{F}(x)$ and probability density function $f(x)$. Recalling the discussions in Section 1.3, the survival function of $X_t = [X|X > t]$, $t > 0$ is given by $\bar{F}_t(x) = \frac{\bar{F}(x)}{\bar{F}(t)}$. Using the density function $f_t(x) = \frac{f(x)}{\bar{F}(t)}$ the entropy of X_t can be expressed from (2.4) as

$$H_F(t) = -\int_t^\infty \frac{f(x)}{\bar{F}(t)} \log\left(\frac{f(x)}{\bar{F}(t)}\right) dx \tag{2.6}$$

$$= -\frac{1}{\bar{F}(t)} \int_t^\infty f(x) \log f(x) dx + \log \bar{F}(t). \tag{2.7}$$

We notice that $H_F(0) = H_F$, the Shannon entropy and that $H_F(t)$ maybe negative or infinite. Also $H_F(t)$ measures the uncertainty in the distribution of X after the device has attained age t and expresses the concentration of probabilities in the distribution of X_t. Accordingly when $H_F(t)$ assumes larger values, the lesser is the concentration and hence the device becomes less reliable. This is the main rationale for the role of information measures in reliability studies along with the fact that the results in this connection are either consistent with or more general than their counterparts in the conventional reliability theory. The residual entropies of some important life distributions are given in Table 2.2, for easy reference.

In terms of mean residual life function $m(x)$ (see Section 1.2.2). Muliere et al. (1993) have shown that

$$H_F(t) = 1 - \int_t^\infty \log h(x) \frac{f(x)}{\bar{F}(t)} dx \tag{2.8}$$

and

$$H_F(t) \leq 1 + \log m(t).$$

The inequality gives an upper bound to the residual entropy which will be finite as long as $m(t)$ is finite. Also

$$H_F'(t) = h(t)[H_F(t) + \log h(t) - 1] \tag{2.9}$$

and

$$H_F''(t) = h'(t)[H_F(t) + \log h(t)] + H_F'(t)h(t). \tag{2.10}$$

Table 2.2. Residual entropies of life distributions.

Distribution	$F(x)$	$H(t)$
1. exponential	$\exp[-\lambda x]$, $x>0$, $\lambda>0$	$1-\log\lambda$
2. Pareto I	$(\frac{x}{\sigma})^{-\alpha}$, $x>\sigma>0$, $\alpha>1$	$\log\alpha-\log t-\frac{1}{\alpha}$
3. Weibull	$\exp[-(\frac{x}{\sigma})^\lambda]$, $x>0$, $\lambda,\sigma>0$	$1-\log(\frac{\lambda}{\sigma})[1+(\lambda-1)\log t+\frac{\lambda-1}{\lambda}e^{(\frac{t}{\sigma})^\lambda}E_1(\frac{t}{\sigma})^\lambda]^{-1}$
4. exponential geometric	$\frac{(1-p)e^{-\lambda x}}{1-pe^{-\lambda x}}$, $x>0$, $0<p<1$, $\lambda>0$	$1-\log\lambda+\frac{1}{p}\log(1-pe^{-\lambda t})+\frac{1-p}{p}(1-pe^{\lambda t})^{-1}$
5. uniform	$\frac{b-x}{b-a}$, $a\le x\le b$	$\log(b-t)$
6. parabolic	$2a(\frac{2b^3}{3}-b^2x+\frac{x^3}{3})$, $0\le x\le b$, $4ab^2=3$	$\log(\frac{2}{3}b^3-b^2t+\frac{t^3}{3})-2\log b$ $-\frac{1}{2}\{\psi(2;\sin^{-1}(\frac{t}{b}))-\psi(\frac{5}{2};\sin^{-1}(\frac{t}{b}))\}$
7. log logistic	$(1+x^\beta)^{-1}$, $\beta>0$, $x>0$	$1-\log\beta-(\beta-1)\log t+(1+t^\beta)^{-1}+\beta\log(1+t^\beta)$
8. Pareto II	$c^\alpha(x+c)^{-\alpha}$, $c,\alpha>0$	$\frac{\alpha+1}{\alpha}+\log\frac{c+t}{\alpha}$
9. rescaled beta	$(1-\frac{x}{R})^c$, $0\le x\le R$, $c>0$	$\frac{c}{c-1}+\log\frac{\alpha}{R-t}$
10. triangular	$\alpha-\frac{x^2}{\alpha}$, $0\le x\le\alpha$, $\frac{(1-x)^2}{1-\alpha}$, $\alpha<x\le1$	$\frac{\alpha}{\alpha-t^2}[t^2\{\log(\frac{2t}{\alpha-t^2})-\frac{1}{2}\}-\frac{1}{2}]-\alpha\{\log\frac{2\alpha}{\alpha-t^2}-\frac{1}{2}\}$, $0\le t\le\alpha$ $\frac{1}{2}+\log(1-t)-\log2$, $\alpha<t\le1$
11. half-logistic	$2[1+\exp\frac{x}{\sigma}]^{-1}$, $x>0$, $\sigma>0$	$2+\log\sigma-e^{-\frac{t}{\sigma}}\log(\frac{1+e^{t/\sigma}}{e^{t/\sigma}})$
12. power	$1-x^c$, $0\le x\le1$, $c>1$	$\frac{c-1}{c}+\log\frac{1-t^c}{c}+\frac{(c-1)t^c\log t}{1-t^c}$
13. generalized Pareto	$(\frac{b}{ax+b})^{\frac{1}{a}+1}$, $x>0$, $b>0$, $a>-1$	$\frac{2a+1}{a+1}+\log\frac{at+b}{a+1}$
14. beta (second kind)	$f(x)=\frac{a+b}{B(a,b)(1+x)}$; $x>0$, $a,c>0$	$\log B(a,b;c(t))-[(b-1)\psi(b,c(t))+(a+1)\psi(a,c(t))]$ $-(a+b)\psi(a+b,c(t))$

Table 2.2. Continued

Distribution	$\bar{F}(x)$	$H(t)$
15 half-normal	$f(x) = \frac{2}{\sqrt{2\pi}} e^{-\frac{x^2}{2}}, x \geq 0$	$\log N(t) + \frac{1}{2}\log\frac{\pi}{2} + \frac{\Gamma(\frac{3}{2},\frac{t^2}{2})}{\sqrt{\pi}N(t)}$ $N(t) = \int_t^\infty \frac{2}{2\pi} e^{-x^2/2} dx$
16 Stacy	$f(x) = \frac{cb^a}{\Gamma(a)} x^{ca-1} e^{-bx^c}$, $x > 0, a, b, c > 0$	$\log\Gamma(a, bt^c) - \log c - \frac{1}{c}\log b$ $+ a - (a - \frac{1}{c})\psi(a, bt^c) + \frac{(bt^c)e^{-bt}}{\Gamma(a,bt^c)}$
17 gamma	$f(x) = \frac{b^a}{\Gamma(a)} x^{a-1} e^{-bx}$, $a, b > 0$	$\log\Gamma(a, bt) - \log b + a - (a-1)\psi(a, bt)$ $+ \frac{(bt)^a e^{-bt}}{\Gamma(a,bt)}$
18 Greenwich	$(1 + \frac{x^2}{b})^{-1}, x \geq 0, b > 0$	$(1 + \frac{t^2}{b^2})\log\frac{t^2}{b^2+t^2} - \log 2t - \log(b^2 + t^2)$
19 modified exponential	$\frac{\theta e^{-\lambda x}}{1-(1-\theta)e^{-\lambda x}}, x > 0, \lambda > 0, \theta > 0$	$1 - \log\theta\lambda - \log[1 - (1-\theta)e^{-\lambda t}]$ $+ (1-\theta)^{-1}e^{\lambda t}[1 - (1-\theta)e^{-\lambda t}]$ $\times [\log(1 - (1-\theta)e^{\lambda t}) - (1 - (1-\theta)e^{-\lambda t})]$
20 Lindley	$\frac{1+\lambda+\lambda x}{1+\lambda} e^{-\lambda x}$	$\log\frac{\lambda^2(1+t)}{1+\lambda+\lambda t} - \frac{1+\lambda+\lambda t}{1+\lambda}e^{-\lambda t}[\frac{te^{-\lambda t}}{1+t} + e^{-\lambda}E_1\lambda(1+t)]$
21 reciprocal Rayleigh	$f(x) = 2\lambda^2 x^{-3} e^{-\frac{\lambda^2}{x^2}}, x > 0, \lambda > 0$	$\log\bar{F}(t) + 1 + \log\frac{\theta}{2} - \frac{3}{2\bar{F}(t)}$ $[\psi(1) - \bar{F}(t)\{\psi(1, \frac{\theta^2}{t^2}) + \frac{2}{3}\frac{\theta^2}{t^2}\}]$
22 log normal	$\frac{1}{x\sqrt{2\pi}} E^{-\frac{(\log x)^2}{2}}, x > 0$	$\log[1 - \phi(\log t) + \frac{1}{2}\log(2\pi e)]$ $-\phi(\log t)(1 + \frac{\log t}{2})$, where $\phi = \frac{1}{\sqrt{2\pi}} e^{-\frac{x^2}{2}}$
23 modified Gompertz	$\exp[-\frac{1}{\alpha}(e^{\alpha x} - \alpha^2 x - 1)]$	$\log(e^{\alpha t} - \alpha^2 t - 1) - \log\alpha$ $+ \exp[\frac{1}{\alpha}(e^{\alpha t} - \alpha^2 t - 1)E_1(\frac{1}{\alpha}e^{\alpha t} - \alpha^2 t - 1)]$ $E_1(x) = \int_0^x \frac{e^{-t}}{t} dt, \psi(p,t) = \frac{\partial}{\partial p}\Gamma(p,t)$

2.4.2. Characterizations. A basic question while considering $H_F(t)$ as a reliability function is whether it can determine $F(x)$ uniquely. Several authors have attempted to answer this problem. A simple solution is available if we assume that the hazard rate is monotone.

THEOREM 2.1. *The residual entropy $H_F(t)$ determines F if $h(t)$ is monotone.*

PROOF. From Kotlarski (1972) if $\phi(\cdot)$ is differentiable and monotone, then $E[\phi(X)|X > t]$ characterizes the distribution of X. Writing (2.8) as

$$H_F(t) = E[1 - \log h(X)|X > t]$$

and noting that $h(x)$ is differentiable under the assumption of absolute continuity of F, it follows that $H_F(t)$ uniquely determines F whenever $h(x)$ is monotone.

A more general result is stated in Muliere et al. (1993).   □

THEOREM 2.2. *If X and Y are non-negative random variables with distribution functions F and G then for all $t > 0$*

$$H_F(t) = H_G(t) \Leftrightarrow F(t) = G(t).$$

The first part of the proof is obvious from the fact that if $F(t) = G(t)$ for all $t > 0$, the hazard rates satisfy $h_F(t) = h_G(t)$ and accordingly from (4.2), $H_F(t) = H_G(t)$. To prove the sufficiency, the equality $H_F(t) = H_G(t)$ for all $t > 0$ implies $H_F'(t) = H_G'(t)$ and $H_F''(t) = H_G''(t)$ and equations (2.9) and (2.10) are utilized. Using only the first part $H_F'(t) = H_G'(t)$, which is equivalent to

$$h_F(t)[H_F(t) + \log h_F(t) - 1] = h_G(t)[H_G(t) + \log h_G(t) - 1] \quad (2.11)$$

Ebrahimi (1996) has stated the following result.

THEOREM 2.3. *Let X be a non-negative continuous random variable with probability density function f and $H_F(t) < \infty$, $t \geq 0$. If f is continuous then $H_F(t)$ uniquely determines $\bar{F}$.*

Belzunce et al. (2004) has argued that proof of Theorem 2.3 given in Ebrahimi (1996) is incorrect. From (2.9) the hazard rate $h(t)$ is a positive solution of the equation

$$g(x) = x[H_F(t) - 1 + \log x] - H_F'(t) = 0, \ x = h(t). \quad (2.12)$$

When $H_F(t)$ is an increasing function $g(x)$ first decreases attains the minimum $x_t = e^{-H_F(t)}$ and then increases, implying a unique $h(t)$ as a solution of (2.12). Thus H_F uniquely determines F. On the other hand if H_F is decreasing and $g(x_t) \neq 0$, there can be two solutions in which either both are hazard rates or one can be a hazard rate and the other may not be a hazard rate. These facts lead to a modification of Theorem 2.3.

THEOREM 2.4. *Let X be a non-negative continuous random variable with distribution function $F(x)$ and residual entropy $H_F(t) < \infty$, $t \geq 0$. Then $H_F(t)$ characterizes the distribution of X.*

Another category of characterization is in terms of the form of H_F and relationships H_F has with various reliability functions that uniquely determine specific life distributions. Most of these are related to the exponential, Pareto II and rescaled beta distributions. These three distributions being subsumed in the Generalized Pareto model (GPD) with survival functions

$$\bar{F}(x) = \left(1 + \frac{ax}{b}\right)^{-\frac{a+1}{a}} \tag{2.13}$$

providing an exponential with mean b when $a \to 0$, Pareto II when $a = (c-1)^{-1}$ and $b = a\alpha$ and the rescaled beta when $a = -(c+1)^{-1}$ and $b = -aR$, we will consider only (2.13). Sankaran and Gupta (1999) proved that each of the relationships

$$H_F(t) = (c + \log m(t))$$

and

$$H_F(t) = a - \log h(t),$$

c and a being real constants, for all $t > 0$ if and only if X follows the GPD. The same result can also be seen in Asadi and Ebrahimi (2000). Belzunce et al. (2004) further extended these to residual entropies of the form

$$H_F(t) = c(t) - \log h(t)$$

for real valued functions $c(t)$ to determine distributions with hazard rate

$$h(t) = K - \int_\alpha^t e^{c(x)}(1 - c(x))dx, \ K = e^{H(\alpha)}, \ \alpha < x < \beta.$$

In particular for

$$h(t) = a\left[\left(2 - b + \frac{a}{b}\right)e^{-at} + b - 2\right]$$

whenever $(2 - b + \frac{b}{a}) \leq 0$ or $(2 - b + \frac{a}{b})e^{-bt} > 0$, the residual entropy has linear form

$$H_F(t) = at + b, \ a, b > 0, \ t > 0.$$

Nair and Rajesh (1998) have reported several characterization results concerning the generalized Pareto distribution using

$$H_F(t) + \log p(t) = \log kt,$$

for distributions satisfying $p(t) = th(t)$, the dynamic proportional hazards models, which is equivalent to the earlier characterizations once we write $\log p(t) = \log t + \log h(t)$

$$H_F(t) + m_1(t) \log q = 1, \ m_1(t) = E(X|X > t)$$

if and only if X has Gompertz law

$$\bar{F}(x) = \exp[pq^x - 1]/\log q, \ x > 0, \ p, q > 0$$

and that

$$H_F(t) + m(t) = 1 - t$$

and

$$H_F(t) + \frac{dm(t)}{dt} = a, \ \text{a constant}$$

for all $t > 0$ respectively are characteristic properties of the extreme value distribution $\bar{F}(x) = \exp[e^{-x}]$, and the logistic distribution $\bar{F}(x) = \frac{Ke^{-cx}}{1+Ke^{-cx}}$. However, notice that the last two models have the entire real line as support and hence do not qualify as life distributions.

A wide variety of life distributions can be involved if we consider the relationship between residual entropy function and the conditional geometric mean function $G(t)$ defined by

$$\log G(t) = E(\log X | X > t) = \frac{1}{\bar{F}(t)} \int_t^\infty \log x f(x) dx$$

and its relationship with the hazard rate,

$$\frac{G'(t)}{G(t)} = h(t)(\log G(t) - \log t). \tag{2.14}$$

A general form of the characteristic property as applied to most of the basic life distributions is available in Nair and Rajesh (2000). It can be further generalized as

$$H_F(t) = A - (c - 1) \log G(t) + K \log \bar{F}(t) \tag{2.15}$$

where A, c and K are real numbers. The values of A, c and K are given in Table 2.3. The proofs are almost along the same lines and therefore those of two typical cases are given below.

THEOREM 2.5. *Let X be a continuous non-negative random variable with survival function $\bar{F}$ and probability density function $f(x)$ satisfying $E(\log X) < \infty$. Then X is distributed as Weibull with*

$$\bar{F}(x) = \exp[-x^\lambda], \ x > 0; \ \lambda > 0$$

if and only if

$$H_F(t) = (1 - \log \lambda) - (\lambda - 1) \log G(t). \tag{2.16}$$

PROOF. By direct calculations from (2.8), we have (2.16). Conversely if (2.16) holds, using (2.9)

$$-\lambda \frac{G'(t)}{G(t)} = h(t)[(1 - \log \lambda) - (\lambda - 1) \log G(t) + \log h(t) - 1]$$

Simplifying $h(t) = \lambda t^{\lambda-1}$, the hazard rate of the Weibull distribution.   □

THEOREM 2.6. *If X is as in Theorem 2.5, then X follows Burr XII distribution with*

$$\bar{F}(x) = (1+x^c)^{-k}, \ x > 0; \ c, k > 0 \qquad (2.17)$$

if and only if

$$H_F(t) = 1 - \log(kc) + \frac{1}{k} - (c-1)\log G(t) - \frac{1}{k}\log \bar{F}(t). \qquad (2.18)$$

PROOF. The hazard rat of (2.17) is

$$h(x) = \frac{kcx^{c-1}}{(1+x^c)}$$

and hence

$$H_F(t) = 1 - \int_t^\infty [\log(kc) + (c-1)\log x - \log(1+x^c)]$$

$$\times \frac{kcx^{c-1}}{(1+x^c)^{k+1}(1+t^c)^{-k}}$$

$$= 1 - \log(kc) - (c-1)\log G(t) + \frac{1}{k} - \frac{1}{k}\log \bar{F},$$

which is (2.18). Conversely if (2.18) is true, (2.9) gives

$$-(c-1)\frac{G'(t)}{G(t)} + \frac{1}{k}h(x) = h(x)[1 - \log(kc) - (c-1)\log G(t)$$

$$+ \frac{1}{k} - \frac{1}{k}\log F + \log h(x) - 1]$$

or

$$[\frac{1}{k} - (c-1)(\log G(t) - \log t)] = [-\log(kc) + \frac{1}{k} - \frac{1}{k}\log \bar{F}$$

$$+ \log h - (c-1)\log G(t)]$$

or

$$t^{c-1} = \frac{f(x)[\bar{F}(x)]^{-\frac{1}{k}-1}}{kc}.$$

Integrating

$$t^c = \bar{F}(x)^{-k} + P,$$

where P is a constant. At $t = 0$, $\bar{F}(x) = 1$, giving $P = -1$. Thus

$$t^c = \bar{F}(x)^{-\frac{1}{k}} - 1,$$

leading to (2.17).   □

The representation (2.15) is not valid for more general life distributions that may involve more than one basic model as special cases. For example in the exponential geometric case we have the identity

$$H_F(t) = 1 - \log \lambda + \frac{1}{p} \log(1-p)e^{\lambda t} - \frac{1}{p} \log \bar{F}.$$

There are some similar results and discussions when X is a discrete random variable. We refer to Nair and Rajesh (1998) and Belzunce et al. (2004) for details.

2.4.3. Classes of life distributions. Ebrahimi (1996) has proposed two classes of life distributions based on the monotonicity of $H_F(t)$ in the same way as such classes have been defined using the reliability functions.

DEFINITION 2.1. *The random variable X or its distribution F has decreasing (increasing) uncertainty of residual life, DURL (IURL) if $H_F(t)$ is decreasing (increasing) in t for all $t \geq 0$. Here, decreasing (increasing) implies non-increasing (non-decreasing).*

The classes are interpreted by Ebrahimi (1996) by saying that intuitively if F belongs to the DURL class then as the device ages the conditional probability density function becomes more informative. If F is DURL (IURL) then $-H'_F(t) \geq 0$ (≤ 0) meaning non-negative local reduction of uncertainty.

Further when $H_F(t) = $ constant, implying F is both IURL and DURL, then $h(t) = $ a constant, so that the distribution is exponential, and it is the only distribution possessing such a property. The following results are proved in Ebrahimi (1996) and Ebrahimi and Kirmani (1996c).

(1) If F is IFR (DFR) then it is DURL (IURL)
(2) If F is DURL (IURL), then $h(t) \leq (\geq) \exp[1 - H_F(t)]$
(3) $H_F(t) \leq (\geq) 1 - \log h(0) = 1 - \log f(0)$ for DURL (IURL) class
(4) If F is DURL then $H_F(t) \leq 1 + \log \mu$ and if F is IURL, $\exp(H_F(0) - 1) \leq m(t)$
(5) If F is DMRL (IMRL) then F is IDURL (IMRL), which is an improvement of the first result above. But the converse is not true. However if $\phi(\cdot)$ is non-negative increasing and convex (concave) then $\phi(X)$ is IURL (DURL)

Towards a better understanding of the nature of the residual entropy function for various life distributions and its relationship with the shape of the hazard rate functions, we consider some examples. For the rescaled beta distribution,

$$H_F(t) = \frac{c-1}{c} + \log \frac{R-t}{c}$$

Table 2.3. Values of A, c and K in the characteristic property (2.15) of life distributions.

	A	c	K
exponential $\bar{F}(x) = e^{-\lambda x}$	$1 - \log \lambda$	1	0
Weibull $\bar{F}(x) = e^{-x^\lambda}$	$1 - \log \lambda$	λ	0
Burr XII $\bar{F}(x) = (1 + x^c)^{-k}$, $x > 0, c > 0, k > 0$	$1 - \log kc + \frac{1}{k}$	c	$-\frac{1}{k}$
Greenwich $\bar{F}(x) = (1 + x^2)^{-a/2}$	$1 - \log a + \frac{2}{a}$	2	$-\frac{2}{a}$
power $\bar{F}(x) = 1 - x^c$	$-\log c$	c	1
log logistic (special case of Burr XII when $k = 1$)			
Generalized Weibull $\bar{F}(x) = (1 - \lambda x^c)^{1/\lambda}$ $0 \le x \le \lambda^{1/c}$	$-\log \lambda$	c	λ
Pareto II $\bar{F}(x) = (1 + \frac{x}{\beta})^{-\alpha}$	$1 + \frac{1}{\alpha} + \log \frac{\beta}{\alpha}$	1	$-\frac{1}{\alpha}$
rescaled beta	$1 - \frac{1}{\beta} + \log \frac{\beta}{\alpha}$	1	$\frac{1}{\alpha}$

and hence

$$H_F'(t) = -\frac{1}{R - t} < 0$$

indicating that $H_F(t)$ is decreasing and F is DURL. Also $h(t) = \frac{cR}{R-t}$ is increasing so that F is IFR.

On the other hand the Pareto II distribution has

$$H_F(t) = \frac{\alpha + 1}{\alpha} + \log \frac{c + t}{\alpha}$$

giving

$$H_F'(t) = \frac{1}{c + t} > 0.$$

Thus $H_F(t)$ is increasing and so F is IURL. The hazard rate $h(t) = \frac{\alpha+1}{\alpha+c}$ is decreasing and hence DFR.

The Weibull distribution is characterized by the relationship

$$H_F(t) = 1 - \log \lambda - (\lambda - 1) \log G(t).$$

Accordingly

$$H_F'(t) = -(\lambda - 1)\frac{G'(t)}{G(t)}.$$

Since $G(t)$ is always an increasing function of t, $H_F'(t) > (<)0$ according as $\lambda < (>)1$. Thus Weibull distribution IURL for $\lambda < 1$ and DURL for

$\lambda > 1$. This is consistent with the hazard rate increasing for $\lambda > 1$ and decreasing when $\lambda < 1$. Notice that the behaviour of $H_F(t)$ depends on the parameter value in contrast with the first two examples.

Now we consider the power distribution for which the residual entropy satisfies

$$H_F(t) = -\log c - (c-1)\log G(t) + \log \bar{F}(t).$$

Differentiating

$$H_F'(t) = -(c-1)\frac{G'(t)}{G(t)} - h(t)$$

or,

$$H_F'(t) = -h(x)[(c-1)(\log G(t) - \log t) + 1].$$

Obviously, when $c > 1$, $H_F'(t) < 0$ and hence the power distribution is DURL for $c > 1$. When $c < 1$, $H_F'(t) = 0$ gives

$$(c-1)(\log G(t) - \log t) = -1. \tag{2.19}$$

By direct calculation,

$$\log G(t) = -\frac{1}{c} - \frac{t^c \log t}{1 - t^c}.$$

Thus equation (2.19) becomes

$$(c-1)\left(\frac{1}{c} + \frac{t^c \log t}{1 - t^c} + \log t\right) = 1$$

which simplifies to

$$\frac{\log t}{1 - t^c} = \frac{1}{c(c-1)},$$

Thus $H_F(t)$ is decreasing initially, reaches a minimum at t_0 and then increasing.

Our last example concerns the log logistic distribution. This model satisfies

$$H_F(t) = 2 - \log c - (c-1)\log G(t) - \log \bar{F}(t)$$

leading to

$$H_F'(t) = h(t)[1 - (c-1)(\log G(t) - \log t)].$$

Then $H_F'(t) = 0$ provides

$$(c-1)(\log G(t) - \log t) = 1. \tag{2.20}$$

Using

$$\log G(t) = \log t - \frac{1 + t^c}{c}\log\frac{t^c}{1 + t^c}$$

(2.20) becomes

$$\frac{c-1}{c}(1 + t^c)\log\frac{t^c}{1 + t^c} = 1.$$

Also $H_F(t)$ increases in $(0, t_0)$ attains a maximum at t_0 and then decreases. To accommodate life distributions of the kind described in the last two examples some additional classes are required.

DEFINITION 2.2. *A distribution F is said to have a bathtub-shaped (BT) residual entropy function $H_F(t)$, if*

$$H_F(t) = \begin{cases} H_F^{(1)}(t), & t \leq t_1 \\ c, \text{ constant}, & t_1 \leq t \leq t_2 \\ H_F^{(2)}(t), & t \geq t_2 \end{cases}$$

where $H_F^{(1)}$ is strictly decreasing and $H_F^{(2)}$ is strictly increasing. When $t_1, t_2 \to 0$. F is IURL and when $t_1, t_2 \to \infty$, F is DURL. We say that t_1 and t_2 are the change points of $H_F(t)$. Similarly if $H_F^{(1)}$ is strictly increasing and $H_F^{(2)}$ is strictly decreasing, $H_F(t)$ has an upside-down bathtub (UBT) shape.

REMARK 2.2. *When $H_F(t)$ has only one change point t_0, it is BT (UBT) if an only if $H_F'(t) < (>)0$ in $(0, t_0)$, $H_F'(t_0) = 0$ and $H_F'(t) > (<)0$ in (t_0, ∞). By the above definition the power distribution has BT residual entropy and the log logistic model has UBT residual entropy.*

It follows from (2.9) that for IURL (DURL) class

$$H_F(t) \geq (\leq)1 - \log h(t) \tag{2.21}$$

and that for BT (UBT) distributions

$$H_F(t_0) = 1 - \log h(t_0)$$

for some $t_0 > 0$ when $H_F(t)$ is assumed to have only one change point. Notice that (2.21) provides a bound for IURL (DURL) distributions. See Nair et al. (2019) for details.

A measure of uncertainty in the interval (a, b) in R^+ can be prescribed as $H_F((a, b)) = H_F(b) - H_F(a)$. It constitutes an increment of uncertainty achieved in the residual life of a device as its lifetime moves from age a to age b. Thus $H_F(t) > H_F(0)$ implies that uncertainty has increased during the age interval $(0, t)$ and hence the conditional distribution $F_t(x)$ becomes less informative as the device ages. Similarly when, $H_F(t) < H_F(0)$, the distribution becomes more informative. Also $H_F(t) = H_F(0)$ if and only if F is exponential, the case of no ageing. These considerations lead to a new classification of life distributions.

DEFINITION 2.3. *Let X be a non-negative random variable with residual entropy $H_F(t)$. Then F is new better (worse) than used in residual entropy, NBURE (NWURE) if $H_F(t) \leq (\geq)H_F(0)$ for all $t > 0$. The distribution F is both NBURE and NWURE iff it is exponential.*

EXAMPLE 2.1. *The rescaled beta distribution has*

$$H_F(t) = \frac{c-1}{c} + \log \frac{R-t}{c}$$

so that $H_F(t) - H_F(0) = \log \frac{R-t}{R} \le 0$ *for all* $0 \le t \le R$ *and hence* F *is* NBURE.

Similarly, for the Pareto II model

$$H_F(t) = \frac{\alpha+1}{\alpha} + \log \frac{c+t}{\alpha}$$

giving

$$H_F(t) - H_F(0) = \log \frac{c+t}{c} \ge 0 \text{ for all } t > 0.$$

Hence Pareto II distribution belongs to the NWURE class.

From the definitions of IURL and NBURE it follows that DURL $\Rightarrow$ NBURE and likewise IURL $\Rightarrow$ NWURE. Since the converse is not true, NBURE (NWURE) class is more general than DURL (IURL) (see example below). Further, the former is more easy to identify than the latter as it is easier to verify $H_F(t) \ge (\le)H_F(0)$ than investigating the monotone behaviour of $H_F(t)$.

EXAMPLE 2.2. *For the power distribution in Table 2.2,*

$$H_F(t) = \frac{c-1}{c} + \log \frac{1-t^c}{c} + \frac{(c-1)t^c}{1-t^c} \log t.$$

Hence

$$H_F(t) - H_F(0) = \log(1-t^c) + \frac{(c-1)\log t}{1-t^c}$$

$$= \log(1-t^c)t^{\frac{(c-1)}{1-t^c}} > 0, \ c < 1.$$

For $0 < c < 1$, *the power distribution is BT, while for* $c > 1$ *it is increasing* *(see Figure 2.1).*

2.5. Order statistics

Recall that if $X_{r:n}$, $r = 1, 2, \ldots, n$ denote the order statistics of a set of independent and identically random variables from a F distribution, then $X_{r:n}$ has probability density function.

$$f_{r:n}(x) = \frac{n!}{(r-1)!(n-r)!}[F(x)]^{r-1}[\bar{F}(x)]^{n-r}f(x)$$

and distribution function

$$F_{r:n}(x) = \sum_{i=r}^{n} \binom{n}{i} [F(x)]^i [\bar{F}(x)]^{n-i}.$$

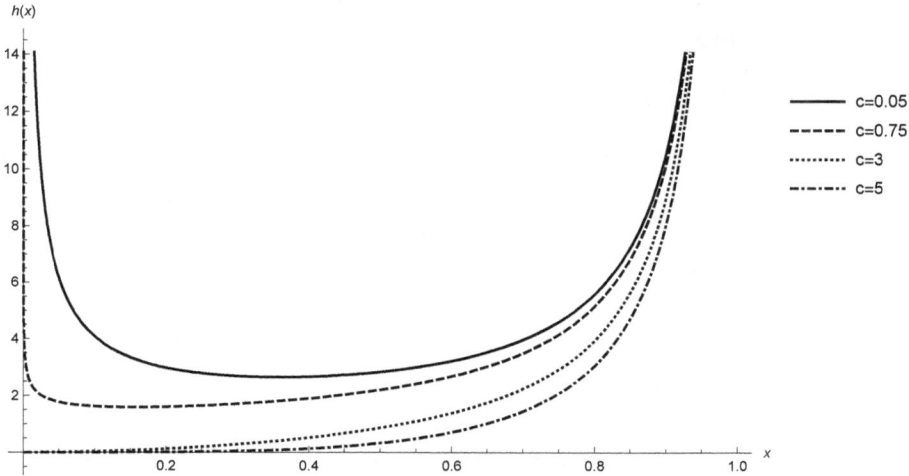

Figure 2.1. $h(x)$ of power distribution for different values of c.

Thus the hazard rate of $X_{r:n}$ is

$$h_{r:n}(x) = \frac{n!}{(r-1)!(n-r)!} \frac{(\frac{F(x)}{\bar{F}(x)})^{r-1}}{\sum_{i=0}^{r-1} \binom{n}{i}(\frac{F(x)}{\bar{F}(x)})^i} h(x).$$

THEOREM 2.7. *Let X and Y be two continuous non-negative random variables with density functions f and g hazard rates $h_F(x)$ and $h_G(x)$, survival functions F and G and residual entropies $H_F(t)$ and $H_G(t)$. If θ be a non-negative increasing function such that $h_G(x) = \theta(x)h_F(x)$, $x \geq 0$ and $0 \leq \theta(x) \leq 1$ and further $\lim_{x \to \infty} \frac{\bar{G}(x)}{\bar{F}(x)}$, then if $H_F(t)$ is decreasing in t, then $H_G(t)$ is also decreasing in t (Asadi and Ebrahimi (2000)).*

THEOREM 2.8. *With the notations of the previous theorem, if $0 \leq \theta(x) \leq 1$, is increasing (decreasing) then if F is IFR (DFR), IFRA (DFRA), NBU (NWU) or DMRL (IMRL) then so is G (Block et al. (1985)).*

When a class of life distribution is introduced based on some reliability property, it is customary to verify whether the property is preserved under certain reliability operations. Important among them are the formulation of coherent structures, convolutions, mixtures, shock models, equilibrium distributions and more. Consider a system of n components whose lifetimes are independent and identically distributed. A system that functions if and only if at least k components of the total of n functions is called a k-out-of-n system. Obviously the lifetime of such a system is $X_{n-k+1:n}$. Two important special cases of a k-out-of-n system are

(i) the series system which functions only when all the components are working, and

(ii) the parallel system which functions if and only if at least one of the components work.

The system lifetime of a series system is $X_{1:n}$, the smallest order statistic and that of a parallel system is $X_{n:n}$, the largest order statistic.

From the above Theorems 2.7 and 2.8, Asadi and Ebrahimi (2000) arrived at the conclusions

(a) If F is DURL, the residual entropy of $X_{n:n}$ is also DURL, that is the DURL property is preserved under the formation of parallel systems

(b) The IURL class is not preserved under the formation of parallel structures

(c) If $X_{k:n}$ is DURL then $X_{k+1:n}, X_{k:n-1}, X_{k+1:n+1}$ are also DURL

The model

$$h_G(x) = \theta(x)h_F(x) \qquad (2.22)$$

where $\theta(x)$ satisfies

(a) $\theta(x) > 0$
(b) $\int_0^\infty \theta(x)h(x)dx = \infty$
(c) $\bar{F}(x) = 0$ for some $x = x_0$ implies

$$\int_0^{x_0} \theta(x)h_F(x)dx = \infty$$

is called the dynamic proportional hazards model and has been investigated by Nanda and Das (2011). For such a model we have the following result.

THEOREM 2.9. *Let $\theta(x)$ be an increasing function and F be IFR. Then G is DURL.*

Returning to order statistics and their entropy we derive some formulas for computing the residual entropy of order statistics. Gupta et al. (2014) and Sunoj et al. (2017) used the fact that the random variable $U = F(X)$ is uniformly distributed over $[0, 1]$ and hence the density function $U_{r:n}$ the rth order statistic based on n independent and identically distributed observations on U can be written as

$$f_{r:n}(u) = \frac{1}{B(r, n-r+1)} u^{r-1}(1-u)^{n-r}, \ 0 \le u \le 1. \qquad (2.23)$$

Hence the entropy of $U_{r:n}$ is

$$H_{r:n} = \log B(r, n-r+1) - (r-1)[\psi(r) - \psi(n+1)]$$
$$- (n-1)[\psi(n-r+1) - \psi(n+1)]$$

where $\psi(p) = \frac{d \log \Gamma(p)}{dp}$, the digamma function. The residual entropy of $X_{r:n}$ is by definition

$$H_{r:n}(t) = 1 - \int_t^\infty \frac{f_{r:n}(\alpha)}{\bar{F}_{r:n}(t)} \log \frac{f_{r:n}(x)}{\bar{F}_{r:n}(t)}$$

$$= 1 - \frac{1}{\bar{F}_{r:n}(t)} \int_t^\infty f_{r:n}(x) \log h_{r:n}(x) dx$$

$$= 1 - \frac{1}{1 - \sum_{i=r}^n \binom{n}{i} [F(t)]^i [\bar{F}(t)]^{n-i}}$$

$$\int_t^\infty \frac{n! [F(x)]^{r-1}}{(r-1)!(n-r)!} [\bar{F}(x)]^{n-r} f(x)$$

$$\log \frac{n!}{(r-1)!(n-r)!} \frac{[F(x)]^{r-1} [\bar{F}(x)]^{n-r} f(x)}{1 - \sum_{i=r}^n \binom{n}{i} [F(x_i)]^r [\bar{F}(x)]^{n-i}}.$$

Using transformation $F(t) = u$,

$$H_{r:n}(t) = 1 - \frac{1}{1 - \sum_{i=1}^r \binom{n}{i} u^i (1-u)^{n-i}} \int_u^1 \frac{n! p^{r-1} (1-p)^{n-r}}{(r-1)!(n-r)!}$$

$$\log \frac{n!}{(r-1)!(n-r)!} \frac{\mu^{r-1} (1-p)^{n-r} f(F^{-1}(p))}{1 - \sum_{i=r}^n \binom{n}{i} p^i (1-p)^{n-i}} dp. \qquad (2.24)$$

The residual entropy of the smallest order static $X_{1:n}$ is obtained by taking $r = 1$ as

$$H_{1:n}(t) = 1 - (1-u)^{-n} \int_u^1 n(1-p)^{n-1} \log \frac{n}{1-p} f(F^{-1}(p)) dp, \ t = F(u) \tag{2.25}$$

and that of $X_{n:n}$ is

$$H_{n:n}(t) = 1 - (1-u^n)^{-1} \int_u^1 np^{n-1} \log \frac{np^{n-1}}{1-p^n} f(F^{-1}(p)) dp. \qquad (2.26)$$

Notice that $X_{1:n}$ and $X_{n:n}$ represent respectively the lifetime of a series and parallel system in the reliability context.

EXAMPLE 2.3. *Let X be distributed as exponential with*

$$\bar{F}(x) = \exp[-\lambda x], \ x > 0, \ \lambda > 0.$$

Then

$$F^{-1}(u) = -\frac{1}{\lambda} \log(1-u)$$

and

$$fF^{-1}(u) = \lambda(1-u).$$

The residual entropy of $X_{1:n}$ is from (2.25)

$$H_{1:n}(t) = 1 - \log n\lambda,$$

independent of t.

T denotes the lifetime of a coherent system of n independent and identically distributed component lifetimes $X_1, X_2, \ldots, X_n$. Chahkandi and Toomaj (2016) have derived the expression for the uncertainty in the conditional density of $T - t$ about the predictability of the remaining life of the system, when all components are alive. They have obtained the expression

$$H(T_t^i) = H(W_i) + \log \bar{F}(t) - E(\log f(F^{-1}(W_i) + t))$$

where W_i has a beta distribution

$$g(w_i) = \frac{\Gamma(n+1)(1-w_i)^{r-1}w_i^{n-1}}{\Gamma(r)\Gamma(n-r+1)}, \ 0 < w_i < 1, \ i = 1, 2, \ldots n$$

$\hat{F}_t(x) = \frac{\bar{F}(x+t)}{\bar{F}(t)}$ and F and f are the common distribution and density functions of the X_i's. They also show that $H(T_t^i) \leq (\geq) H_{X_{i:n}}$ according as X is IFR (DFR).

2.6. Quantile-based residual entropy

A major departure from the discussions so far made on residual entropy in terms of the distribution function of lifetimes is accomplished by introducing the notion of quantile functions in the definitions and in the analysis. The role of the quantile function in statistical analysis, its properties and the relative advantages it has over the distribution function approach were explained in Section 1. We assume throughout this section that $F(x)$ is continuous and strictly increasing so that its quantile function

$$F^{-1}(u) = Q(u) = \inf_x [x|F(x) \geq u]$$

is such that $F(x) = u$ implies $x = Q(u)$ and the inverses involved are unique. The quantile density function $q(u) = \frac{dQ(u)}{du}$ is such that

$$f(Q(u))q(u) = 1.$$

This transforms the definition of differential entropy in (2.4) to (Vasicek (1976))

$$H_Q = \int_0^1 \log q(p)dp. \tag{2.27}$$

Also the transformation $x = Q(u)$ in (2.7) gives the quantile version of (2.7) as

$$\xi(u) = H_F(Q(u)) = \log(1-u) + (1-u)^{-1}\int_u^1 \log q(p)dp, \tag{2.28}$$

proposed by Sunoj and Sankaran (2012) as an alternative to $H_F(t)$ and presented different properties of the measure $\xi(u)$. Since the hazard quantile function $H_Q(u)$ is directly related to $q(u)$ by

$$h_Q(u) = [(1-u)q(u)]^{-1}$$

we can express (2.28) in the alternative form

$$\xi(u) = 1 - (1-u)^{-1} \int_u^1 \log h_Q(p)dp. \tag{2.29}$$

Table 2.4 contains the quantile forms of residual entropies of some distributions in common use. Note that the entries 1 through 10 are cases in which the quantile functions have closed form distribution functions, while 11 through 14 do not enjoy this property.

Three important identities emerge from (2.28). These are

$$\log h_Q(u) = (1-u)\xi'(u) + (1-\xi(u)) = -\frac{d}{du}\left((1-u)(1-\xi(u))\right) \tag{2.30}$$

$$h_Q(u) = \exp[(1-\xi(u)) + (1-u)\xi'(u)] \tag{2.31}$$

and

$$h_Q'(u) = h_Q(u)\left[(1-u)\xi''(u) - 2\xi'(u)\right]. \tag{2.32}$$

An explicit expression for the hazard rate function in terms of $H_F(t)$ is not available in the distribution function approach whereas its equivalents are well connected in equation (2.31). Similarly, while establishing the unique determination of F from $H_F(t)$, various authors have chosen the path that if two random variables have the same residual entropy, the variables are stochastically equal. As a result there was no inversion formula that provides F in terms of $H_F(t)$. A remarkable feature of the quantile approach is that one can specify the distribution once the functional form of $\xi(u)$ is known. To see this, we first differentiate (2.28) to get

$$(1-u)\xi'(u) = -1 + \xi(u) - \log(1-u) - \log q(u)$$

and thus

$$q(u) = \exp[\xi(u) - (1-u)\xi'(u) - \log(1-u) - 1]. \tag{2.33}$$

Since

$$h_Q(u) = [(1-u)q(u)]^{-1}$$

we can obtain (2.33) from (2.31) also and further

$$q(u) = \frac{1}{1-u}\exp\left[-\frac{d}{du}\left((1-u)(1-\xi(u))\right)\right]. \tag{2.34}$$

Table 2.4. Quantile forms of residual entropy.

Distribution	Quantile function	Residual entropy
1. exponential	$\lambda^{-1}(-\log(1-u))$	$1-\log\lambda$
2. Pareto II	$\alpha[(1-u)^{-1/c}-1]$	$\log\frac{\alpha}{c}+\frac{c+1}{c}-\frac{1}{c}\log(1-u)$
3. log logistic	$\alpha^{-1}\left(\frac{u}{1-u}\right)^{1/\beta}$	$2-\log(\alpha\beta)+\frac{\beta-1}{\beta}\frac{u\log u}{1-u}-\frac{1}{\beta}\log(1-u)$
4. beta	$R(1-(1-u)^{1/c})$	$\log\frac{R}{c}+\frac{c-1}{c}+\frac{1}{c}\log(1-u)$
5. half logistic	$\sigma\log\left(\frac{1+u}{1-u}\right)$	$2+\log(2\sigma)-\frac{2\log 2}{1-u}+\frac{1+u}{1-u}\log(1+u)$
6. power	$\alpha u^{1/\beta}$	$\log\frac{\alpha}{\beta}+\left(\frac{\beta-1}{\beta}\right)+\log(1-u)-\frac{1}{\alpha}\log(1-u)$
7. Pareto I	$\sigma(1-u)^{-1/\alpha}$	$\log\frac{\sigma}{\alpha}+\frac{\alpha+1}{\alpha}-\frac{1}{\alpha}\log(1-u)$
8. exponential geometric	$\frac{1}{\lambda}\log\frac{1-pu}{1-u}$	$2+\log\left(\frac{1-p}{\lambda}\right)+p^{-1}(1-u)^{-1}[(1-p)\log(1-p)$ $-(1-pu)\log(1-pu)]$
9. generalized Pareto	$\frac{b}{a}\left[(1-u)^{-a/a+1}-1\right]$	$\log\frac{b}{a+1}+\frac{2a+1}{a+1}-\frac{a}{a+1}\log(1-u)$
10. residual entropy model I	$\frac{e^{a-b}}{2b}\left(e^{2bu}-1\right)$	$\log(1-u)+a+bu$
11. linear hazard quantile function	$(a+b)^{-1}\log\frac{a+bu}{a(1+u)}$	$2+\log\frac{b-a}{a+b}+\log(1-u)-\frac{a+b}{b(1-u)}\log(a+b)$ $+\frac{a+bu}{a+b}\log(a+bu)$ $+\frac{1-u}{1-u}-\frac{2\log 2}{1-u}+\frac{1+u}{1-u}\log(1+u)$
12. residual entropy model II	$q(u)=cu^{\alpha}(1-u)^{\beta-1}e^{\alpha+\beta-1}e^{-\alpha/u}$	$\log cu^{\alpha}(1-u)^{\beta}$
13. Govidarajulu	$\sigma[(\beta+1)u^{\beta}-\beta u^{\beta+1}],\ \sigma,\beta>0$	$2\log(1-u)+\log[\sigma\beta(\beta+1)]-\frac{(\beta-1)}{1-u}$ $(1+u(\log u-1))$

Table 2.4. Continued

Distribution	Quantile function	Residual entropy
14. power Pareto	$cu^{\lambda_1}(1-u)^{-\lambda_2}$ $c > 0, \lambda_1, \lambda_2 > 0$ with one of λ_1, λ_2 may be zero	$\log c + \lambda_2 - \lambda_1 + 1 - (1-u)^{-1}(\lambda_2 - \lambda_1)^{-1}\lambda_2 \log \lambda_2$ $-\lambda_2 \log(1-u) - (1-u)^{-1}(\lambda_1-1)u\log u - (\lambda_2-\lambda_1)^{-1}$ $(1-u)^{-1}(\lambda_1(1-u) + \lambda_2 u) - (1-u)^{-1}(\lambda_1-1)u\log u$ $-(\lambda_2 - \lambda_1)u\log u - (\lambda_2-\lambda_1)^{-1}(1-u)^{-1}(\lambda_1(1-u) + \lambda_2 u)$ $\log(\lambda_1(1-u) + \lambda_2 u)$
15. linear mean residual	$-(a+b)\log(1-u) - 2bu$ $a > 0, a+b > 0$	$(1-u)^{-1}\{\frac{a-b+2}{2}[\log(a-b+2)-1]$ $-\frac{a-b+2u}{2}(\log(a-b+2u)-1)\} + 1$
16. Nair et al. (2011b)	$q(u) = Ku^\alpha(1-u)^{A+\alpha}$	$K + A + \alpha - (1-u)^{-1}[\alpha + \alpha u(\log u - 1)]$ $+(A + \alpha - 1)\log(1-u)$

The advantage (2.33) or equivalently (2.34) is manifold. In the first place it enables us to propagate new lifetime models by specifying the functional form of $\xi(u)$. Most of the existing life distributions are obtained by assuming convenient forms of various reliability functions. When uncertainty is used as a criterion to assess the reliability of a device, such characteristics of a reliability function may not always be appropriate and will be difficult to interpret in terms of entropy. A more direct approach is therefore to model lifetimes on the basis of the properties of $H_F(t)$ or its counterpart $\xi(u)$ derived from the evidence supplied by the observations in the sample. We now present some simple functional forms of $\xi(u)$ and derive the corresponding life distributions.

2.6.1. Residual entropy model I. Let $\xi(u) = \log(1 - u) + a + bu$, $b > 0$. Then from (2.33)

$$q(u) = \exp[a - b + 2bu], \tag{2.35}$$

and the corresponding quantile function is

$$Q(u) = \frac{e^{a-b}}{2b} \left(e^{2bu} - 1\right). \tag{2.36}$$

It is easy to see that (2.36) is quantile function, since $Q(u)$ is continuous and increasing. Also $Q(0) = 0$ and $Q(1) = \frac{e^{a-b}(e^{2b}-1)}{2b}$ so that X is a random variable with support $\left(0, \frac{e^{a+b}-e^{a-b}}{2b}\right)$. To understand the nature of $\xi(u)$, we note that

$$\xi'(u) = -\frac{1}{1-u} + b,$$

so that X is DURL whenever $0 < b < 1$. When $b > 1$, $\xi(u)$ is initially decreasing, reaches a minimum at $u = \frac{b-1}{b} < 1$ for $b > 1$ and then increases. This is illustrated in Figure 2.2 for different values of a and b.

Thus $\xi(u)$ is BT with change point $\frac{b-1}{b}$. The hazard quantile function of X is

$$h_Q(u) = (1 - u)^{-1} \exp[b - a - 2bu]$$

which is also BT with change point $\frac{2b-1}{2b}$, $b > \frac{1}{2}$. We notice that both the residual entropy and hazard function are bathtub-shaped but with different change points. When $b \le 1$, $\xi'(u) < 0$ and so X is DURL. Since

$$h'_Q(u) = e^{b-a-2bu}(2bu + 1 - 2b),$$

the sign of h' depends $(2bu + 1 - 2b)$ which is positive so that X is IHR whenever $b < \frac{1}{2}$.

The quantile function (2.36) admits a tractable distribution function

$$F(x) = (2b)^{-1} \log \left(1 + 2e^{b-a}bx\right)$$

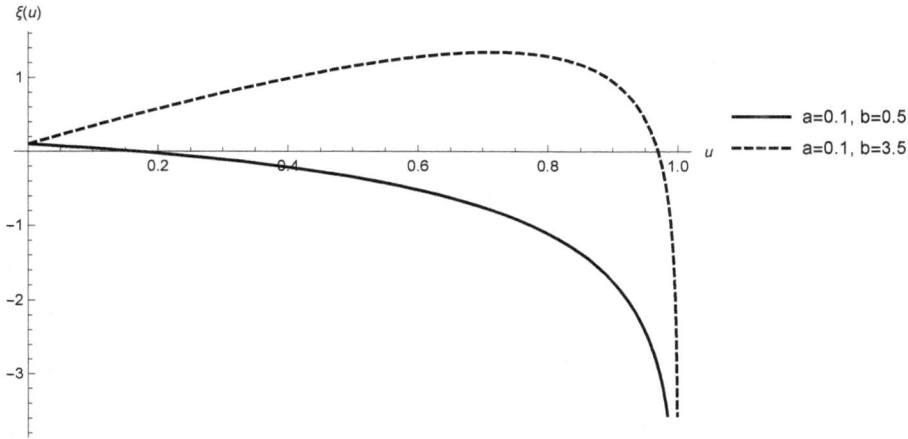

Figure 2.2. $\xi(u)$ of Model I for different values of a and b.

and density function

$$f(x) = \left(e^{a-b} + 2bx\right)^{-1}.$$

In terms of the distribution function

$$H_F(t) = \log \frac{a}{2b}\left(1 + 2be^{b-a}t\right)^{\frac{1}{2}} + \log\log\left(1 + 2be^{b-a}t\right).$$

EXAMPLE 2.4. *In order to validate Model I in a real situation, we consider the data on the failure times of 50 devices given in Aarset (1987) and estimate the parameters by the method of percentiles by equating the 20th and 80th percentile points of the sample with those of the population. Solving the resulting equations give the estimates of a and b as*

$$\hat{a} = 4.96 \text{ and } \hat{b} = 1.8.$$

The data is distributed into five bins of ten observations each in ascending order of magnitude. When model I holds for the data, the estimated value of $Q(u)$ at $u = \frac{1}{5}, \frac{2}{5}, \frac{3}{5}$ and $\frac{4}{5}$ and the fact that if U is uniform over $[0,1]$, then X and $Q(U)$ are identically distributed enable us to find the observed frequencies with $Q\left(\frac{1}{5}\right) = 6.77, Q\left(\frac{2}{5}\right) = 20.8, Q\left(\frac{3}{5}\right) = 4.49$ and $Q\left(\frac{4}{5}\right) = 80.08$, the observed frequencies turn out to be, 10, 8, 8, 11 and 13 in the five classes as against the expected frequency of 10 in each. These give a chi-square value of 1.75, justifying the appropriateness of the model. Notice that since $b > 1$, $\xi(u)$ is UBT shaped.

2.6.2. Residual entropy model II. The choice

$$\xi(u) = \alpha \log u + \beta \log(1 - u) + \log c, \quad \alpha, \beta \text{ real}, \ c > 0$$

leads to

$$q(u) = \exp\left[\log cu^\alpha(1-u)^\beta - (1-u)(\alpha u^{-1} - \beta(1-u)^{-1}) - \log u - 1\right]$$

$$= cu^\alpha(1-u)^{\beta-1}\exp\left[\beta - 1 - \frac{\alpha(1-u)}{u}\right]$$

$$= cu^\alpha(1-u)^{\beta-1}e^{\alpha+\beta-1}\exp\left[-\frac{\alpha}{u}\right], \quad 0 < u \leq 1. \tag{2.37}$$

We have

$$\xi'(u) = \frac{\alpha}{u} - \frac{\beta}{1-u}. \tag{2.38}$$

When $\alpha = 0$ and $\beta > 0$, $\xi'(u) < 0$ in which case X is DURL and similarly $\alpha = 0$, $\beta < 0$ provides IURL distribution. The case $\alpha = 0$, $\beta = 0$ corresponds to $\xi(u) = \log c$ and X is exponential. Notice that the above three special cases arise when X is distributed respectively as rescaled beta, Pareto II and exponential. A solution to (2.38) is

$$u_0 = \frac{\alpha}{\alpha + \beta}, \quad 0 < u_0 < 1, \quad \alpha, \beta > 0.$$

Thus $\xi(u)$ is increasing, reaches its maximum value $\alpha\log\frac{\alpha}{\alpha+\beta} + \beta\log\frac{\beta}{\alpha+\beta} + \log C$ at u_0 and then decreases. Hence (2.37) provides a model that has UBT shaped residual entropy. Arguing in the same manner $\alpha > 0$ and $\beta > 0$ shows that X is IURL and so is when $\alpha > 0$ and $\beta = 0$. Similarly $\alpha < 0$ and $\beta \geq 0$ gives a DURL model. When α and β are < 0, u_0 is in $(0,1)$ and $\xi''(u) \geq 0$ implies that $\xi(u)$ attains the minimum at u_0 and therefore $\xi(u)$ is BT. The entire analysis can be read from following representation of the parameter space.

$\beta\backslash\alpha$	$= 0$	> 0	< 0
$= 0$	Exponential	IURL	DURL
> 0	Rescaled beta (DURL)	UBT	DURL
< 0	Pareto II (IURL)	IURL	BT

Thus our model is simple and yet flexible capable of assuming different shapes according to various regions in the parameter space. The family is capable of modelling quantile residual entropy functions of different forms, which is evident from the plots of $\xi(u)$ through Figures 2.3 through 2.9.

If the model is analyzed from a reliability point of view, the hazard quantile function is given by

$$\log h_Q(u) = -\frac{\alpha}{u} + \frac{\beta}{1-u} - \frac{\alpha}{u^2}$$

with derivative

$$h'_Q(u) = \left[-\frac{\alpha}{u} - \frac{\alpha}{u^2} + \frac{\beta}{1-u}\right]h_Q(u).$$

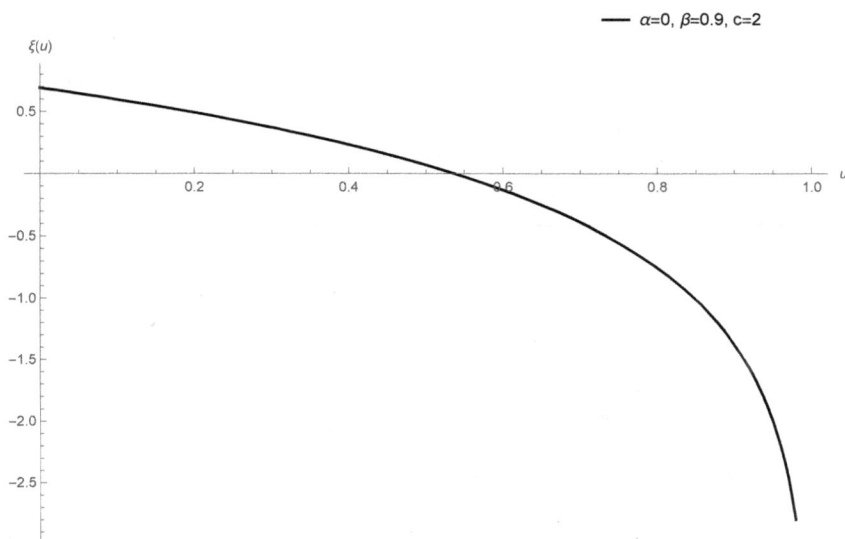

Figure 2.3. $\xi(u)$ of Model II for $\alpha = 0, \beta = 0.9$ and $c = 2$.

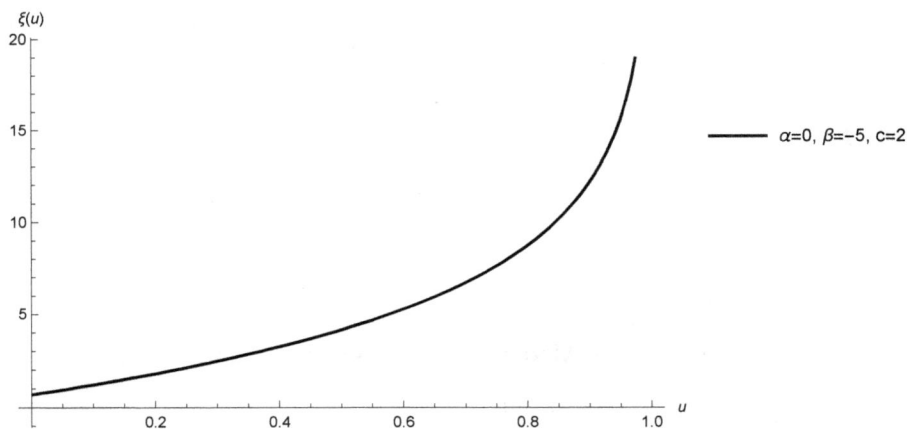

Figure 2.4. $\xi(u)$ of Model II for $\alpha = 0, \beta = -5$ and $c = 2$.

The various possibilities are X is IFR for $\alpha > 0, \beta = 0; \alpha = 0, \beta > 0$, DFR for $\alpha = 0, \beta < 0$.

The empirical verification of the shape of the $\xi(u)$ for a given data can be accomplished in the following manner. It is shown that X is DURL (IURL) if $h(t)$ is increasing (decreasing). Similarly X is BT (UBT) if $h(t)$ or $h_Q(u)$ is UBT (BT). Thus using a nonparametric estimate of $h(t)$ available in literature, a plot of $\hat{h}(t)$ can suggest the nature of $H_F(t)$. Alternatively one can use the scaled total time on test transform (TTT) plot. Recall

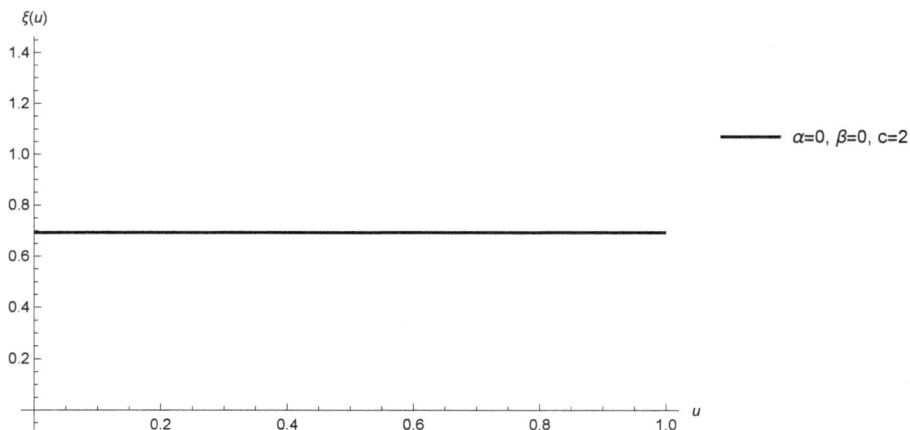

Figure 2.5. $\xi(u)$ of Model II for $\alpha = 0, \beta = 0$ and $c = 2$.

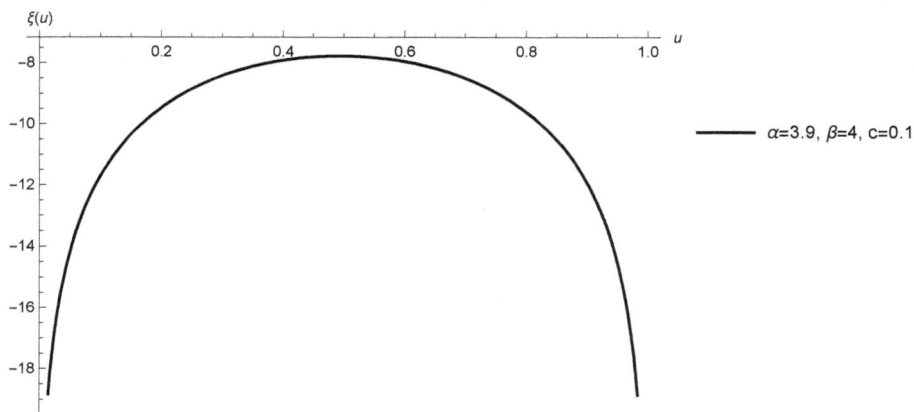

Figure 2.6. $\xi(u)$ of Model II for $\alpha = 3.9, \beta = 4$ and $c = 0.1$.

that the scaled TTT is defined as

$$\phi(u) = \frac{\int_0^u (1-p)q(p)dp}{\mu}, \ \mu = E(X)$$

and its empirical version

$$\phi_{r:n} = \frac{\sum_{j=1}^{r} (n-j+1)(X_{j:n} - X_{j-1:n})}{\sum_{j=1}^{n} (n-j+1)(X_{j:n} - X_{j-1:n})}$$

is called scaled total time on test statistic. Since X is IFR (DFR, BT, UBT) if and only if $\phi(u)$ is concave (convex, convex in $(0, t)$ and concave in

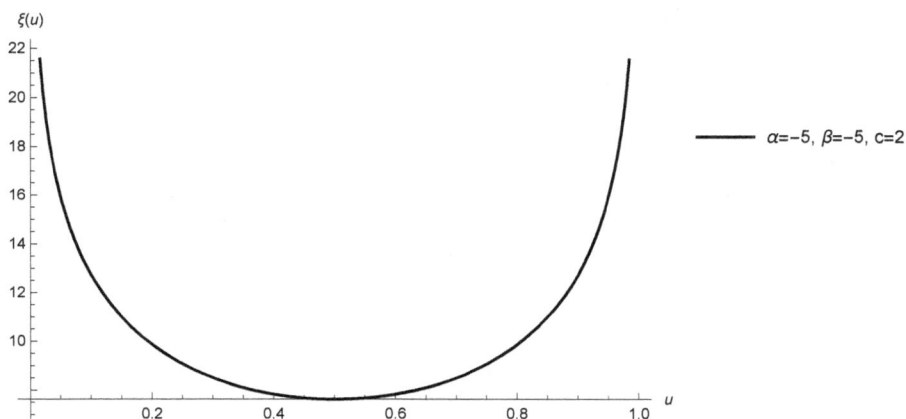

Figure 2.7. $\xi(u)$ of Model II for $\alpha = -5, \beta = -5$ and $c = 2$.

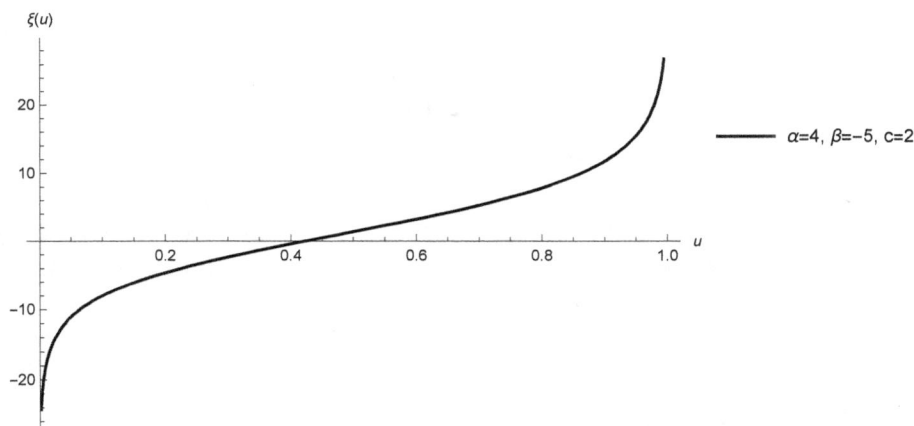

Figure 2.8. $\xi(u)$ of Model II for $\alpha = 4, \beta = -5$ and $c = 2$.

(t, ∞), concave in $(0, t)$ and convex in (t, ∞)), the $\phi_{r:n}$ plot can indicate whether X is DURL, IURL, BT or UBT. Lastly one can use the kernel residual entropy function proposed by Belzunce et al. (2001).

In view of the above examples we formally define BT and UBT models when the distribution is specified by a quantile function.

DEFINITION 2.4. *A random variable X with quantile function $Q(u)$ is said to have a bathtub-shaped (BT) residual entropy quantile function $\xi(u)$ if*

$$\xi(u) = \begin{cases} \xi_1(u), & u \leq u_1 \\ c, & u_1 \leq u \leq u_2 \\ \xi_2(u), & u \geq u_2 \end{cases}$$

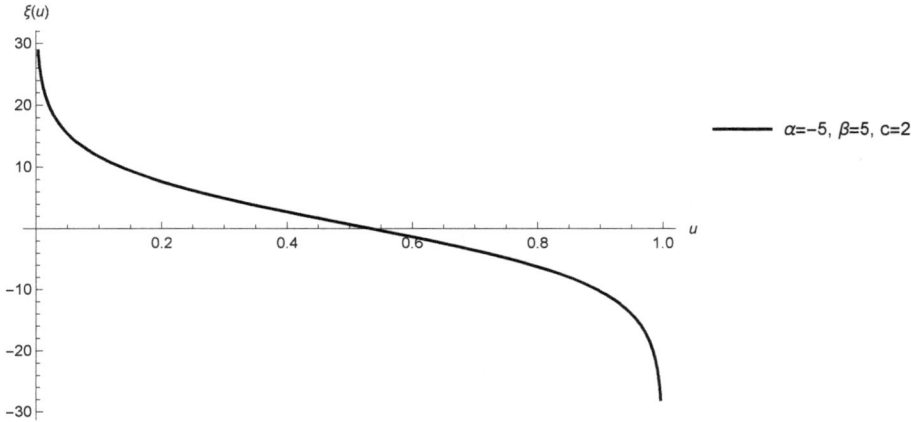

Figure 2.9. $\xi(u)$ of Model II for $\alpha = -5, \beta = 5$ and $c = 2$.

where c is a constant, $\xi(u)$ is strictly decreasing and $H^{(2)}(u)$ is strictly increasing. If $\xi_1(u)$ is strictly increasing and $\xi_2(u)$ is strictly decreasing an upside-down bathtub-shaped residual entropy quantile function is formed.

Assuming differentiability of $\xi(u)$, X is BT (UBT) if and only if $\xi'(u) < (>)0$ for u in $(0, u_0)$, $\xi'(u_0) = 0$ and $\xi'(u)(<)0$ in $(u_0, 1)$ with a single change point u_0. When $H_F(t)$ is BT (UBT) so is $\xi(u)$ with change points of $H_F(t)$ and u_0 of $\xi(u)$ satisfying $u_0 = F(t_0)$. Some further properties of $\xi(u)$ are

(a) If $Y = aX + b$, $\xi_Y(u) = \xi_X(u) + \log a$, $a > 0$
(b) If $Q(u) = Q_1(u) + Q_2(u)$ where Q_1 and Q_2 are quantile functions, then $Q(u)$ is also a quantile function with

$$\xi_Q(u) = \log(1 - u) + (1 - u)^{-1} \int_u^1 \log(q_1(p) + q_2(p))dp.$$

(c) If $Y = T(X)$ is a non-decreasing function of X, then the residual entropy quantile function of T is given by

$$\xi_T(u) = \log(1 - u) + (1 - u)^{-1} \int_u^1 \log t(p)dp$$

where $t(p) = \frac{dT(Q(p))}{dp}$.

Sunoj and Sankaran (2012) have defined the increasing (decreasing) residual quantile entropy class, IRQE (DRQE) according as $\xi(u)$ is increasing (decreasing). Observe that from the above discussions,

$$\frac{dH_F(t)}{dt} = \frac{dH_F(Q(u))}{dQ(u)} = \frac{d\xi(u)}{du}\frac{du}{dQ} = \frac{d\xi(u)}{du}\frac{1}{q(u)}$$

which shows that X is IURE (DURE) if and only if X is IRQE (DRQE). Thus the classes in the two approaches are the same and hence the original nomenclature IURE (DURE) will be used through out this work irrespective of whether $F(x)$ or $Q(u)$ is used.

Besides the properties mentioned in Section 2.4.3, the IURE (DURE) classes satisfy some additional properties as well.

THEOREM 2.10. *The class DURL (IURL) does not imply IHR (DHR). However if X is DURL (IURL) and $\xi(u)$ is convex (concave) then X is IHR (DHR).*

PROOF. Let $\xi(u) = \log(1 - u) + \frac{3}{4}u$. Then

$$\xi'(u) = -\frac{1}{1-u} + \frac{3}{4} < 0$$

showing X is DURL. At the same time

$$h'_Q(u) = e^{3/4 - 3/2u}\left(\frac{3}{2}u - \frac{1}{2}\right)$$

which is positive for $u = \frac{1}{2}$ and negative for $u = \frac{1}{8}$, and hence not IHR. The second part of the theorem follows from equation (2.32). □

(2) X is DURL (IURL) if and only if

$$\xi(u) \leq (\geq)1 + \log(m_Q(u) - (1 - u)m'_Q(u)).$$

PROOF. Equation (2.30) shows that X is DURL if and only if

$$(1 - \xi(u)) \geq \log h_Q(u)$$

or

$$\xi(u) \leq 1 - \log h_Q(u)$$
$$= 1 + \log \frac{1}{h_Q(u)}$$
$$= 1 + \log(m_Q(u) - (1 - u)m'_Q(u)), \text{ using (1.19).}$$

□

The result also provides a bound for $\xi(u)$ in terms of mean residual quantile function.

(3) If X is IHR (IHRA, NBU) and $f(0) \geq \lambda$, then $\xi(u) \leq 1 - \log \lambda$.

(4) If X has a decreasing density function such that $f(0) \leq 1$ then $\xi(u) \geq \log(1 - u)$.

(5) The DURL is not closed under the formation of mixtures

(6) If F_1 and F_2 are IURL their convolution is not IURL nor the parallel system formed by them. The results (3) through (6) are discussed in Qiu et al. (2019).

A function that is closely associated with $H_F(t)$ is given by

$$A_F(t) = 1 - \int_t^\infty \log \frac{h(x)}{h(t)} dF_t(x). \tag{2.39}$$

Using quantile functions,

$$\xi_1(u) = A_F(Q(u)) = 1 - \frac{1}{1-u} \int_u^1 \log \frac{h_Q(p)}{h_Q(u)} dp$$

$$= 1 - \frac{1}{1-u} \int_u^1 \log h_Q(p) dp + \log h_Q(u)$$

$$= \xi(u) + \log h_Q(u). \tag{2.40}$$

By virtue of (2.40), the identities (2.30), (2.31) and (2.32) reduce to

$$\xi_1(u) = 1 + (1-u)\xi'(u) \tag{2.41}$$

$$h_Q(u) = \exp[\xi_1(u) - \xi(u)] \tag{2.42}$$

$$h'_Q(u) = h_Q(u)(\xi'_1 - \xi') = h_Q(u) \left(\xi'_1(u) - \frac{\xi_1(u) - 1}{1-u} \right). \tag{2.43}$$

Equations (2.41) through (2.43) enable the investigation of properties of $\xi(u)$ in terms of $\xi_1(u)$ which is simpler to apply analytically. There is a statistical interpretation to (2.39). We have

$$\int_t^\infty \log \frac{h(x)}{h(t)} dF_t(x) = \int_t^\infty [\log h(x) - \log h(t)] f_t(x) dx$$

$$= \int_t^\infty \log h(x) f_t(x) dx - \log h(t)$$

$$= E[\log h(X) - \log h(t)|X > t]. \tag{2.44}$$

If $G_1(t)$ is the geometric mean of the random variable $h(X)$ given $X > t$,

$$\log G_1(t) = E(\log h(X)|X > t)$$

and hence (2.43) gives the geometric mean of residual hazard rates of those who have survived age t. When a device is subject to positive ageing $h(x) > h(t)$, (2.43) is positive while for negative ageing it is negative. Thus (2.39) or its counterpart $\xi_1(u)$ in (2.40) is an indicator of the ageing property of the device with life distribution F.

Like $\xi(u)$ the functional form of $\xi_1(u)$ can also determine the life distribution. We consider some simple forms of $\xi_1(u)$.

THEOREM 2.11. *The random variable X will follow a generalized Pareto distribution if and only if $\xi_1(u)$ is a constant for all $0 < u < 1$.*

PROOF. When X has a generalized Pareto distribution with quantile function

$$Q(u) = \frac{b}{a}\left[(1-u)^{-\frac{a}{(a+1)}} - 1\right], \ b > 0, \ a > -1 \quad (2.45)$$

then

$$h_Q(u) = b^{-1}(a+1)(1-u)^{\frac{a}{a+1}}$$

$$\xi(u) = 1 - \log b^{-1}(a+1) - \frac{a}{a+1}(\log(1-u) - 1)$$

and

$$\xi_1(u) = \xi(u) + \log h_Q(u) = \frac{2a+1}{a+1},$$

a constant. Conversely assuming $\xi_1(u) = C$, a constant we have from (2.41)

$$\xi(u) = K - (C-1)\log(1-u)$$

and hence

$$\log h_Q(u) = \xi_1(u) - \xi(u) = (C - K) + (C - 1)\log(1 - u).$$

Thus the hazard quantile function is of the form

$$h_Q(u) = A(1-u)^{a/(a+1)}, \ A = C - K, \ C = \frac{a}{a+1}$$

which characterizes the generalized Pareto distribution. $\qquad\square$

A much larger class of distributions which subsumes the result of Theorem 2.11 can be characterized by a simple functional form of $\xi_1(u)$.

THEOREM 2.12. *A random variable X follows the distribution with quantile density function of the form*

$$q(u) = Cu^\alpha(1-u)^{-(A+\alpha)}, \ c > 0, \ A, \alpha \ real \quad (2.46)$$

if and only if

$$\xi_1(u) = A + B\frac{\log u}{1-u} \quad (2.47)$$

for all $0 < u < 1$ and real constants A and B.

PROOF. If we assume (2.47), we have on using (2.47),

$$\xi'(u) = \frac{A-1}{1-u} + \frac{B\log u}{(1-u)^2}.$$

Integrating with respect to u,

$$\xi(u) = -(A-1)\log(1-u) + \frac{Bu\log u}{1-u} + B\log(1-u) + BK$$

$$= BK + \frac{Bu\log u}{1-u} + (B - A + 1)\log(1-u),$$

where K is the constant of integration. Then

$$\log h_Q(u) = \xi_1(u) - \xi(u)$$
$$= (A - BK) + B \log u + (A - B - 1) \log(1 - u)$$

or

$$h_Q(u) = e^{A-BK} u^B (1 - u)^{A-B-1}.$$

Finally

$$q(u) = [(1 - u)h_Q(u)]^{-1}$$
$$= \begin{cases} e^{BK} u^{-B}(1 - u)^{B-1}, & A = 1 \\ e^{-(A-BK)} u^{-B}(1 - u)^{B-A} & A \neq 1 \end{cases}$$

which has form (2.46) with $C_1 = e^{BK-A}$ and $\alpha = -\beta$. Conversely if X has the quantile density function stated in the theorem,

$$\xi(u) = C_1 + A + \alpha - \quad (1 - u)^{-1}[\alpha + \alpha u(\log u - 1)]$$
$$+ (A + \alpha - 1) \log(1 - u)$$

and

$$h_Q(u) = C_1^{-1} u^{-\alpha}(1 - u)^{A+\alpha-1}.$$

Substituting in (2.40) we have (2.47) and the proof is complete.   □

There are several special cases of Theorem 2.12 which are of interest. The random variable X is distributed as

- exponential $(A = 1, B = 0)$
- Pareto II $(A = \frac{c+1}{c} > 1, B = 0)$
- rescaled beta $(A = \frac{c-1}{c} < 1, B = 0)$
- power $(A = \frac{\beta-1}{\beta} = B)$
- log logistic $(A = 2, B = \frac{\beta-1}{\beta})$
- Govindarajulu $(A = 1 - \beta = B)$

Corollary. The family (2.46) is also characterized by $\xi(u)$ with its form given above.

THEOREM 2.13. *A non-negative random variable X is distributed as the residual entropy model*

$$q(u) = \exp[a - b + 2bu]$$

if and only if

$$\xi_1(u) = b(1 - u) \text{ for all } 0 < u < 1.$$

The function $\xi_1(u)$ can be used to generate new models of lifetimes. We give one example.

EXAMPLE 2.5. *Assume the simple linear form*

$$\xi_1(u) = 1 + a + bu, \ a + b + 1 \geq 0; \ b \geq 0$$

so that

$$\xi'(u) = \frac{a + bu}{1 - u}.$$

Integrating

$$\xi(u) = -(a + b)\log(1 - u) - bu + K.$$

The hazard quantile function is calculated from

$$\log h_Q(u) = \xi_1(u) - \xi(u)$$
$$= (1 - K) + a + bu + (a + b)\log(1 - u) + bta$$

as

$$h_Q(u) = e^{1-K+a}e^{2bu}(1 - u)^{a+b}.$$

Thus the quantile density function of X becomes

$$q(u) = e^{K-a-1}e^{-2bu}(1 - u)^{-(a+b+1)}$$
$$= Ce^{-2bu}(1 - u)^{-(a+b+1)}, \ C = e^{K-a-1}. \qquad (2.48)$$

The special cases of the model are the generalized Pareto when $b = 0$ and $A = \frac{1}{a}$, $a > -1$ and the model (2.35) when $b + a + 1 = 0$.

REMARK 2.3. *The distribution (2.48) is characterized by the residual entropy functions*

$$\xi(u) = -(a + b)\log(1 - u) - bu + K.$$

An equivalent of the definition of the NBURE (NWURE) class in terms of quantile functions is that $\xi(u) \leq (\geq)\xi(0)$ for all u.

THEOREM 2.14. *DURL (IURL) $\Rightarrow$ NBURE (NWURE), but the converse is not true.*

PROOF. Since X is DURL the function $\xi(u)$ is decreasing and hence $\xi(u) \leq \xi(0)$ so that X is NBURE. The proof of NWRUE is similar. Assume that X follows distribution (2.35) with $a > 0$ and $b > 1$. Then $\xi(u) \geq \xi(0)$ and hence X is NWURE. However, $\xi(u)$ has bathtub shape for the above parameter values and so NWURE does not imply IURL. $\qquad \square$

REMARK 2.4. *X is DURL (IURL) if $\xi_1(u)$ is $< (>)1$.*

2.7. Weighted residual entropy

The role of weighted distributions in reliability modelling and analysis was discussed in Section 1.12. In this section we define and study the properties of the residual entropy of weighted distributions by recalling that a weighted distribution with weight function $w(x)$, of a random variable X_w is defined as

$$f_w(x) = \frac{w(x)f(x)}{Ew(X)}, \quad E\,w(X) < \infty \tag{2.49}$$

Belis and Guiasu (1968) proposed the weighted entropy of X_w as

$$H_w = -\int_0^\infty f_w(x) \log f_w(x) dx.$$

See also Guiasu (1986) for various results on the properties of H_w and its applications. As a natural extension of (2.50) to the residual life X_t, one can define the residual entropy of X_t as

$$H_w(t) = -\int_0^\infty f_{w,t}(x) \log f_{w,t}(x) dx \tag{2.50}$$

where

$$f_{w,t} = \frac{w(x)f(x)}{\bar{F}(t)}.$$

Two special cases of interest in the context of reliability studies are the length-biased model when $w(x) = x$ and the equilibrium distribution when $w(x) = [h(x)]^{-1}$. In the first case

$$H^L(t) = -\int_t^\infty x\frac{f(x)}{\bar{F}(t)} \log \frac{f(x)}{\bar{F}(t)} dx \tag{2.51}$$

which was investigated in detail by Di Cresenzo and Longobardi (2006).

$$H^L = -\frac{1}{\bar{F}(t)} \int_t^\infty xf(x) \log f(x)dx + \frac{\log \bar{F}(t)}{\bar{F}(t)} \int_t^\infty xf(x)dx$$

$$= tH_F((t) + \frac{1}{\bar{F}(t)} \int_t^\infty \bar{F}(y) \left[H_F(y) + \log \frac{\bar{F}(t)}{\bar{F}(y)} \right] dy.$$

It is shown that

(i) $H^L_{aX+b} = a[H_L + E(X) \log a] + b[H_F + \log a]$, $H^L_X = -\int_0^\infty xf(x) \log f(x)dx$

(ii) $H^L_{X,Y} = E(Y)H^L_X + E(X)H^L_Y$, where $H^L_{X,Y} = E[XY \log f(X,Y)]$ and X and Y are independent random variables.

(iii) Two random variables can have the same weighted entropy and that H_L may be $-\infty$. When the support is finite $[0, c]$,

$$H^L \leq \mu \log \frac{c^2}{2\mu}$$

(iv) The expressions for weighted entropy when X is distributed as
  (a) exponential, $H^L = (2 - \log \lambda)\lambda^{-1}$
  (b) uniform $[a, b]$, $H^L = \frac{a+b}{2} \log(b - a)$
  (c) gamma, $f(x) = \frac{x^{\alpha-1}e^{-x/\beta}}{\beta^\alpha \Gamma(\alpha)}$,

$$H^L = \alpha\beta \log(\beta^\alpha \Gamma(\alpha)) - \alpha(\alpha - 1)\beta\{\log \beta + \psi_0(\alpha + 1)\}$$
$$+ \alpha(\alpha + 1)\beta.$$

  (d) beta, $f(x) = \frac{1}{B(\alpha,\beta)} x^{\alpha-1}(1 - x)^{\beta-1}$,

$$H^L = \frac{\log B(\alpha, \beta)\Gamma(\alpha + 1)\Gamma(\beta)}{B(\alpha, \beta)\Gamma(\alpha + \beta + 1)}$$

$$-\frac{(\alpha - 1)\Gamma(\alpha + 1)\Gamma(\beta)}{B(\alpha, \beta)\Gamma(\alpha + \beta - 1)}\{\psi_0(\alpha + 1) - \psi_0(\alpha + \beta + 1)\}$$

$$-\frac{(\beta - 1)\Gamma(\beta)}{B(\alpha, \beta)\Gamma(\alpha + \beta + 1)}\{\psi_0(\beta)\Gamma(\alpha + 1) - \alpha\Gamma(\alpha)\psi_0(\alpha + \beta + 1)\}$$

  where $\psi_0(t) = \frac{d \log \Gamma(t)}{dt}$
  (e) When $H_F(t)$ is increasing (decreasing) $H^L(t)$ is also increasing (decreasing)
  (f) If $h(t)$ is increasing

$$H^L(t) \geq E(X|X > t) \log h(t)$$

  (g) $H_{aX}^L(t) = aH_X^L(\frac{t}{a}) + M(\frac{t}{a})a \log a$, $M(t) = E(X|X > t)$

$$H_{X+b}^L(t) = H_X^L(t - b) + bH_X^b(t - b)$$

  (h) If $Y = \phi(X)$ is strictly monotone continuous and differentiable

$$H_Y^L(t) = H^L(\phi^{-1}(t)) + E[\phi(X)\log \phi'(X)|X > \phi^{-1}(t)],$$
$$\text{if } \phi(t) \text{ is increasing}$$
$$= H^L(\phi^{-1}(t)) + E[\phi(X)\log(-\phi'(X))|X > \phi^{-1}(t)],$$
$$\text{if } \phi(t) \text{ is decreasing.}$$

Belzunce et al. (2004) have given an alternative expression for H_L in the form

$$H^L(t) = 1 - \frac{1}{(t + m(t))\bar{F}(t)} \int_t^\infty xf(x) \log\left(\frac{xh(x)}{x + m(x)}\right) dx.$$

They have observed that the result of Oluyede (1999) that $H_F(t) \leq H^L(t)$ for all $t > 0$ is not true and gave examples to show that $H_L(t) \leq H_F(t)$ and also $H_F(t) \geq H^L(t)$. Introducing the additional condition X or X_L have DFR then the modified result X has smaller residual entropy than X_L. In addition if X has larger residual entropy than X_L then (i) X is DURL

$\Rightarrow X_L$ is DURL and (ii) X is IURL $\Rightarrow X$ is DURL. However X is IURL in general does not mean that X_L is also IURL.

When a general weight function $w(x)$ is used as in (2.50), the hazard rate becomes

$$h_w(x) = \frac{w(x)}{E(w(X)|X > x)} h(x)$$

and further if $E(w(X)|X > t)$ is increasing (decreasing) $E(w(X)|X \geq t)/w(t)$ is decreasing (increasing) and $\lim_{x \to \infty} w(x) < \infty \ (> 0)$ and X is DURL (the weighted random variable X_W is DURL) then X_w is DURL (X ie. DURL). The corresponding result for the IURL class is that X IURL (X_W is IURL) then X_W is IURL (X is IURL) provided that $E(w(X)|X \geq t)$ is increasing (decreasing), $\frac{E(w(X)|X \geq t)}{w(t)}$ is increasing (decreasing) and $0 < \lim_{x \to \infty} w(x) < \infty$.

The entropy of the equilibrium random variable X_E with density function $f_E(x) = \frac{\bar{F}(x)}{\mu}$, $\mu = E(X)$ is

$$H_E = \log \mu - \frac{1}{\mu} \int_0^\infty \bar{F}(x) \log \bar{F}(x) dx.$$

EXAMPLE 2.6.   *(1) When X is exponential X_E is also exponential with the same parameter. Hence $H_E = 1 - \log \lambda$*
(2) When X is generalized Pareto with $\bar{F}(x) = (\frac{a}{a+bt})^{1/b}$,

$$H_F(t) = \frac{a}{a+1} + \log(at+b) - \log(a+1),$$

$$H_E(t) = \frac{a+1}{a} - \log \frac{a}{at+b}$$

In general it is difficult to find a simple relationship between H_F and H_E, but there exists some for specific distributions. As seen from the above example when X is exponential $H_E(t) = H_F(t)$ and for the generalized Pareto $H_E(t) = H_F(t) + C$, where C is a constant.

Note that the equilibrium residual entropy is

$$H_E(t) = 1 - \frac{1}{F_E(t)} \int_t^\infty \log h_E(t) f_E(x) dx$$

$$= 1 - \frac{1}{\int_t^\infty \bar{F}(x) dx} \int_t^\infty \log \frac{1}{m(t)} \bar{F}(x) dx$$

$$= 1 + \frac{1}{m(t)\bar{F}(t)} \int_t^\infty \log m(x) \bar{F}(x) dx.$$

Differentiation leads

$$H_E'(t) = \frac{1}{m(t)} [H_E(t) - 1 - \log m(t)].$$

Using the above formulas for $H_E(t)$ and $H'_E(t)$ one can derive various properties of $H_E(t)$ in the same manner as we have done for $H_F(t)$.

2.8. Stochastic orders

The major stochastic orders employed in reliability theory were presented in Section 1.13, and we take a similar approach with reference to residual entropy. Our main objective to compare life distributions with reference to the measures of residual entropy is to say which of two models F and G have lesser $H(t)$. These orders also help in deriving some theoretical results. Ebrahimi and Pellerey (1995); Ebrahimi and Kirmani (1996c); Gupta et al. (2014) have provided several basic contributions in this connection.

Let X and Y be two non-negative random variables with absolutely continuous distribution functions F and G, hazard rates h_F and h_G and residual entropy functions $H_F(t)$ and $H_G(t)$.

DEFINITION 2.5. *We say that lifetime X is smaller than Y in residual entropy ordering if $H_F(t) \leq H_G(t)$ for all $t > 0$ and denote it as $X \leq_{RE} Y$.*

With the aid of examples in each case Ebrahimi and Pellerey (1995) have proved the following results.

(a) If for any $t \geq 0$,

$$\frac{1}{F(t)} \log h_F(t) \geq \frac{1}{G(t)} \log h_G(t)$$

for all $x \geq t$ and $h_F(x)$ is a non-decreasing function of x, then $X \leq_{RE} Y$.

(b) Let $\frac{h_F(x)}{h_G(x)}$ be non-decreasing, $h_F(x)$ is non-increasing and $X \leq_{hr} Y$. Then $X \leq_{RE} Y$.

(c) If $X \leq_{LR} Y$ and $h_F(x)$ or $h_G(x)$ be increasing then $X \leq_{RE} Y$.

(d) Let $z_1 = a_1 X + b_1$ and $Z_2 = a_2 Y + b_2$, $a_1, a_2 > 0$ and $b_1, b_2 \geq 0$ be linear transformations of X and Y. Given $X \leq_{RE} Y$, $a_1 = a_2$ and $b_1 = b_2$, we have $Z_1 \leq_{RE} Z_2$. On the other hand $a_1 \leq a_2$, $b_1 \leq b_2$ and either $H_F(t)$ or $H_G(t)$ is non-increasing then also $Z_1 \leq_{RE} Z_2$.

(e) Given two devices with lifetimes X and Y, X is very strongly better than Y if $X \leq_{RE} Y$ and $X \geq_{LR} Y$, strongly better if $X \leq_{RE} Y$ and $X \geq_{hr} Y$, weakly better if $X \leq_{RE} Y$ and $X \geq_{mrl} Y$ and X is better than Y if $X \geq_{RE} Y$ and $X \geq_{st} Y$. With the above definitions we have the implications.

very strongly better $\Rightarrow$ strongly better $\Rightarrow$ better

and, very strongly better $\Rightarrow$ strongly better $\Rightarrow$ weakly better

Another set of results is given in Ebrahimi and Kirmani (1996c) using the hazard rate order and dispersive order.

(i) If $X \leq_{\text{hr}} Y$, $h_F(x)$ or $h_G(x)$ is decreasing then $X \leq_{\text{RE}} Y$.

(ii) If $X \leq_{\text{disp}} Y$ and $h_F(x)$ or $h_G(x)$ is increasing then $X \leq_{\text{RE}} Y$.

(iii) If If $\phi(x)$ is non-negative and increasing satisfying $\phi'(x) \geq 1$ for all $x > 0$ and F is DURL, then $\phi(X) \geq_{\text{RE}} X$. Similarly if $\phi(x)$ satisfies the above conditions except $\phi'(x) \leq 1$ (instead of $\phi'(x) \geq 1$) then $\phi(x) \leq_{\text{RE}} X$.

(iv) Given $X \leq_{\text{RE}} Y$, $X \leq_{\text{hr}} Y$ and $\phi(x)$ is non-negative and increasing then $\phi(X) \leq_{\text{RE}} Y$.

(v) Let $Y = \phi(X, Z)$, X and Z being independent. Under the conditions $\phi(x, z) \geq 0$. $\phi(x, z)$ is increasing in x for all $z \geq 0$, $\frac{\partial}{\partial x}\phi(x, z) \geq 1$ and F is DURL, the ordering $\phi(x) \leq_{\text{RE}} \phi(Y)$ holds.

The result (c) above was further strengthened by Gupta et al. (2014) by showing that

(vi) When X is DHR $X \leq_{\text{RE}} X_{n:n}$ and $X_{r:n} \leq_{\text{RE}} X_{s:n}$, $s > r$. Also for $(n+1)$ independent and identically distributed random variables their order statistics satisfy

$$X_{n:n} \leq_{\text{RE}} X_{n+1:n+1} \text{ and } X_{1:n+1} \leq_{\text{RE}} X_{j:n}$$

Various conditions under which two random variables X and Y have identical distributions were investigated by Ebrahimi (2001). His main results are

- if $H_F(t) \leq H_G(t)$, $X \leq_{\text{st}} Y$ and $g(x)$ is decreasing ($f(x)$ is increasing) then $X =_{\text{st}} Y$, where f and g are the probability density functions of X and Y.
- if $H_F(t) = H_G(t)$, $X \leq_{\text{icx}} Y$ and $\log f(x)$ is increasing and convex ($\log g(x)$ is decreasing and convex) then $X =_{\text{st}} Y$.
- if $H_F(t) = H_G(t)$, $X \leq_{\text{cx}} Y$ and $\log f(x)$ is convex ($\log g(x)$ is concave) then $X =_{\text{st}} Y$.

When dealing with quantile functions, Sunoj and Sankaran (2012) offered the following definition.

DEFINITION 2.6. *Let X and Y be continuous non-negative random variables with quantile functions $Q_X(u)$ and $Q_Y(u)$. Then X is said to be smaller than Y is residual quantile entropy denoted by*

$$X \leq_{RQE} Y \text{ if } \xi_X(u) \leq \xi_Y(u). \tag{2.52}$$

They have two results concerning the above ordering.

-

$$X \leq_{\text{HQ}} Y \Rightarrow X \leq_{\text{RQE}} Y \tag{2.53}$$

- If $X \leq_{\text{HQ}} Y$ and ϕ is non-negative increasing and convex, then $\phi(X) \leq_{\text{RQE}} \phi(Y)$. Of these, the first result directly establishes a link between hazard quantile function ordering and residual quantile entropy ordering, which is different from the result (vi) above and offers a less stringent condition.

One basic question that arises between the residual entropy order $\leq_{RE}$ and the residual quantile entropy order $\leq_{RQE}$ is whether they are identical. Our next example clarifies that $\leq_{RE}$ nether implies nor is implied by $\leq_{RQE}$.

EXAMPLE 2.7. *Let* $Q_X(u) = [(1 - u)^{-1/2} - 1]$ *and* $Q_Y(u) = [(1 - u)^{-1/3} - 1]$. *Then*

$$\xi_X(u) = -\log 2 + \frac{3}{2} - \frac{1}{2}\log(1 - u)$$

and

$$\xi_Y(u) = -\log 3 + \frac{4}{3} - \frac{1}{3}\log(1 - u)$$

showing that

$$\xi_X(u) \geq \xi_Y(u) \text{ or } X \geq_{RQE} Y.$$

The survival functions of X *and* Y *are*

$$\bar{F}_X(x) = (1 + x)^{-2} \text{ and } \bar{F}_Y(x) = (1 + x)^{-3}$$

with residual entropies

$$H_X(t) = 2 + \log(2 + t)$$

and

$$H_Y(t) = 2 + \log(3 + t).$$

Obviously $H_X(t) \leq H_Y(t)$ *or* $X \leq_{RE} Y$. *Thus* $X \leq_{RE} Y \not\Rightarrow X \leq_{RQE} Y$. *Interchanging the roles of* X *and* Y, *it is easy to see that* $X \leq_{RQE} Y \not\Rightarrow X \leq_{RE} Y$.

Result (2.53) states that if $h_{Q_X}(u) \geq h_{Q_Y}(u)$ then $\xi_X(u) \leq \xi_Y(u)$. The stochastic order defined by $h_{Q_X}(u) \geq h_{Q_Y}(u)$ is the hazard quantile function order defined in Vinesh Kumar et al. (2015) which is mathematically equivalent to $X \leq_{disp} Y$, the dispersive order, although $\leq_{HQ}$ and $\leq_{disp}$ has different interpretations. First we examine whether the converse of (2.53) holds good.

EXAMPLE 2.8. *Let* $q_X(u) = \exp[-\frac{1}{4} + \frac{1}{2}u]$ *and* $q_Y(u) = \exp[-\frac{1}{2} + u]$ *both following the residual entropy model* (2.35). *Then*

$$\xi_X(u) = \log(1 - u) + \frac{u}{4} \text{ and } \xi_Y(u) = \log(1 - u) + \frac{u}{2}$$

confirming that $\xi_X(u) \leq \xi_Y(u)$ *or* $X \leq_{QE} Y$. *Now*

$$h_{Q_X}(u) = (1 - u)^{-1} \exp\left[\frac{1}{4} - \frac{1}{2}u\right] \text{ and } h_{Q_Y}(u) = (1 - u)^{-1} \exp\left[\frac{1}{2} - u\right]$$

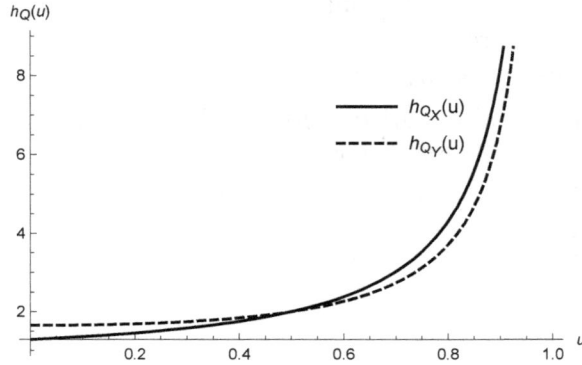

Figure 2.10. Plot of $h_{Q_X}(u)$ and $h_{Q_Y}(u)$ for different values of u.

It can be seen that $h_{Q_Y}(u) \geq h_{Q_X}(u)$ for $u = \frac{1}{4}$ and $h_{Q_X}(u) > h_{Q_Y}(u)$ for $u = \frac{5}{8}$ showing that $X \not\leq_{HQ} Y$.

In view of the reverse implications in (2.54) being not true it is interesting to know the additional conditions that renders some implication between the two orders.

THEOREM 2.15. *Let X and Y be absolutely continuous non-negative random variables such that $\frac{1-\xi_X(u)}{1-\xi_Y(u)}$ is non-decreasing (non-increasing). Then $X \leq_{RQE} (\geq_{RQE})Y \Rightarrow X \geq_{HQ} (\leq_{HQ})Y$.*

PROOF. First we observe that when $X \leq_{RE} (\geq_{RQE})Y$, we have $\xi_X(u) \leq (\geq)\xi_Y(u)$ for all u and $\frac{1-\xi_X(u)}{1-\xi_Y(u)}$ is non-decreasing

$$\Rightarrow \frac{d}{du}\frac{(1-u)(1-\xi_X(u))}{(1-u)(1-\xi_Y(u))} \geq 0$$

$$\Rightarrow (1-u)(1-\xi_Y(u))\frac{d}{du}(1-u)(1-\xi_X(u))$$

$$- (1-u)(1-\xi_X(u))\frac{d}{du}(1-u)(1-\xi_Y(u)) \geq 0$$

$$\Rightarrow \frac{\frac{d}{du}(1-u)(1-\xi_X(u))}{\frac{d}{du}(1-u)(1-\xi_Y(u))} \geq \frac{1-\xi_X(u)}{1-\xi_Y(u)} \geq 1$$

$$\Rightarrow \frac{d}{du}(1-u)(1-\xi_X(u)) \geq \frac{d}{du}(1-u)(1-\xi_Y(u))$$

$$\Rightarrow \log h_{Q_Y}(u) \geq h_{Q_X}(u) \Leftrightarrow h_{Q_X}(u) \leq h_{Q_Y}(u)$$

The proof of $\frac{1-\xi_X(u)}{1-\xi_Y(u)}$ decreasing is similar. □

EXAMPLE 2.9. *Let*

$$\xi_X(u) = \log(1 - u) + a_1 + bu$$

and

$$\xi_Y(u) = \log(1 - u) + a_2 + bu$$

where $a_1 \geq a_2$ and $b < 1$, both X and Y following distribution (2.35) with parameters (a_1, b) and (a_2, b). In this case $X \geq_{RQE} Y$ and also

$$\frac{1 - \xi_X(u)}{1 - \xi_Y(u)} = \frac{1 - \log(1 - u) - a_1 - bu}{1 - \log(1 - u) - a_2 - bu}$$

has derivative $(a_1 - a_2)(-b + (1 - u)^{-1}) \geq 0$ since $b < 1$. The hazard quantile functions

$$h_{Q_X}(u) = (1 - u)^{-1} \exp[-b - a_1 - 2bu]$$

and

$$h_{Q_Y}(u) = (1 - u)^{-1} \exp[-b - a_2 - 2bu]$$

satisfy $h_{Q_X}(u) \geq h_{Q_Y}(u)$ or $X \leq_{HQ} Y$.

There are some stochastic orders based on ageing concepts that can be related to residual quantile entropy ordering.

THEOREM 2.16. *The random variables X and Y are such that $F_X(0) = F_Y(0) = 0$ and $F_Y(x)$ is strictly increasing in an interval support, satisfy convex transform order $X \leq_c Y$, (X is more IHR than Y if and only $(1 - u)(\xi_X(u) - \xi_Y(u))$ is convex.*

PROOF. By virtue of convex transform order, $F_Y^{-1}(F_X(x))$ is a convex function implies that

$$\frac{d}{dx} F_Y^{-1}(F_X(x)) = \frac{f_X(F_X^{-1}(x))}{f_Y(F_Y^{-1}(x))} = \frac{q_Y(u)}{q_X(u)}$$

is increasing. Thus

$$\frac{q_Y(u)}{q_X(u)} = \frac{(1 - u)q_Y(u)}{(1 - u)q_X(u)} = \frac{h_{Q_X}(u)}{h_{Q_Y}(u)} \text{ is increasing.}$$

$$\Rightarrow \log h_{Q_X}(u) - \log h_{Q_Y}(u) \text{ is increasing}$$

$$\Rightarrow \frac{d}{du}(1 - u)(1 - \xi_Y(u)) - \frac{d}{du}(1 - u)(1 - \xi_X(u)) \text{ is increasing}$$

$$\Rightarrow \frac{d}{du}(1 - u)(\xi_X(u) - \xi_Y(u)) \text{ is increasing}$$

$$\Rightarrow (1 - u)(\xi_X(u) - \xi_Y(u)) \text{ is convex.}$$

$\square$

THEOREM 2.17. *Let X and Y be absolutely continuous random variables with $F(0) = 0 = G(0)$ and $g(0) \geq f(0) > 0$. Then*

$$Y \leq_* X \Rightarrow Y \leq_{RQE} X.$$

PROOF. Under the conditions on X and Y stated in the Theorem, Deshpande and Kochar (1983) have proved that

$$Y \leq_* X \Rightarrow Y \leq_{disp} X.$$

Since dispersive ordering is the same as hazard quantile function ordering the result follows from (2.53). □

REMARK 2.5. *The star order $X \leq_* Y$ is also interpreted as X is less IHRA than Y.*

THEOREM 2.18. *If $Y \leq_{su} X$ and either $Y \leq_{st} X$ or $\lim \frac{F^{-1}(G(x))}{x} \geq 1$, then*

$$Y \leq_{su} X \Rightarrow Y \leq_{RQE} Y.$$

PROOF. When the given conditions in Theorem are satisfied $Y \leq_{disp} X$ (Ahmed et al. (1986)) and hence the result. □

2.9. Estimation and testing

The estimation problem can be resolved in two ways: one by inferring the distribution by means of either a parametric or non-parametric methods and the second by directly estimating the residual entropy function. In this section we consider a brief review of the work done in these contexts. Belzunce et al. (2001) employed the kernel approach, a non-parametric method for estimating the probability density function to estimate the residual entropy function. For a random sample of n observations $X_1, X_2, \ldots, X_n$ from $F(x)$, the kernel estimate of the differential entropy is

$$\hat{H} = -\frac{1}{n} \sum_{i=1}^{n} \log \hat{f}_i(X_i)$$

where

$$\hat{f}_i(X_i) = \frac{1}{n-1} \sum_{j \neq i} K_h(X_i - X_j)$$

in which K_h is a non-negative function that integrates to unity and $h > 0$ is a smoothing parameter called bandwidth. For sample values greater than t, the estimate of $H_F(t)$ is written similarly as

$$H_1(t) = -\frac{1}{m_t} \sum_{i=1}^{m_t} \log \hat{f}_{t,i}(X_i)$$

where

$$m_t = I(X_i \geq t) \text{ is the indicator function}$$

$$\hat{f}_{t,i}(X_i) = \frac{1}{m_t - 1} \sum_{j=i} K_h(X_i - X_j),$$

$$K_h(t) = \frac{K(t/h)}{h}$$

and $K(\cdot)$ is a symmetric probability density function with finite variance. Based on the complete sample the estimate is

$$H_1^*(t) = -\frac{1}{m_i} \sum_{i=1}^n \frac{\log \hat{f}_i(X_i)}{\hat{R}(t)} I(X_i > t)$$

where $\hat{R}(t)$ is the empirical or kernel estimate of the survival function and $\hat{f}_i(X_i) = \frac{1}{n-1} \sum_{j \neq i} K_h(X_i - X_j)$. Writing $g(x) = f(x) \log f(x)$, two estimates of g are

$$g_1 = \frac{1}{n} \sum_{i=1}^n K_h(x - X_i) \log(\frac{1}{n} \sum_{i=1}^n K_h(x - X_i))$$

and

$$g_2 = \frac{1}{n} \sum_{i=1}^n K_h(x - X_i) \log(\frac{1}{n-1} \sum_{j \neq i} K_h(X_i - X_j)).$$

Thus estimations of H are

$$H_i(t) = \log \hat{R}(t) - \frac{1}{\hat{R}(t)} \int_t^\infty g_i(x) dx, \ i = 1, 2.$$

If $\hat{R}$ is the empirical estimator, then $H_2(t) = H_1^*(t)$. The residual entropy function is estimated by using

$$G_i(t) = \int_t^\infty g_i(x) dx.$$

Belzunce et al. (2001) has used the exponential and power distributions for illustration and the results are demonstrated for two real data sets.

Jeevanand and Abdul-Sathar (2009) have developed a parametric approach for estimation of the residual entropy function of the exponential model. They used the double censoring and progressive censoring sampling schemes to derive the Bayes and maximum likelihood estimators. We assume that n units are placed on test at time zero with m observed failures. At the occurrence of the first failure r_1 units are randomly selected and removed, and at the second failure, r_2 units are randomly selected and removed and so on and finally when the mth failure takes place all

$r_m = n - r_1 \cdots - r_{m-1}$ surviving sampling units are removed. The likelihood function is

$$l(\underline{x}|\lambda) = K\lambda^m \exp[-\lambda \sum_{i=1}^{m} x_i(1 + r_i)], \quad \underline{x} = (x_1, \ldots, x_n)$$

leading to the conjugate prior

$$g(\lambda|c) = C_1 \lambda^{p-1} e^{-\tau\lambda}$$

and the posterior density

$$f(\lambda|\underline{x}) = C_2 \lambda^{N-1} e^{-\lambda T}$$

with $N = m + p$ and $T = r + \sum_{i=1}^{m}(1 + r_i)x_i$.

Thus the posterior distribution of $H_F(t)$ becomes

$$f(H_F|\underline{x}) = [C_3(0)]^{-1} e^{(1-H)N - T \exp(1-H)}$$

where

$$C_3(d) = \int_{-\infty}^{\infty} H^d \exp[(1 - H)_N - T \exp(1 - H)]dH.$$

The Bayes estimator under the squared error loss function becomes

$$\hat{H}_1 = 1 + \log T - \text{polygamma}\,[0, N].$$

On the other hand if we use the LINEX loss function

$$L(\delta) \sim e^{c\delta} - c\delta - 1, \ c \neq 0$$

the estimator is

$$\hat{H}_2 = \frac{1}{a} \log G_1$$

where

$$G_1 = [C_3(0)]^{-1} \int_{-\infty}^{\infty} \exp[aH + (1 - H)N - T \exp(1 - H)]dH$$

which requires numerical evaluation.

We have the maximum likelihood estimator as

$$\hat{H}_3 = 1 - \frac{\log m}{\sum_{i=1}^{m} x_i(1 + r_i)}.$$

In the double censoring scheme r times are censored on the left and s on the right giving the observations $x_{r+1}, \ldots, x_{n-s}$, $r = [nq_1] + 1$, $s = [nq_2] + 1$, $0 < q_1, q_2 < 1$. The Bayes estimator under squared error loss is

$$\hat{H}_4 = \frac{C_6(1)}{C_6(d)}$$

where

$$C_6(d) = \int_{-\infty}^{\infty} H^d (1 - e^{-y_{r+1}(e^{1-H})})^r e^{N(1-H) - Ze^{1-H}},$$

$$N = p + n - r - s \text{ and } Z = \tau^{n-s} + \sum_{i=r+1}^{n-s} [x_i + s y_{n-s}]$$

while with LINEX loss,

$$\hat{H}_5 = \frac{1}{a} \log G_2,$$

$$G_2 = [C_6(0)]^{-1} \int_{-\infty}^{\infty} (1 - e^{-y_{r+1}(e^{1-H})} - z e^{1-H}).$$

The maximum likelihood estimator is

$$\hat{H}_6 = 1 - \log \lambda_{ml}$$

where λ_{ml} is the solution of

$$\frac{r y_{r+1} e^{-\lambda y_{r+1}}}{1 - e^{-\lambda y_{r+1}}} + \frac{n - (r+s)}{\lambda} - 2 = 0.$$

A simulation study to compare the performance of the estimators is also carried out by the authors.

Rajesh et al. (2015) proposed a non-parametric estimator for the residual entropy for censored dependent observations. Let $X_1, \ldots, X_n$ be a random sample F in which X_i is right censored y_i, where the y_i's are independent and identically distributed as $G(\cdot)$ and also independent of the X_i's. If $Z_i = X_i \wedge Y_i$ then Z_i has distribution function $L(\cdot)$ satisfying

$$1 - L(t) = (1 - F(t))(1 - G(t)), \ 0 < t < \infty.$$

The non-parametric estimator of $f(x)$ under the sampling scheme is

$$f_n(x) = \frac{1}{h_n} \int_0^\infty \frac{K(x-u)/h_n}{1 - G_n} dL_n^*(u), \quad L^*(t) = P(z_1 \le t, \delta_1 = 1).$$

The Kaplan-Meier estimate of F is

$$1 - F_n(t) = \prod_{s \le t} \left(1 - \frac{dN_n(s)}{Y_n(s)} \right)$$

in which

$$N_n(t) = \sum_{i=1}^{n} I(z_i \le t, \delta_i = 1), \quad \delta_i = P(X_i \le Y_i))$$

the number of uncensored observations not exceeding t and $y_n = \sum_{i=1}^{n}(z_i \geq t)$, the number of censored or uncensored values $\geq t$ and $dN_n(s) = N(s) - N(s-1)$. A non-parametric estimator of $H_F(t)$ based on central data is

$$\hat{H}_F = -\frac{1}{n}\sum_{i=1}^{n}\log\frac{f_n(Z_i)}{1-F_n(t)}I(z_i > t)$$

where $f_i(z_i) = \frac{1}{n-1}\sum_{j\neq 1}^{n}\frac{1}{h_n}K\left(\frac{Z_i - Z_j}{h_n}\right)$ is the kernel estimator obtained from the sample without Z_i and the estimator under censoring is

$$H_n(t) = \log(1 - F_n(t)) - \frac{1}{1-F_n(t)}\int_t^\infty f_n(x)\log f_n(x).$$

The work also established asymptotic properties of the estimates, a simulation study to compare the two estimators and illustrated the methodology for real data sets.

Ebrahimi (1997) has proposed a procedure for testing the null hypothesis H_0 that X is exponential with $F = 1 - \exp(-\lambda x)$ against the alternative H_1 that X is DURL. Under the consideration that H_1 holds if $H_F(t) - H_F(s) \geq 0$ for all $s \geq t$ and hence under H_0, $\Delta F = 0$ or $\frac{1}{2} - \Delta F$ is small, when

$$\Delta F = \int_0^\infty \int_\alpha^\infty [H_F(t) - H_F(s)]f(t)f(s)dtds$$

and

$$= \frac{1}{2} - \int_0^\infty [\log \bar{F}(x) + 2F(x)](\log f(x))f(x)dx.$$

For a random sample $t_1, t_2, \ldots, t_n$ define a partition of $t_{1:n}, \ldots, t_{n:n}$ in to

$$I_i = (t_{a_i-1:n}, t_{a_i:n}), \; i = 1, 2, \ldots, K < n.$$

and $a_i = 1 + \sum_{j=1}^{i}\lambda_j$. The proposed test statistic is

$$J_i(k) = \frac{1}{n-1}\sum_{i=1}^{k}(A_n(i))\log\frac{(n-1)d_i(k)}{\lambda_i}$$

where

$$d_i(k) = t_{a_i:n} - t_{a_i-1:n}$$

and

$$A_n(i) = n - a_{i-1}\log(n - a_{i-1}) - (n - a_1)\log(n - a_1)$$
$$+ \frac{1}{n-1}\lambda_i(\lambda_i + 2\sum_{j=1}^{i-1}\lambda_j).$$

Large values of $J_n(k)$ favours H_1 and reject H_0 in favour of H_1 at significance level α if $J_n(k) \geq C_{k,n}(\alpha)$ where $C_{K,n}(\alpha)$ is determined by the $(1-\alpha)$th quantile of the distribution of $J_{n,1}(k)$ static under H_0. The tabled values of $C_{k,n}(\alpha)$ are provided by the author. For the asymptotic properties of the test and its application to real data sets we refer to Ebrahimi (1997).

2.10. Residual extropy

The Shannon's measure of entropy in the discrete case

$$H_X = -\sum_{i=1}^{H} p_i \log p_i$$

where $p_i = P(X = x_i)$ has a complementary dual

$$J_X = -\sum_{i=1}^{n} (1 - p_i) \log(1 - p_i)$$

which is called the extropy of the random variable X. It originated in environmental investigations by Ayres and Martinas (1995) in the name of M-potential and the term extropy was coined in Martinas (1997). The two measure entropy and extropy address measurement of uncertainty in contrasting styles. A detailed discussion on the motivation, importance and properties of extropy in the context of thermodynamics and statistical mechanics is given in Martinas (1997) and Martinas and Frankowicz (2000). They argue that both entropy and extropy share similar mathematical properties and that the latter has some conceptual superiority in certain situations. Lad et al. (2015) points out that as a measure of uncertainty extropy is invariant under permutations and monotonic transformations, maximum extropy distribution is uniform and satisfies Shannon's first and second axioms. As X increases in such a way that $\sum p_i$ decreases to zero, $\sum (1 - p_i) \log(1 - p_i)$ is well approximated by

$$-\sum (1 - p_i)(-p_i) = 1 - \sum p_i^2. \tag{2.54}$$

Also

$$\lim_{\Delta \to 0} \left(\frac{J_X - 1}{\Delta x} \right) = -\frac{1}{2} \int f^2(x) dx.$$

The discrete approximation (2.54) has implications as repeat rate of a distribution (Good (1989)) and as Gini index of homogeneity widely used in data base applications. It is also an indicator of the degree of skewness of the data. If p_i denotes the proportion of instances belonging to class i, $i = 1, 2, \ldots, n$, the Gini index is $1 - \frac{1}{n}$ when the elements are heterogeneous ($p_i = \frac{1}{i}$) and zero in the case of homogeneity, and if we use entropy the counterpart values are $\log n$ and zero.

Qiu (2017) has obtained several results on the extropy of order statistics and records. If $X \leq_{\text{disp}} Y$, then $J_{X_{i:n}} \leq J_{Y_{i:n}}$. Also if $f(0) \geq g(0) > 0$, where $f(\cdot)$ and $g(\cdot)$ are the probability density functions of X and Y, and $X \leq_{\text{su}} Y$ ($X \leq_* Y$ or $X \leq_c Y$) then also $J_X \leq J_Y$. An upper bound to entropy can be prescribed in terms of the ageing properties as, if

$$X \in \text{IFR (IFRA, NBU) and } f(0) \geq \mu = E(X), \text{ then } J_X \leq -\frac{\mu}{4}.$$

Further if X and Y have a common lower boundary b, then for any fixed m $(1 \leq m \leq n)$

$$X \overset{d}{=} Y \Leftrightarrow J_{X_{m:n}} = J_{Y_{m:n}} \text{ for every } n \geq m.$$

Some characterization results can also be established.

THEOREM 2.19. (i) *The random variable X follows the exponential law if and only if* $J_{X_{1:n}} = nJ_X$, $n \geq 1$
(ii) $F(x) = e^{\mu x}$, $x < 0$, $\mu > 0$ *iff* $J_{X_{n:n}} = nJ_X$, $n \geq 1$.

In terms of ageing concepts we have some properties.

THEOREM 2.20. (a) *If X is DFR then $J_{X_{i:n}}$ is decreasing in n for fixed i, $1 \leq i \leq n$*
(b) *If X is IRHR then $J_{X_{i:n}}$ is increasing in n for fixed i and decreasing in i for fixed n.*
    If $f(\mu + x) = f(\mu - x)$, $x \geq 0$.
  (i) $J_{X_{i:n}} = J_{X_{n-k+1:n}}$
  (ii) $\Delta J_{X_{i:n}} = -\Delta J_{X_{n-i:n}}$, Δ *being the forward difference operator*
  (iii) $J_{Y_{i:n}} = \frac{1}{a}J_{X_{i:n}}$, *where* $Y = \frac{x-\mu}{a}$.
A special feature of (ii) above is that the extropy of $X_{i:n}$ at the median always provides a local minimum or maximum. Again, the distribution of X is symmetric if and only if $J_{X_{1:n}} = J_{X_{n:n}}$ *for all $n \geq 1$.*

A system is coherent if it does not have irrelevant components and the structure function is monotone. Let T denote the lifetime of a mixed system born out of a mixture of coherent systems. Then the density function of T is

$$f(x) = \sum_{i=1}^{n} s_i f_{i:n}(x) = \frac{F^{i-1}(x)\bar{F}^{n-i}(x)f(x)}{B(i, n-i+1)}$$

with F as the common distribution function of the component lifetimes X_i which are independent and identically distributed. The vector $s = (s_1, \ldots, s_n)$ is called the signature of the mixed system. Qiu et al. (2019) obtained the extropy of the system as

$$J_X = -\frac{1}{2} \int_0^1 g^2(v)f(F^{-1}(v))dv$$

where

$$g(v) = \sum_{i=1}^{n} s_i \frac{v^{i-1}(1-v)^{n-1}}{B(i, n-i+1)}.$$

They have also obtained same ordering relations.

(a) If T_X and T_Y be lifetimes of two mixed systems and $X \leq_{\text{disp}} Y \Rightarrow$ $J_{T_X} \leq J_{T_Y}$

(b) Either X or Y is DFR and $X \leq_{\text{hr}} Y$ then also $J_{T_X} \leq J_{T_Y}$

(c) $J_{X_{i:n}} \leq J_{T_X}$.

Some bounds for the extropy of mixed systems are

$$M^2 J_X \leq J_T \leq m^2 J_X, \ m = \inf g(v), \ M = \sup g(v)$$

$$J_T \geq \sum_{i=1}^{n} s_i J(X_{i:n})$$

$$J_{X+Y} \geq \max(J_X, J_Y), \ X \text{ and } Y \text{ being independent.}$$

Qiu and Jia (2018a) have also provided two estimators for extropy of an absolutely continuous random variable using spacings. They have defined the residual extropy of X as

$$J_F(t) = -\frac{1}{2\bar{F}^2(t)} \int_t^{\infty} f^2(x)dx \tag{2.55}$$

and showed that,

$$J_F'(t) = -\frac{1}{2}(h^2(t) + 4h(t)J_F(t)). \tag{2.56}$$

Solving the last different equation

$$J_F(t) = \{\exp[2 \int h(t)dt]\}\frac{1}{2} \int h^2(t)e^{-2\int h(t)dt}dt + C \tag{2.57}$$

where $C = J_X$. A characterization of the generalized Pareto distribution is available based on the identity $J_F(t) = kh(t)$.

THEOREM 2.21. *The identity $J_F(t) = -kh(t)$ where k is a non-negative constant, holds iff. X has rescaled beta distribution ($k > \frac{1}{4}$) or exponential distribution ($k = \frac{1}{4}$) or Pareto II distribution ($k < \frac{1}{4}$).*

A main limitation of residual extropy is that it is not defined for certain regions of the parameter space. For example when X has rescaled beta distribution $\bar{F}(x) = (1 - \frac{x}{R})^c, 0 \leq X \leq R, R, c > 0$.

$$J_F(t) = -\frac{c^2}{(4C-2)(R-t)}$$

so that $J_F(t)$ being negative is not defined for $c < \frac{1}{2}$. Similarly for the power distribution $F(x) = (\frac{x}{\beta})^\alpha$, $0 \le x \le \beta$, $\alpha, \beta > 0$.

$$J_F(t) = -\frac{\alpha^2(1 - (\frac{t}{\beta})^{2\alpha-1})}{(2 - 4\alpha)(1 - (\frac{t}{\beta})^\alpha)^2}$$

is valid only for $\alpha > \frac{1}{2}$. Thus the parameter values become a crucial aspect when discussing properties of residual extropy. Qiu and Jia (2018b) in Theorem 2.20 (their Theorem 2.9) prove the result for $k > \frac{1}{4}$ by taking the distribution of X as $\bar{F}(x) = (1 - x)^c$ where $c > 1$. One can consider a more flexible range for the parameter as seen from the next theorem which modifies Theorem 2.20.

THEOREM 2.22. *The relationship $J_P(t) = -kh(t)$ where k is a non-negative constant holds for all $t > 0$ if and only if X has*

  (1) rescaled beta distribution

$$\bar{F}(x) = \left(1 - \frac{x}{R}\right)^c, \ 0 \le x \le R;\ R > 0,\ C > \frac{1}{2}$$

  if $k > \frac{1}{4}$
  (2) exponential law,

$$\bar{F}(x) = e^{-\lambda x}, \ x > 0;\ \lambda > 0$$

  if $k = \frac{1}{4}$ and
  (3) Pareto II distribution,

$$\bar{F}(x) = \left(1 + \frac{x}{\alpha}\right)^{-c}, \ x > 0;\ c, \alpha > 0$$

  if $k < \frac{1}{4}$.

PROOF. It is enough to prove (i) only. When X has the given distribution

$$J_F(t) = \frac{-c^2}{4c - 2(R - t)} \text{ and } h(t) = \frac{c}{R - t}$$

provides

$$J_F(t) = -kh(t),\ k = \frac{c}{4c - 2}.$$

Thus $k > \frac{1}{4}$ when $c > \frac{1}{2}$. Conversely the relationship between $J_F(t)$ and $h(t)$ assumed above leads to

$$h(t) = (pt + d)^{-1},\ p = \frac{1 - 4k}{2k},\ d > 0.$$

When $K < \frac{1}{4}$, p is negative and hence

$$\bar{F}(x) = \left(1 + \frac{pt}{d}\right)^{-p^{-1}}$$

which is rescaled beta with $c = -\frac{1}{p}$ and $R = -\frac{p}{d} > 0$.   □

Equation (2.56) needs further investigation by treating it as a quadratic in $h(t)$,

$$h^2(t) + 4h(t)J(t) - 2J'_F(t) = 0. \qquad (2.58)$$

From Descartes rule of signs it is seen that (2.58) has only one positive root when $J'_F(t) > 0$ or when $J'_F(t) = 0$ and two positive root when $J'_F(t) < 0$. In the first two cases $J'_F(t)$ determines $h(t)$ and the distribution of X uniquely. However the last case shows that, in general, $J(t)$ does not characterize the distribution if we adopt (2.58) as the basis of argument. From (2.58),

$$h(t) = -2J_F(t) \pm (rJ_F^2(t) + 2J'_F(t))^{1/2}. \qquad (2.59)$$

EXAMPLE 2.10. *Let X follow the power distribution $F(x) = x^2$, $0 \le x \le 1$,*

$$J_F(t) = -\frac{2}{3}\frac{(1 - t^3)}{(1 - t^2)^2} \text{ and } J'_F(t) = -\frac{2}{3}\frac{t(1 - t)(t^2 + t + 4)}{(1 - t^2)^3}$$

showing that both $J_F(t)$ and $J'_F(t)$ are less than zero.

After some algebra we get

$$4J_F^2(t) + 2J'_F(t) = \frac{16(1 - t)^4(t + 2)^2}{36(1 - t^2)^4}.$$

Thus we have two solutions for $h(t)$ given by (2.59)

$$h_1(t) = \frac{2}{(1 - t)^2} \text{ and } h_2(t) = \frac{2(t^2 + t + 4)}{3(1 - t)(1 + t)^2}.$$

Writing $h_2(t)$ as

$$h_2(t) = \frac{3}{2(1 - t)} + \frac{1}{2(1 + t)} + \frac{2}{(1 + t)^2},$$

it is easy to see that $h_2(t) \ge 0$ and $\int_0^1 h(t)dt$ diverge to ∞ proving it to be a hazard rate. Also $h_1(t)$ is the hazard rate of the power distribution mentioned at the beginning of this example and $h_2(t)$ corresponds to

$$\bar{F}(x) = (1 - x)^{3/2}(1 + x)^{-1/2}e^{-2x/(1+x)}, \ 0 \le x \le 1.$$

If X is assumed to be a uniform distribution in $[0, b]$, still there are two positive roots, but they are equal to $(b-t)^{-1}$. Hence the uniform distribution is characterized by its residual extropy function.

The random variable X is said to have increasing (decreasing) residual extropy, IRE (DRE) class is defined by the property $J_F(t) \ge (\le) - \frac{h(t)}{4}$ for all t. From (2.58), the equality $J_F(t) = -\frac{h(t)}{4}$ holds for all $t > 0$ if and only if X is exponential. Since rescaled beta is DRE and Pareto II is IRE the two classes are not empty. There is a clear distinction between monotonic entropy and extropy. While the monotonicity of the hazard rate implies monotone residual entropy, the magnitude of the hazard rate is the determining factor for residual entropy. The rescaled beta is DRE where $h(t)$ is increasing, power distribution is DRE when in $\frac{1}{2} < a < 1$, $h(t)$ is bathtub shaped.

Within the framework of quantile functions we can write the residual quantile extropy as

$$J_Q(u) = J_F(Q(u)) = -\frac{1}{2(1-u)^2} \int_u^1 \frac{dp}{q(p)}. \qquad (2.60)$$

Differentiation of (2.60) yields

$$q(u) = [-2(1-u)^2 J_Q'(u) + 4(1-u) J_Q(u)]^{-1}. \qquad (2.61)$$

The utility of (2.61) is too-fold. Firstly it shows that the distribution of X is characterized in terms of $J_Q(u)$ and secondly it helps to generate new quantile functions based on assumed functional forms of $J_Q(u)$. The hazard quantile function of X now becomes

$$h_Q(u) = [(1-u)q(u)]^{-1} = [4 J_Q(u) - 2(1-u) J_Q'(u)] \qquad (2.62)$$

Accordingly $J_Q(u)$ determines $h_Q(u)$ and the distribution through $Q(u)$. Equations (2.61) and (2.62) exhibit the advantage of quantile approach over the distribution function counterpart in the sense that the former gives a unique representation of the distribution where as this could not be accomplished in the latter as was shown in the above example.

EXAMPLE 2.11. *Consider the linear mean residual quantile function family of distribution (Midhu et al. (2013)) specified by*

$$Q(u) = -(c + \mu)\log(1 - u) - 2cu, \quad \mu > 0; \ -\mu \le c < \mu$$

which contains the exponential and uniform distributions and closely approximate several continuous distributions. It does not have a tractable distribution to study the properties of $J_F(t)$ using F. In this case, the quantile residual entropy is

$$J_Q(u) = \frac{1}{2(1-u)} - \frac{c+\mu}{8c^2(1-u)^2} \log \frac{\mu+c}{\mu-c+2u}.$$

EXAMPLE 2.12. *The example shows how to construct new quantile functions that conform to known functional forms of $J(u)$. Let $J(u) = -e^{bu}$. Then from (2.60)*

$$\int_u^1 \frac{dp}{q(p)} = 2(1-u)^2 e^{bu}, \ 0 < b < 2$$

or

$$q(u) = [2(1-u)e^{bu}(2 - b(1-u))]^{-1}$$

which is a quantile density function.

Equation (2.62) can be employed to characterize probability distributions.

THEOREM 2.23. *Let X be an absolutely continuous non-negative random variable. The $J_Q(u)$ is linear if and only if $h_Q(u)$ is linear.*

PROOF. Let $J_Q(u) = -(A + Bu)$, $A < 0$, $A + B > 2A$. Then from (2.62).

$$h_Q(u) = 2B - 4A - 6BU = a + bu, \quad a = 2B - 4A > 0, \ b = -6B > 0$$

which is a hazard quantile function whenever $B > 4A$. Conversely let $h_Q(u) = a + bu$, $a, b > 0$. Then

$$a + bu = 4J_Q(u) - 2(1-u)J_Q(u) \tag{2.63}$$

which reduces to a linear differential equation

$$J_Q'(u) - \frac{2J_Q(u)}{(1-u)} = -\frac{(a+bu)}{2(1-u)}$$

with integrating factor $(1-u)^2$. Thus

$$\frac{d}{du}(1-u^2)J_Q(u) = -\frac{(a+bu)(1-u)}{2}$$

giving the general solution

$$J_Q(u) = \frac{3a + b - 2bc}{12} + \frac{c}{(1-u)^2}.$$

It satisfies (2.63) only when $c = 0$. Thus

$$J_Q(u) = \frac{3a + b - 2bu}{12} = -(A + Bu).$$

□

REMARK 2.6. *Distribution with linear hazard quantile function has been discussed in Nair et al. (2013a) and Midhu et al. (2013) where its properties and applications are given. It subsumes the exponential, half-logistic, exponential-geometric and Marshall-Olkin type distributions. Thus Theorem 2.23 proves a characterization of the linear hazard quantile function family.*

REMARK 2.7. *Polynomial hazard rates $h(t) = c_0 + c_1 t + \cdots + c_n t^n$ were used from early days (Bain (1978), Gore et al. (1986)) to represent various shapes and there is a recent revival of interest in such models, see for example Bebbington et al. (2009). One can also consider polynomial hazard quantile functions to accommodate shapes like $\vee, \wedge, N, M$ and W for $h_Q(t)$ to derive appropriate models. Theorem 2.23 can be further extended with the same method of proof to show that $J_Q(u)$ is polynomial equivalent to $h_Q(u)$ also polynomial with appropriate conditions than the coefficient that renders $h_Q(u)$ positive. A special case of interest is quadratic $h_Q(u)$ that provides both BT and UBT shapes.*

Qiu and Jia (2018b) have presented several results on residual extropy of order statistics. If $J_{X_{i:n}}(t)$ and $J_{Y_{i:n}}(t)$ are residual entropies of order statistics of two non-negative random variables X and Y, then X and Y have identical distributions if and only if $J_{X_{i:n}}(t) = J_{Y_{i:n}}(t)$ for all $t \geq 0$ and $n \geq 1$. In particular if $J_{X_{i:n}}(t) = J_{Y_{i:n}}(t)$ then $X \overset{d}{=} Y$. Also if X has a decreasing probability density function then $J_{X_{1:n}}(t)$ is decreasing in $t \geq 0$ and in n. They have also given a characterization of some distributions which needs modification in the following manner.

THEOREM 2.24. *The relationship $J_{X_{1:n}}(t) = -kh(t)$ holds for all $t \geq 0$ and a non-negative constant k if and only if X has*

(i) *exponential distribution when $k = n/4$*
(ii) *Pareto II distribution when $k < \frac{n}{4}$ and*
(iii) *rescale beta distribution when $k > \frac{n}{4}$ and $c > \frac{1}{2n}$.*

With the aid of (2.62) we can derive some ageing properties of X by means of $J_Q(u)$. Differentiating

$$h'_Q(u) = J'_Q(Q) - 2(1-u)J''_Q(u).$$

Consequently X is IFR if J_Q is increasing and concave, while X is DFR when J_Q is decreasing and convex. Also we have some stochastic orders connecting two lifetimes X and Y with residual extropies $J_{Q_X}(u)$ and $J_{Q_Y}(u)$.

DEFINITION 2.7. *We say that X has less residual quantile extropy than Y, $X \leq_{RQE} Y$ if $J_{Q_X}(u) \leq J_{Q_Y}(u)$ for all $0 < u < 1$.*

THEOREM 2.25.

$$X \leq_{HQ} Y \Rightarrow X \leq_{RQE} Y$$

where $\leq_{HQ}$ is the hazard quantile function order.

PROOF.

$$X \leq_{\text{HQ}} Y \Leftrightarrow \frac{1}{(1-u)q_X(u)} \geq \frac{1}{(1-u)q_Y(u)}$$

$$\Rightarrow \int_u^1 \frac{dp}{q_X(p)} \geq \int_u^1 \frac{dp}{q_Y(p)} dp$$

$$\Rightarrow -\frac{1}{2(1-u)^2} \int_u^1 \frac{dp}{q_X(p)} \leq \frac{1}{2(1-u)^2} \int_u^1 \frac{dp}{q_Y(p)}$$

$$\Leftrightarrow X \leq_{\text{RQE}} Y.$$

The implication of this theorem is that when $X \leq_{\text{HQ}} Y$, the device with life-time X is less reliable than that with lifetime Y which also means that the information content in the residual life distribution of X is smaller than the content in Y. The converse of Theorem 2.24 may not be true and therefore the additional requirements that ensure the converse is stated in the next theorem. $\square$

THEOREM 2.26. *If* $\frac{J_{Q_Y}(u)}{J_{Q_X}(u)}$ *is increasing in* u, *then* $X \leq_{RQE} Y \Rightarrow$ $X \leq_{HQ} Y.$

PROOF. $\frac{J_{Q_Y}(u)}{J_{Q_X}(u)}$ is increasing

$$\Leftrightarrow \frac{\int_u^1 \frac{dp}{q_Y(p)}}{\int_u^1 \frac{dp}{q_Y(p)}}$$

$$\Leftrightarrow \frac{1}{q_X(u)} \int_u^1 \frac{dp}{q_Y(p)} - \frac{1}{q_Y(u)} \int_u^1 \frac{dp}{q_X(p)} \geq 0$$

$$\Leftrightarrow \frac{q_Y}{q_X} \geq \frac{\int_u^1 \frac{dp}{q_Y(p)}}{\int_u^1 \frac{dp}{q_X(p)}} \geq 1$$

$$\Leftrightarrow (1-u)q_Y \geq (1-u)q_X$$

$$\Leftrightarrow H_X(u) \geq H_Y(u) \Rightarrow X \leq_{\text{HQ}} Y.$$

Since $X \leq_{\text{HQ}} Y \Leftrightarrow X \leq_{\text{RHQ}} Y$ we also have

$$X \leq_{\text{RHQ}} Y \Rightarrow X \leq_{\text{RQE}} Y.$$

$\square$

EXAMPLE 2.13. *For the uniform distribution* $Q_X(u) = u$, $J_{Q_X}(u) = -\frac{2}{1-u}$ *and for* $Q_Y(u) = u^{\frac{1}{2}}$, $J_{Q_Y}(u) = -\frac{4}{6}\frac{(1-u^{\frac{3}{2}})}{(1-u)^2}$ $0 < u < 1$, *so that* $X \geq_{RQE} Y$. *From Figure 2.11 it is clear that* X *and* Y *are not in hazard quantile order. However, Figure 2.12 shows that* $\frac{J_{Q_Y}(u)}{J_{Q_X}(u)}$ *is decreasing in* u, *so that the condition given in Theorem 2.26 cannot be relaxed.*

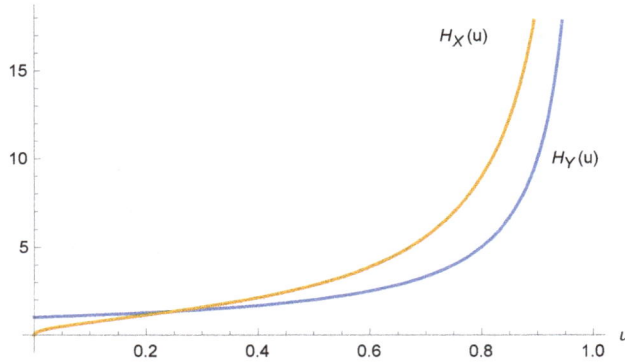

Figure 2.11. Plot of $H(u)$ against u.

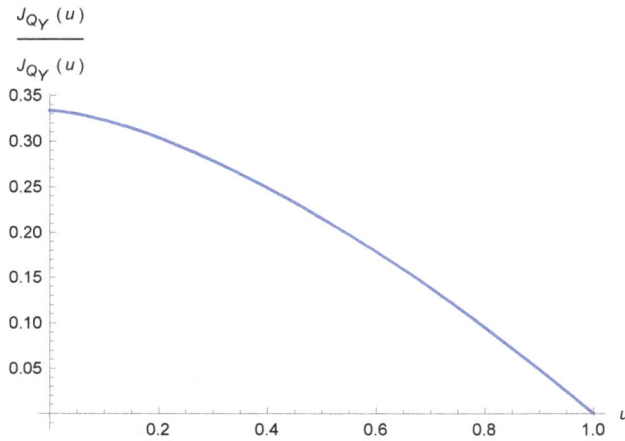

Figure 2.12. Plot of $\frac{J_{Q_Y}(u)}{J_{Q_X}(u)}$ against u.

DEFINITION 2.8. *The random variable X is less than the random variable Y in residual extropy if $J_{F_X}(t) \leq J_{F_Y}(t)$ for all $t \geq 0$ and is denoted by $X \leq_{RE} Y$.*

EXAMPLE 2.14. *For the uniform distribution $F_X(x) = x$, $0 \leq x \leq 1$, $J_{F_X}(t) = -\frac{2}{1-t}$ and for $f_Y(x) = x^2$, $J_{F_Y}(t) = -\frac{2}{3}\frac{1+t+t^2}{(1-t)(1+t)^2}$ so that $X \geq_{RE} Y$.*

We examine if there exist any implications among the two orderings $\leq_{RE}$ and $\leq_{RQE}$. The following example shows that $\leq_{RE}$ does not imply $\leq_{RQE}$.

EXAMPLE 2.15. *Consider the distribution in Example 2.11. We have* $J_{Q_X}(u) = -\frac{2}{1-u}$ *and* $J_{Q_Y}(u) = -\frac{4}{6}\frac{(1-u)^{3/2}}{(1-u)^2}$. *Then* $X \leq_{RQE} Y$.

For further properties and results on extropy we refer to the recent papers Noughabi and Jarrahiferi (2019), Krishnan et al. (2020, 2021), Abdul-Sathar and Nair (2021).

CHAPTER 3

Entropy of Past Life

3.1. Past entropy

The differential entropy of past lifetime usually abbreviated as past entropy is the dual of the residual entropy in which we consider the random variable $X_{(t)} = [X|X \leq t]$. By the similarity to X_t discussed in the previous chapter most of the results in past entropy run parallel to those of residual entropy. However as an uncertainty measure its relevance and application to various scientific disciplines, where there are occasions to analyze data which is left truncated at some point, cannot be discounted. In the reliability framework $X_{(t)}$ is the lifetime of a device which cannot survive age t with distribution function $F_{(t)}(x) = \frac{F(x)}{F(t)}$, $x \leq t$. The random variable $(t - X|X \leq t)$ called the reversed residual lifetime or inactivity time, is the period between the time of failure and t. The mean, variance, percentile life among others, of the inactivity time has been discussed extensively in the literature.

Continuing with the notations in the previous chapter, we assume that F is absolutely continuous with probability density function $f(x)$, reversed hazard rate $\lambda(x) = \frac{f(x)}{F(x)}$ and mean inactivity time $\mu(t) = E(t - X|X \leq t)$. Introduced by Keilson and Sumita (1982), $\lambda(x)$ has found applications in estimation and modelling of censored data, comparison of distributions through stochastic orders, characterization problems and in evolving repair and maintenance strategies of equipments in reliability engineering. Block et al. (1998) have shown that there does not exist a non-negative random variable having increasing or constant reversed hazard rate, a result that has wide implications in past entropy considerations.

Muliere et al. (1993) defined the entropy of $(X|X \leq t)$ as

$$
\begin{aligned}
\bar{H}_F(t) &= \int_0^t \frac{f(x)}{F(t)} \log \left(\frac{f(x)}{F(t)} \right) dx \\
&= 1 - \frac{1}{F(t)} \int_0^t f(x) \log \lambda(x) dx.
\end{aligned}
\tag{3.1}
$$

Though definition (3.1) bears a similarity to residual entropy, the fact that the properties of $\bar{H}_F(t)$ cannot be observed from those of $H_F(t)$ makes a study of $\bar{H}_F(t)$ meaningful. The function $\bar{H}_F(t)$ satisfies the differential equations

$$\frac{d}{dt}\bar{H}_F(t) = \lambda(t)[1 - \bar{H}(t) - \log \lambda(t)], \tag{3.2}$$

$$\frac{d^2}{dt^2}\bar{H}_F(t) = (\lambda'(t) - \lambda(t))\frac{d\bar{H}(t)}{dt} - \lambda'(t). \tag{3.3}$$

Let $E = I(X > t)$, $I(\cdot)$ be the indicator functions. For the bivariate random variable (X, E) conditioned on E, its entropy is

$$H_{X,E}(t) = H_E + F(t)\bar{H}_F(t) + \bar{F}(t)H_F(t),$$

where

$$H_E = -F(t)\log F(t) - \bar{F}(t)\log \bar{F}(t).$$

Similarly by conditioning on X, $H(E|X) = 0$, so that $H(X, E) = H_X$. In terms of the residual entropy function, $H_X = H_F(0)$ or

$$H_F(0) = H_E + F(t)\bar{H}(t) + \bar{F}(t)H(t)$$

and

$$H_F(0) - H_F(t) = H_E + F(t)[\bar{H}(t) - H(t)]. \tag{3.4}$$

Muliere et al. (1993) pointed out an application of the above result in search problems where it is desired to find a partition point t for which both outcomes of E are equally informative. This is obtained by solving $H_F(t) = \bar{H}_F(t)$. When $H_F(t)$ is increasing no solution exists, while with decreasing $H_F(t)$,

$$t = H^{-1}(H_F(0) - H(E)).$$

Equation (3.4) also gives a relationship between $H_F(t)$ and $\bar{H}_F(t)$.

EXAMPLE 3.1. *Let X follow power distribution with $F(x) = x^a$, $0 \le x \le 1$, $a > 0$. Then $\lambda(x) = \frac{a}{x}$ and so from (3.1).*

$$\bar{H}_F(t) = 1 - \frac{1}{t^a}\int_0^t \log\left(\frac{a}{x}\right)\left(\frac{d}{dx}x^a\right)dx$$

$$= \frac{a-1}{a} + \log\frac{t}{a}.$$

Values of $\bar{H}_F(t)$ for some standard distributions can be seen in Table 3.1.

Table 3.1. Past entropy of distributions.

Distribution	$F(x)$	$\lambda(x)$	$\bar{H}(t)$
uniform	$\frac{x-a}{b-a}$ $a \leq x \leq b$	$\frac{1}{x-a}$	$\log(t-a)$
reciprocal exponential	$e^{-\frac{\lambda}{x}},$ $x > 0, \lambda > 0$	$\frac{\lambda}{x^2}$	$\log \frac{\lambda}{t^2} + e^{\frac{\lambda}{t}} E_1(\frac{\lambda}{t})$
exponential	$1 - e^{-\lambda x}$	$\frac{\lambda e^{-\lambda x}}{1-e^{-\lambda x}}$	$1 + \log(1 - e^{-\lambda t})$ $- \frac{\lambda t e^{-\lambda t}}{1-e^{-\lambda t}}$
reversed exponential	$\exp[x-b],$ $0 \leq x \leq b$	a	$1 - \log a$
power	$(\frac{x}{\alpha})^\beta,$ $0 \leq x \leq \alpha, \beta > 0$	$\frac{\beta}{x}$	$1 - \log \beta - \frac{1}{\beta} + \log t$

3.2. Properties of past entropy

The past entropy $\bar{H}_F(t)$ satisfies the following properties

(a) $\bar{H}_F(t)$ can take values in $(-\infty, \infty)$

(b) Even if $H_F(t)$ is the same for two distributions their past entropies can be different

(c) If $f(x)$ is decreasing then $\bar{H}_F(t)$ is decreasing

(d) If $\lambda(x)$ is decreasing for all $x > 0$ then $\bar{H}_F(t)$ is increasing for all $t > 0$ and further $\bar{H}_F(t) \leq 1 - \log \lambda(t)$

(e) If $\bar{H}_F(t)$ is increasing then $\lambda(t) \leq \exp[1 - \bar{H}_F(t)]$

(f) Let $Y = \phi(X)$ be strictly monotonic, continuous and differentiable

$$\bar{H}_Y(t) = H_X(\phi^{-1}(t)) + E(-\log \phi'(X)|X > \phi^{-1}(t)), \ \phi \text{ is strictly increasing}$$
$$= H_X(\phi^{-1}(t)) + E(-\log \phi'(X)|X < \phi^{-1}(t)), \ \phi \text{ is strictly decreasing}$$

(g) $\bar{H}_Z(t) = \bar{H}_X\left(\frac{t-b}{a}\right) + \log a$, where $Z = aX + b, a > 0; b \geq 0$.

(h) $\bar{H}_F(t) = 1 - E(\log \lambda(X)|X \leq t)$

(i) If X is a non-negative random variable with absolutely continuous distribution function F and past entropy function $\bar{H}_F(t)$, then $\bar{H}_F(t)$ uniquely determine F. The results (a) through (f) are discussed in Di Crescsenzo and Longobardi (2002), (g) in Nanda and Paul (2006a) and (i) in Gupta (2009).

There are several properties of $\bar{H}_F(t)$ available for specific distributions and families of distributions. Kundu et al. (2010) have obtained several such results concerning absolutely continuous random variables with support $-\infty \leq a < b < \infty$. If X has a finite mean then

$$\bar{H}_F(t) = k + \log r(t), \tag{3.5}$$

where $r(t) = E(t - X|X \leq t)$ is the mean inactivity time, with $k = -\log r(b) = H_X$ if an only if $r(t)\lambda(t) = p$, a constant. A more special property that excludes $\lambda(t)$ in the above is that (3.5) is satisfied if and only if X is distributed as one of

$$F_1(x) = \exp[(x - b)/b - \mu] \text{ for } p = 1,$$

$$F_2(x) = \left(\frac{bp - \mu + x(1 - p)}{b - \mu}\right)^{\frac{p}{1-p}}$$

where $x \in \left(\frac{\mu - bp}{1 - p}, b\right)$ for $p < 1$ and $x \in (-\infty, b)$ if $p > 1$. Further

$$\bar{H}_F(t) = k + \log(\alpha + \beta t), \ \alpha > 0$$

characterize the above distributions at the values $p = 1 - \beta, \beta \geq 0$ and $\beta > 0, < 0$. Two further special cases of interest are the necessary and sufficient condition for X to be (a) exponential is $\bar{H}_F(t) + \log \lambda(t) + \frac{r(t)}{\mu} = 0$ and (b) power is $\bar{H}_F(t) + \log p = k$. The truncated extreme value distribution

$$F(x) = \exp[-(e^{-x} - e^{-b})], \ -\infty < x < b$$

has the unique property $\bar{H}_F(t) + \mu(t) = 1 + t$ and the truncated logistic model

$$F(x) = k[1 + (k - 1)e^{c(b-x)}], \ -\infty < x < b, \ c > 0, \ k > 1$$

is characterized by $\bar{H}_F(t) - r(t) = p$, Asha and Rajeesh (2015) have proved characteristic properties of certain distributions (with support $(-\infty, b)$).

Given below are $\bar{H}_F(t)$ values shown against them.

Distribution ($F(t)$)	Property
$e^{c(t-b)}$	$1 - \log \lambda(t)$
$\left(\frac{t}{b}\right)^c$	$\frac{c-1}{c} - \log \lambda(t)$
$(1 - t)^{-c}$	$\frac{c+1}{c} - \log \lambda(t)$
$e^{-b} - e^{-t}$	$1 - \log \lambda(t) - r(t)$

When viewed in the context of lifetime data analysis, the above results are not convenient for applications since their support extends to negative values. The main reason for the choice of support as $(-\infty, b)$ is that many results in reversed hazard rates will be similar to their counterpart hazard rates only in the above interval, if we insist on the absolute continuity of $F(x)$. By dropping the assumption of absolute continuity of F and restricting X to be non-negative $\bar{H}_F(t)$ can be shown to possess some interesting properties. Since the calculation of $\bar{H}_F(t)$ requires the reversed hazard rate, some properties of $\lambda(x)$ will be presented first.

THEOREM 3.1. *A non-negative random X taking values in $[0, b]$, $0 < b < \infty$, will have a constant reversed hazard rate if and only if it has a distribution function*

$$F(x) = e^{a(x-b)}, \ 0 \le x \le b, \ b > 0. \tag{3.6}$$

PROOF. When the distribution function of X is (3.6), it is observed that $x = 0$ is a point of discontinuity satisfying $F(0) = e^{-ab}$. It's density function is

$$f(x) = \begin{cases} e^{-ab}, & x = 0 \\ ae^{a(x-b)}, & 0 < x \le b \end{cases}$$

so that the reversed hazard function is

$$\lambda(x) = \begin{cases} 1, & x = 0 \\ a, & 0 < x \le b \end{cases}$$

which is a constant. We have the case of a mixed distribution. Conversely, if $\lambda(x) = a$, a constant,

$$\frac{d \log F(x)}{dx} = a, \ 0 < x \le b.$$

Integrating with respect to x, $F(x) = e^{ax+K}$ and $F(b) = 1$ gives (3.6) in $0 < x < b$. The density function

$$f(x) = ae^{a(x-b)}, \ 0 < x < b$$

gives a proper distribution if X has jump e^{-ab} at zero and this gives the required $F(x)$.   □

REMARK 3.1. *The distribution (3.6) will be referred to as the reversed exponential distribution. Its mean inactivity time function is $r(t) = a^{-1}$, a constant in $(0, b]$ and unity when $x = 1$ and satisfies $\lambda(t)r(t) = 1$ for all $t \ge 0$. Later it will be seen that the constancy of $\mu(t)$ and the reciprocal relationship between $\lambda(t)$ and $r(t)$ are infact characteristic properties of the model. We shall denote by F_c, f_c, λ_c, r_c, $\bar{H}_c$ the distribution function, density function, reversed hazard rate, mean inactivity time and past entropy of the continuous part of the random variable X.*

A more general class of distributions can be arrived at as seen in the following theorem.

THEOREM 3.2. *Let $g(x)$ be a non-negative increasing function differentiable in $(0, b]$, $b < \infty$ satisfying $g(0) = 0$. Then*

$$F(x) = \exp[g(x) - g(b)], \ 0 \le x \le b \tag{3.7}$$

is the distribution function of a random variable X in $[0, b]$.

REMARK 3.2. *The distribution (3.7) has a discrete mass point at $x = 0$ with $P(X = 0) = \exp[-g(b)]$ and an absolutely continuous part*

$$F_c(x) = \exp[g(x) - g(b)]$$

in $0 < x \leq b$ with density function $f_c(x) = g'(x) \exp[g(x) - g(b)]$. The reversed hazard rate is $\lambda(x) = g'(x)$ in $0 <\leq b$ and $\lambda(x) = 1$ at $x = 0$. The distribution function can be decomposed as the sum of a step function and of a continuous function as

$$F(x) = e^{-g(b)} + \left(1 - e^{-g(b)}\right) exp[g(x) - g(b)].$$

REMARK 3.3. *In the above representation X denotes the lifetime of a device which may fail immediately on installation (generally attributed to a manufacturing defect) with probability $e^{-g(b)}$ or live upto age x with probability $\left(1 - e^{-g(b)}\right) exp\left[g(x) - g(b)\right]$. Here $F(0-) = 0$, $F(0) = e^{-g(b)}$ and $F(b) = 1$. The kind of distributions presented here have appeared in literature under the name semi-continuous models in several contexts including insurance, rainfall analysis, epidemiology, radio audience data, marine science and labour economics. For example, because of loss of bonus in the next year, many insured persons do not claim losses and so policies with zero claims will be substantial. The claims depend on the extent of loss, insured amount and follow a continuous distribution with positive support. Modelling data of this nature requires a distribution of the above type. In epidemiology, patient compliances which measure the extent to which a subject's behaviour coincides with medical or health advices is recorded on a $[0, 100]$ scale as percentages. The proportion of people with zero compliance is important and is usually large enough to be neglected. The percentage of compliance is taken as continuous and those at zero is a jump. For details we refer to Nanjundan and Pasha (2018) and Chang and Pocock (2000).*

The construction of (3.7) requires a method to choose an appropriate $g(x)$. This is considerably simplified with the aid of the next result.

THEOREM 3.3. *Let X be as in Theorem 3.2 and Y be a non-negative random variable with absolutely continuous distribution function $F_Y(x)$. Then a necessary and sufficient condition that X has distribution (3.7) is that $\lambda_X(x) = h_Y(x)$ for all x in $(0, b]$, where $h_Y(\cdot)$ is the hazard rate of Y.*

PROOF. Assuming $\lambda_X(x) = h_Y(x)$, the absolutely continuous part of the distribution of X is

$$F_{c,X}(x) = \exp\left[-\int_x^b \lambda(t)dt\right]$$

$$= \exp\left[-\int_x^b h(t)dt\right] \qquad (3.8)$$

$$= \frac{\exp[H_Y(x)]}{H_Y(b)} = \frac{\bar{F}_Y(b)}{\bar{F}_X(x)}$$

where $H_Y(x) = \int_0^x h(t)dt$ is the cumulative hazard rate of Y. Now using $P(0 \leq X \leq b) = 1$, we have (3.8). Conversely if (3.8) is true, $g(x) = H(x)$ satisfies all the conditions of Theorem 3.1 and $\lambda_Y(x) = g'(x) = h(x)$. □

In any of the life distributions available in the literature the choice of $H_Y(x) = g(x)$ leads to a $F_X(x)$ whose reversed hazard rate has the same functional form as the hazard rate of the chosen distribution. Thus the reversed hazard rate of X will be increasing, decreasing BT, UBT or periodic according as $h_Y(x)$ is so.

The following definitions will be used to classify various life distributions.

DEFINITION 3.1. *We say that X is*
  (i) *increasing (decreasing) reversed hazard rate, IRHR (DRHR) if $g(x)$ is convex (concave) in $(0, b]$*
  (ii) *BT (UBT) if $g(x)$ is concave (convex) in $(0, t_0)$ and convex (concave) in $(t_0, b]$ for some $0 < t_0 < b$, and*
  (iii) *periodic reversed hazard rate if $\lambda(x+c) = \lambda(x)$ for all x and some c in $0 < x \leq b$.*

A list of some distributions arrived at in this fashion and the nature of their reversed hazard rates are given in Table 3.2.

To demonstrate the applications of Theorems 3.2 and 3.3 we present in Table 3.2 some useful distributions that have a variety of shapes for their reversed hazard rate functions. To our knowledge, the distributions except 4 are new. The first distribution is derived from the Burr type XII with $\bar{F}(x) = (1 + x^c)^{-k}$ using (3.8) whose reliability properties are discussed in Zimmer et al. (1988). Its reversed hazard rate is decreasing for $c \leq 1$ and UBT for $c > 2$ (see Figure 3.1). Similarly the models 2 through 10 in Table 3.2 in that order were generated by the same methods from the power distribution $\bar{F}(x) = \left(\frac{x}{\alpha}\right)^\beta, 0 \leq x \leq \alpha$, Hjorth model $\bar{F}(x) = e^{-\frac{\alpha x^2}{2}} (1 + \beta x)^{-\frac{\theta}{\beta}}, x > 0$ (Hjorth (1980)), exponential distribution, Dhillon's model (Dhillon (1981)), $\bar{F}(x) = exp\left[-(\log(\lambda x + 1))^{\beta+1}\right]$,

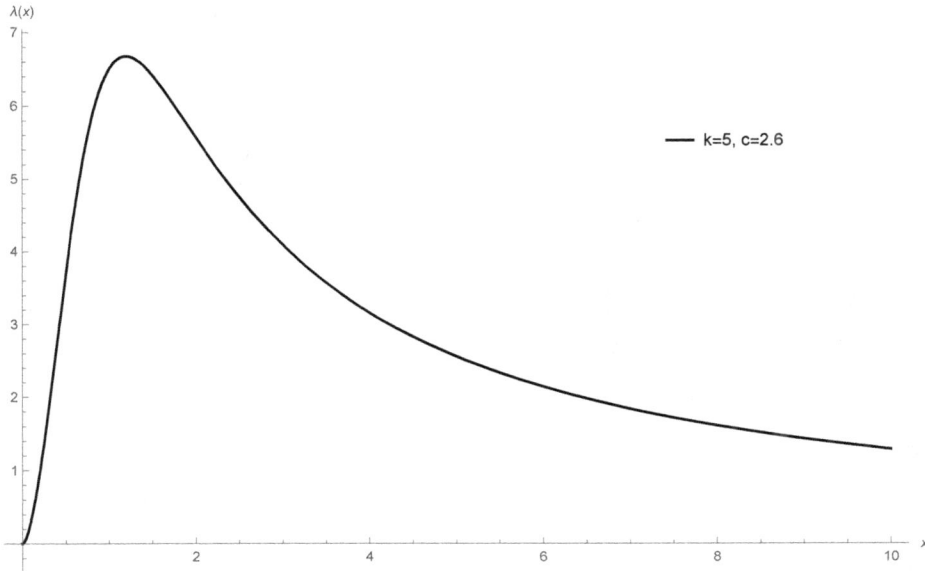

Figure 3.1. Shape of $\lambda(x)$ for Model 1 with $(k = 5, c = 2.6)$.

$x > 0$, exponential power distributions $\bar{F}(x) = exp\left[1 + e^{-x^{\alpha}}\right], x > 0$ (Smith and Bain (1975)), the beta law $\bar{F}(x) = (1 - x)^c, 0 \le x \le 1$, a new model $\bar{F}(x) = exp\left[-ax - \frac{\theta}{c}\sin cx\right]$ with periodic hazard rate and the generalized Pareto $\bar{F}(x) = \left(1 + \frac{ax}{c}\right)^{-\frac{a+1}{a}}$. We have chosen a few representative life distributions for illustration and the methods apply to other distributions as well. The shapes of the reversed hazard rates can be seen in the last column. For instance, in the case of model 5, we have decreasing (increasing) reverse hazard rates for $\lambda < 1(> 1)$ as evident from Figure 3.2. In the case of model 7, $\lambda(x)$ is BT as seen in Figure 3.3 and for model 9, $\lambda(x)$ is periodic, see Figure 3.4. The relevance of these and other distributions generated by (3.8) is that they are candidate models for data which possess reversed hazard rates of a particular shape evidenced from the observations. They are useful for modelling semi-continous data in the other applications mentioned above.

THEOREM 3.4. *The random variable X satisfies the property $\lambda(t)\mu(t) = C$ for all $0 < t \le b$, where C is a constant if and only if its distribution function is of the form*

$$F(x) = \left[\frac{K + (1 - c)x}{K + (1 - c)b}\right]^{\frac{c}{1-c}}, \quad 0 \le x \le b. \quad (3.9)$$

Table 3.2. Life distributions and classifications by nature of reversed hazard rates.

Distribution	$g(x)$	$\lambda(x)$	Classification
1. $\left(\frac{1+x^c}{1+b^c}\right)^k$, $k, c > 0$	$k\log(1+x^c)$	$\frac{kcx^{c-1}}{1+x^c}$	UBT
2. $\frac{\alpha^\beta - b^\beta}{\alpha^\beta - x^\beta}$, $\alpha, \beta > 0$, $x < \alpha < b$	$\log(1 - \frac{x^\beta}{b^\beta})$	$\frac{\beta x^{\beta-1}}{\alpha^\beta - x^\beta}$	increasing for $\beta > 1$ BT for $\beta < 1$
3. $\exp[\frac{\alpha}{2}(x^2 - b^2)](\frac{1+\beta x}{1+\beta b})^{\theta/\beta}$	$\alpha\frac{x^2}{2} + \frac{\theta}{\beta}\log(1 + \frac{x}{\beta})$	$\alpha x + \frac{\theta}{1+\beta x}$	$\begin{cases}$ decreasing $(\alpha < 0)$ increasing $\alpha > \theta\beta$; $\theta = 0$ constant, $\alpha = \beta = 0$ BT, $0 < \alpha < \theta\beta$
4. $\exp[a(x - b)]$, $a > 0$	ax	a	constant
5. $\exp[\sigma(x^\lambda - b^\lambda)]$, $\sigma, \lambda > 0$	σx^λ	$\sigma\lambda x^{\lambda-1}$	increasing, $\lambda > 1$ decreasing, $\lambda < 1$ constant $\lambda = 1$
6. $\exp[\{\log(1+x)\}^{\beta+1}$ $-\log\{(1+b)\}^{\beta+1}]$, $c > 0, b < 1$	$[\log(1+x)]^{\beta+1}$	$(\beta+1)\log(1+x)$	UBT
7. $(\frac{1-b}{1-x})^c$, $c > 0, b < 1$	$c\log(1-x)$	$\frac{c}{1-x}$	increasing
8. $\exp[e^{x^\lambda} - e^{b^\lambda}]$	$e^{x^\lambda} - 1$	$\lambda x^{\lambda-1} e^{x^\lambda}$	BT
9. $(\frac{ax+c}{ab+c})^{\frac{a+1}{a}}$, $c > 0$, $a > -1$	$\frac{a+1}{a}\log(1 + \frac{ax}{c})$	$\frac{a+1}{ax+c}$	$\begin{cases}$ decreasing $(-1 < a < 0)$ increasing $a > 0)$ constant, $a = 0$
10. $\frac{e^{-\alpha b - \frac{\theta}{c}\sin b}}{e^{-\alpha x - \frac{\theta}{c}\sin x}}$, $\alpha, \theta, c > 0$	$\alpha x - \frac{\theta}{c}\sin cx$	$\alpha + \theta\cos cx$	periodic with period $\frac{2\pi}{c}$

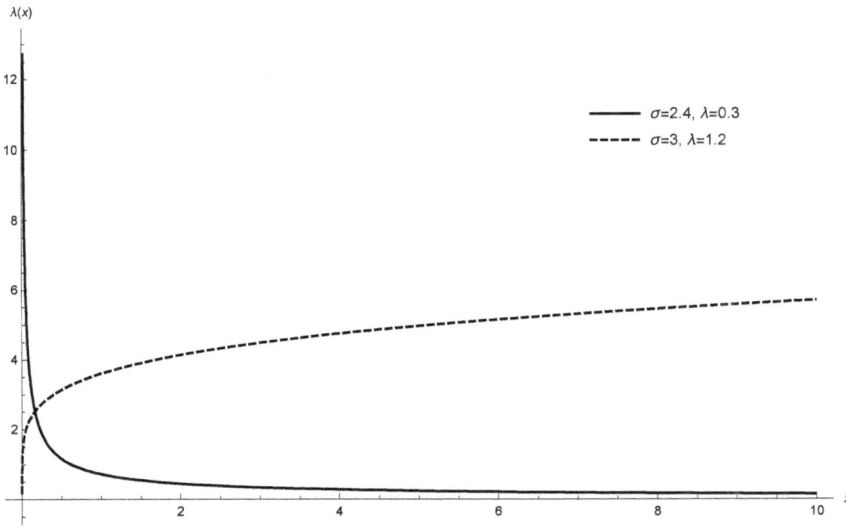

Figure 3.2. Shapes of $\lambda(x)$ for Model 5 with $(\sigma = 2.4, \lambda = 0.3)$ and $(\sigma = 3, \lambda = 1.2)$.

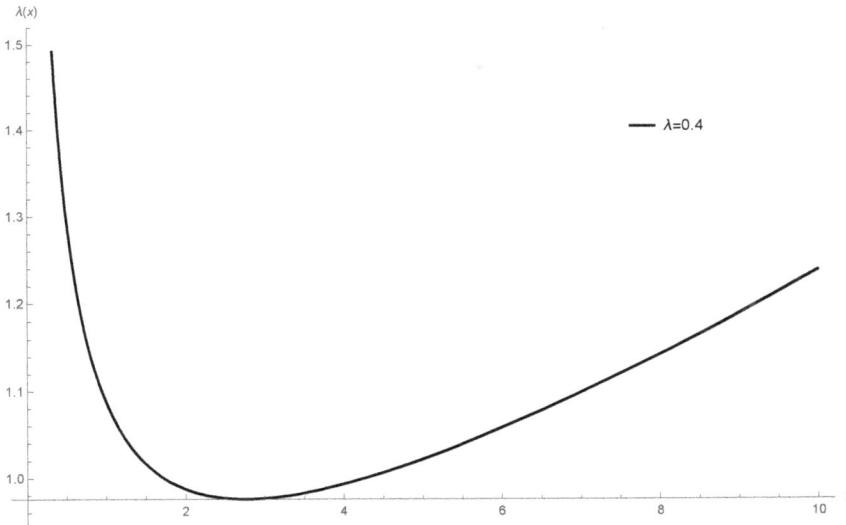

Figure 3.3. Shape of $\lambda(x)$ for Model 7 with $\lambda = 0.4$.

PROOF. Assuming the distribution (3.9) of X,

$$f(x) = \frac{c}{K} \frac{\left(1 + \frac{1-c}{K}x\right)^{\frac{c}{c-1}-1}}{\left(1 + \frac{1-c}{K}b\right)^{\frac{c}{c-1}}}$$

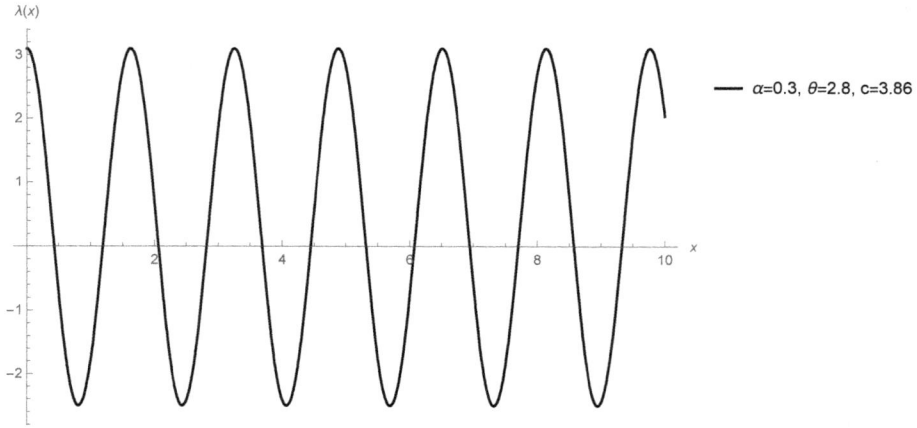

Figure 3.4. Periodic shape of $\lambda(x)$ for Model 9 with $\alpha = 0.3$, $c = 3.86$ and $\theta = 2.8$.

giving

$$\lambda(x) = \frac{c}{K + (1 - c)x}.$$

Also

$$\int_0^t F(x)dx = \frac{K\left(1 + \frac{1-c}{K}x\right)^{\frac{c}{1-c}+1}}{c\left(1 + \frac{1-c}{K}\right)^{\frac{c}{1-c}}}$$

leads to

$$r(t) = \frac{1}{F(t)} \int_0^t F(x)dx = K + (1 - c)x, \ 0 < x \le b$$

and the property, $\lambda(t)r(t) = c$. On the other hand if $\lambda(t)r(t) = c$ for $0 < t \le b$, from (1.11)

$$1 - r'(t) = 1 - c$$

which gives,

$$r(t) = (1 - c)t + K.$$

With the help of (1.12) we find

$$F(x) = \exp\left[-\int_x^b \frac{cdt}{(1 - c)t + k}\right]$$

$$= \left[\frac{(1 - c)x + K}{(1 - c)b + K}\right]^{\frac{c}{1-c}}, \ 0 < x \le b.$$

Computing $f(x)$ in $(0, b]$ and then requiring $P[a \le x \le b] = \phi$, we get (3.9).   □

In fact, (3.9) represents a class of distributions similar to the generalized Pareto distribution. We call it the reversed generalized Pareto. As $c \to 1$ it reduces to the reversed exponential (3.6),

$$F_E(x) = e^{\frac{(x-b)}{K}}$$

$$F_B(x) = \left(\frac{1 + \frac{c-1}{K} x}{1 - \frac{c-1}{K} b} \right)^{-\frac{c}{c-1}}, \quad c > 1 \tag{3.10}$$

is the reversed beta and for $0 < c < 1$

$$F_P(x) = \left(\frac{1 + \frac{1-c}{K} x}{1 - \frac{1-c}{K} b} \right)^{\frac{c}{1-c}}, \quad 0 < c < 1$$

is called the reversed Pareto. Notice that for the class (3.9) the reversed mean residual life is linear, $r(x) = K + (1 - c)x$, and the reversed hazard rate is reciprocal linear $\lambda(x) = c(K + (1 - c)x)^{-1}$ in $0 < x \le b$ and $\lambda(x) = r(x) = 1$ at the point of discontinuity $x = 0$ so that $\lambda(x)\mu(x) = 1$. This is always the case for all distributions with the origin as the discontinuity point. Block et al. (1998) have shown that there does not exist a non-negative random variable with increasing reversed hazard rate (decreasing mean inactivity time) whenever F is absolutely continuous. When $c > 1$, (3.10) has

$$\lambda(x) = c(K + (1 - c)x)^{-1}$$

which is increasing for $x > 0$. Thus when distributions with a discontinuity point are considered, we have non-negative random variables for which $\lambda(x)$ is increasing and similarly mean inactivity time is decreasing. At all continuity points of $F(x)$, we have $\lambda(x)r(x) = c$.

THEOREM 3.5. *A random variable X taking values in $[0, b]$ has a mean inactivity time*

$$r(x) = \alpha e^{-\beta x}, \quad \alpha, \beta > 0, \ x > 0 \tag{3.11}$$

if and only if

$$F(x) = \exp \left[\beta(x - b) + \frac{1}{\alpha\beta} \{\exp(\beta x) - \exp(\beta b)\} \right], \quad 0 \le x \le b. \tag{3.12}$$

PROOF. Given (3.11),

$$F(x) = \exp \left[-\int_x^b \frac{1 - r'(t)}{r(t)} dt \right]$$

leads to (3.11). Conversely for the distribution function (3.12) the use of formula (1.11) gives $r(x)$. □

Some remarks on Theorem 3.5 are in order. In the first place the reserved hazard function of (3.11) is

$$\lambda(x) = \beta + \frac{1}{\alpha}e^{\beta x}, \ x > 0$$

and the relationship

$$r(x)\left(\lambda(x) - \beta\right) = 1$$

characterizes the model. The point of discontinuity of (3.12) is $x = 0$ with jump $e^{1/\alpha\beta}\left[e^{\beta b} + (\frac{e}{\alpha\beta})^{\beta b}\right]^{-1}$. Further the mean inactivity time is decreasing while reversed hazard rate is increasing.

In the light of the above discussion some new aspects of past entropy need investigation and the relevant results are summarized in a few theorems that follow.

THEOREM 3.6. *If X is a non-negative random variable taking values in $[0, b]$, $b < a$ with distribution function $F(x)$, then*

$$\bar{H}_F(t) = c - \log \lambda(t), \text{ for all } t \text{ in } [0, b]$$

if and only if

$$F(x) = \left[\frac{k + (1-c)x}{k + (1-c)b}\right]^{\frac{c}{1-c}}, \ 0 \le x \le b, \ c > 0. \qquad (3.13)$$

PROOF. Assume $F(x)$ as in (3.13). Then by definition

$$\bar{H}_F(t) = 1 - \left[0 + \frac{1}{F_c(t)}\int_0^t \log\frac{c}{k+(1-c)x}\right]\frac{d}{dx}F(x)dx$$

$$= 1 - \log\left(\frac{c}{k+(1-c)t}\right) - \frac{1-c}{F_c(t)}\int_0^t f_c(x)dx$$

$$= c - \log\left(\frac{c}{k+(1-c)t}\right) = c - \log \lambda(t).$$

To prove the 'only if' part, we write

$$c - \log \lambda_c(t) = 1 - \frac{1}{F_c(t)}\int_0^t \log \lambda_c(x)f_c(x)dx$$

$$[1 - c - \log \lambda_c(t)]F_c(t) = \int_0^t \log \lambda_c(x)f_c(x)dx.$$

Differentiating and simplifying

$$\frac{\lambda'(t)}{\lambda(t)} = (c-1)\lambda(t).$$

Solving

$$\lambda_c(t) = \frac{1}{k + (1 - c)t}.$$

This gives

$$F_c(x) = \left[\frac{k + (1 - c)x}{k + (1 - c)b}\right]^{\frac{c}{1-c}}, \ 0 < x \le b$$

and $P[0 \le X \le b] = 1$ ensures that

$$F(0) = \left[\frac{k}{k + (1 - c)b}\right]^{\frac{c}{1-c}}.$$

Thus $F(x)$ is as stated above.   □

REMARK 3.4. *Since $r(t)\lambda(t) = 1$ is a characteristic property of (3.12)*
X satisfies

$$\bar{H}_F(t) = B + \log r(t)$$

for all t and some $B > 0$, if and only if $F(x)$ is as in (3.12).

THEOREM 3.7. *The following statements are equivalent*

(a) *X follows reversed Gompertz law*

$$F(x) = \exp\left[\frac{1}{\alpha}(e^{\alpha x} - e^{\alpha b})\right], \ 0 \le x \le b, \ \alpha > 0 \qquad (3.14)$$

(b)

$$\bar{H}_F(t) = 1 - \alpha t + \alpha r(t). \qquad (3.15)$$

PROOF. To prove (a) ⇒ (b) we first note that (3.14) has density function
in $0 < x \le b$,

$$f(x) = e^{\frac{1}{\alpha}}(e^{\alpha x} - e^{\alpha b})e^{\alpha x}$$

so that $\lambda(x) = \alpha x$ and $\log \lambda(x) = \alpha x$. Thus

$$\bar{H}_F(t) = 1 - \left[0 + \frac{1}{F_c(t)}\int_0^t \alpha x f(x)dx\right]$$

$$= 1 - \left[\alpha t - \frac{\alpha}{F_c(t)}\int_0^t F_c(x)dx\right]$$

$$= 1 - \alpha t + \alpha r_c(t).$$

Now if $\bar{H}_F(t)$ is as above

$$1 - \frac{1}{F_c(t)}\int_0^t \log \lambda(x)dF(x) = 1 - \alpha t + \frac{\alpha}{F_c(t)}\int_0^t F_c(x)dx$$

or

$$\alpha \int_0^t F_c(x)dx + \int_0^t \log \lambda(x)dF(x) = \alpha t F(t).$$

Differentiating and simplifying, $\log \lambda_c(b) = \alpha t$ and so

$$F_c(x) = \exp\left[\frac{1}{\alpha}(e^{\alpha x} - e^{-\alpha b})\right].$$

Using $P[0 \leq x \leq b] = 1$, we have (3.14).   □

REMARK 3.5. *The distribution has a jump $P(X = 0) = \exp\left[-\frac{1}{\alpha}e^{\alpha b}\right]$.*

Some other interesting properties of $\bar{H}_F(t)$ can be derived if we look at its relationship with the reversed geometric mean life function $G(t)$ given by

$$\log \bar{G}(t) = E(\log X | X \leq t).$$

THEOREM 3.8. *Let X be a non-negative random variable with absolutely continuous distribution function $F(x)$. Then*

$$\bar{H}_F(t) = C_1 + C_2 \log \bar{G}(t) \tag{3.16}$$

for all $t > 0$ if and only if X is distributed as

$$F(x) = e^{-\frac{K}{C_2-1}x^{1-C_2}}, \quad x > 0, \ K = e^{1-C_1}. \tag{3.17}$$

PROOF.

$$\bar{H}_F(t) = C_1 + C_2 \log \bar{G}(t)$$

implies

$$1 - \frac{1}{F(t)}\int_0^t \log \lambda(x)f(x)dx = C_1 + \frac{C_2}{F(t)}\int_0^t \log x f(x)dx$$

and

$$(1-c_1)F(t) = C_2 \int_0^t \log x f(x)dx + \int_0^t \log \lambda(x)f(x)dx.$$

Differentiating and simplifying

$$\lambda(x) = Kx^{-c_2}, \ K = e^{1-c_1}$$

from which the form of $F(x)$ is recovered. Conversely from (3.17)

$$\bar{H}_F(t) = 1 - \frac{1}{F(t)}\int_0^t (\log K - C_2 \log x)f(x)dx$$
$$= 1 - \log K + C_2 \log \bar{G}(t).$$

Since $K = e^{1-C_1}$, we have the required result in (3.16).   □

REMARK 3.6. *(i) When $C_2 = 1$, $F(x) = x^K$, the power distribution in $[0, 1]$.*
(ii) When $C_2 = \beta + 1$, $\beta > 0$

$$F(x) = \exp\left[-\frac{K}{\beta}x^{-\beta}\right], \quad x > 0, K, \ \beta > 0$$

is the inverse Weibull distribution discussed in Erto (1989). A further specialization when $\beta = 1$ is the inverted exponential model.

A still larger set of distributions can be generated if the relationship (3.16) can be further extended to the form

$$\bar{H}_F(t) = C_1 + C_2 \log \bar{G}(t) + C_3 \log F(t). \tag{3.18}$$

Relationship (3.18) means that

$$1 - \frac{1}{F(t)} \int_0^t \log \lambda(x) f(x) dx = C_1 + \frac{C_2}{F(t)} \int_0^t \log x f(x) dx + C_3 \log F(t)$$

or

$$(1 - C_1) F(t) = \int_0^t \log \lambda(x) f(x) dx + C_2 \int_0^t \log x f(x) dx + C_3 F(t) \log F(t).$$

Differentiating and simplifying

$$1 - C_1 - C_3 = \log \left(\lambda(t) t^{C_2} [F(t)]^{C_3} \right)$$

which reduces to

$$[F(t)]^{C_3 - 1} f(t) = K t^{-C_2}, \quad K = e^{1 - C_1 - C_3}.$$

Integrating with respect to t

$$F(t) = \left[BC_3 + \frac{C_3 K t^{-c_2 + 1}}{1 - C_2} \right]^{\frac{1}{C_3}}$$

where B is the constant of integration. As $t \to \infty$, $B = \frac{1}{C_3}$ or

$$F(t) = \left[1 + \frac{C_3 K t^{1 - c_2}}{1 - C_2} \right]^{1/c_3}, \quad C_2 > 1. \tag{3.19}$$

When X is distributed as (3.19) by direct calculation one can verify (3.18). Thus we have proved the following theorem.

THEOREM 3.9. *If X is a non-negative random variable with an absolutely continuous distribution satisfying $E(\log X) < \infty$, then*

$$F(x) = \left(1 + \frac{c_3 K t^{1 - c_2}}{1 - c_2} \right)^{\frac{1}{c_3}}$$

for real constants C_1, C_2 and C_3 if and only if

$$\bar{H}_F(t) = C_1 + C_2 \log \bar{G}(t) + C_3 \log F(t).$$

3.2.1. Special cases.

(i) When $C_3 \to 0$, we have the inverse Weibull and inverse exponential distributions as seen in Theorem 3.5

(ii) $C_1 = 1 - \frac{1}{c} - \log \frac{c}{\alpha}$, $c, \alpha > 0$, $C_2 = 2$ and $C_3 = -\frac{1}{c}$ give the reciprocal Pareto law

(iii) $C_1 = 1 - \frac{1}{c} - \log \frac{c}{R}$, $c, R > 0$, $C_2 = 2$ and $C_3 = \frac{1}{c}$ leads to the reciprocal rescaled beta distribution

(iv) $C_1 = 1 - \log \beta\theta - \frac{1}{\theta}$, $C_2 = 1 + \beta$, $C_3 = \frac{1}{\theta}$ gives the generalized power distribution $F(x) = (1 - x^{-\beta})^\theta$.

After a some what detailed investigation of the properties of the reversed hazard we return to the past entropy. The fact $\lambda(x)$ can assume different shapes, than decreasing along the non-negative real line, leaves scope for extending the domain of application to models in which $\bar{H}_F(t)$ can also have characteristics corresponding to the extended behaviour of $\lambda(x)$. From (3.2) we find that whenever $\bar{H}_F(t)$ is decreasing it has a lower bound prescribed as

$$\bar{H}'_F(t) = \lambda(t)[1 - \bar{H}_F(t) - \log \lambda(t)] \leq 0$$

or

$$\bar{H}_F(t) \geq 1 + \log \lambda(t).$$

Corresponding to Definition 3.1, we propose the following.

DEFINITION 3.2. *The random variable X is said to be (i) increasing (decreasing) uncertainly in past entropy, IUPL (DUPL) if $\bar{H}_F(t)$ is increasing in t for all $t > 0$ (all t in $(0, b]$) (ii) BT (UBT) if*

$$\bar{H}_F(t) = \begin{cases} \bar{H}_F^{(1)}(t), & t \leq t_1 \\ c, & a \ constant, t_1 \leq t \leq t_2 \\ \bar{H}_F^{(2)}(t), & t \geq t_2 \end{cases}$$

where $\bar{H}_F^{(1)}(t)$ is strictly decreasing (increasing) and $\bar{H}_F^{(2)}(t)$ is strictly increasing (decreasing). The points t_1 and t_2 are called the change points of $\bar{H}_F$. When $\bar{H}_F$ is differentiable and has only one change point it is BT (UBT) if $\bar{H}'_F(t) < (>0)$ in $(0, t_0]$. $\bar{H}'_F(t_0) = 0$ and $\bar{H}'_F(t) > (<)0$ in $(t_0, b]$.

We now state a sufficient condition for X to be DUPL.

THEOREM 3.10. *If X is IRHR then it is DUPL.*

PROOF. From (3.2),

$$\bar{H}'_F(t) = -\lambda(t) \log \lambda(t) + \frac{\lambda(t)}{F(t)} \int_0^t \log \lambda(x) f(x) dx.$$

Since X is IRHR $\lambda(x) \leq \lambda(t)$ for all $x \leq t$ and so

$$\bar{H}_F'(t) \leq -\lambda(t) \log \lambda(t) + \lambda(t) \log \lambda(t) = 0,$$

proving the result. □

The above condition is necessary as illustrated by the following example.

EXAMPLE 3.2. *Let*

$$F(x) = \frac{1 - b^{1/2}}{1 - x^{1/2}}, \ 0 \leq x \leq b < 1.$$

We have

$$\bar{H}_F(t) = 1 + \log 2 + 2 \log(1 - t^{1/2}) + \frac{1}{2} t^{1/2} \log t - t^{1/2} \log(1 - t^{1/2}) + t^{1/2}$$

and

$$\bar{H}_F'(t) = \log \frac{t^{1/2}}{1 - t^{1/2}} - \frac{t^{1/2}}{1 - t^{1/2}} < 0$$

for all $t > 0$. Thus $\bar{H}_F'(t)$ is decreasing or DULP (see Figure 3.5). However

$$\lambda(t) = [2t^{1/2}(1 - t^{1/2})]^{-1}$$

is BT shaped (see Figure 3.6) with change point $t = \frac{1}{2^{1/2}}$.

It was seen that increasing (decreasing) $\lambda(x)$ implies decreasing (increasing) $H_F(t)$. Can BT (UBT) reversed hazard rate produce UBT (BT) past entropy? We show that this is not the case.

EXAMPLE 3.3. *Assume that X is distributed as*

$$F(x) = \left(\frac{x + d}{b + \alpha}\right)^c, \ 0 \leq x \leq b, \alpha, c > 0, c > \alpha.$$

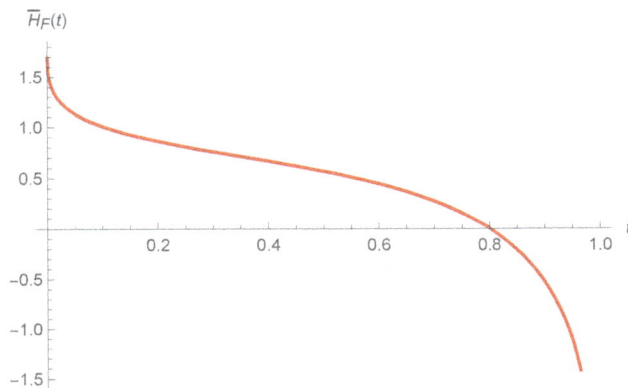

Figure 3.5. Plot of $\bar{H}_F(t)$.

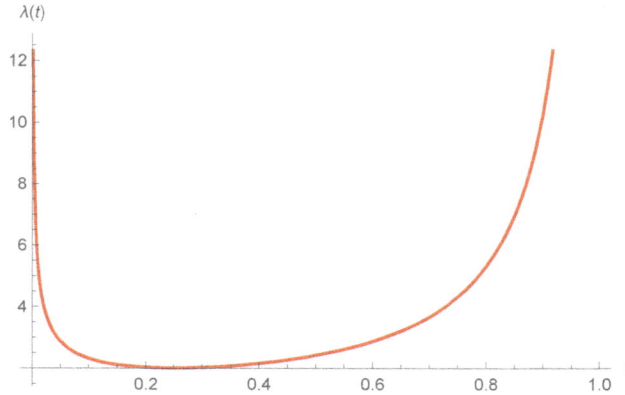

Figure 3.6. Plot of $\lambda(t)$.

Then $\lambda(x) = \frac{c}{x+d}$ is decreasing. At the same time

$$\bar{H}_F(t) = 1 - \frac{1}{c} - \log\left(\frac{c}{t+\alpha}\right) + \left(\frac{1}{c} + \log\frac{c}{\alpha}\right)\frac{\alpha^c}{(t+\alpha)^c}.$$

Differentiating

$$\bar{H}'_F(t) = \frac{1}{t+\alpha}\left[1 - \left(1 + c\log\frac{c}{\alpha}\right)\frac{\alpha^c}{(t+\alpha)^c}\right].$$

It is easy to see that there is a solution to $\bar{H}'_F(t) = 0$ given by

$$t_0 = \alpha\{\left(1 + c\log\frac{c}{\alpha}\right)^{1/c} - 1\}$$

so that $\bar{H}_F(t)$ is BT. However, $\lambda(t)$ is not BT, but decreasing.

EXAMPLE 3.4. *In Example 3.2, $\lambda(t)$ is BT but $\bar{H}_F(t)$ is decreasing.*

3.3. Past quantile entropy

By means of the transformation $x = Q(u)$ in (3.1). Sunoj et al. (2013) proposed a quantile oriented measure of past entropy as

$$\bar{\xi}(u) = \bar{H}_F(Q_X(u)) = \log u + u^{-1}\int_0^u \log q(p)dp \qquad (3.20)$$

$$= 1 - u^{-1}\int_0^u \log \lambda_Q(p)dp \qquad (3.21)$$

where

$$\lambda_Q(u) = [uq(u)]^{-1}$$

is the reversed hazard quantile function defined in Section 1.3.3. They have also obtained the formulas

$$q(u) = \exp[u\xi^{-1}(u) + \xi(u) - \log u - 1] \tag{3.22}$$

and

$$\xi_X = \xi(u, 1 - u) + u\bar{\xi}(u) + (1 - u)\xi(u).$$

The relationship between residual quantile entropy and past quantile entropy with entropy of X is given by $\xi(X) = \xi(u, 1 - u) + u\bar{\xi}(u) + (1 - u)\xi(u)$, where $\xi(u, 1 - u) = -u\log u - (1 - u)\log(1 - u)$ denotes the entropy of the Bernoulli distribution. Equation (3.22) shows that the past quantile entropy function $\bar{\xi}(u)$ determines the distribution of X uniquely.

EXAMPLE 3.5. *Let X be distributed as exponential with $\bar{F}(x) = e^{-\lambda x}$, $x > 0$; $\lambda > 0$. Then $Q_X(u) = -\frac{1}{\lambda}\log(1 - u)$ and*

$$\lambda_Q(u) = \frac{1}{uq(u)} = \frac{\lambda(1 - u)}{u}.$$

Thus

$$\hat{\xi}(u) = 1 - u^{-1}\int_0^u \log \lambda_Q(p)dp$$

$$= 1 - \log \lambda + \log u + \frac{1 - u}{u}\log(1 - u).$$

We refer to Table 3.3 for more examples of distributions and their past quantile entropies.

Equation (3.22) provides the opportunity for generating new quantile functions that do not correspond with the existing distribution function. This is done by assuming functional forms for $\xi(u)$ with desirable properties.

EXAMPLE 3.6. *Let $\bar{\xi}(u) = \beta \log(1 + u)$. Then*

$$q(u) = \exp\left[\frac{\beta u}{1 + u} + \beta \log(1 + u) - \log u - 1\right], \beta > 0.$$

We have $\lambda_Q(u) = (1 + u)^{-\beta} \exp\left[1 - \frac{\beta u}{1 + u}\right]$, which is decreasing for $\beta > 0$ and as is easily seen $\bar{\xi}(u)$ is increasing.

EXAMPLE 3.7. *Taking $\bar{\xi}(u) = au(1 - u)$, $a > 0$*

$$q(u) = \frac{1}{u}\exp[2au - 3au^2 - 1].$$

The corresponding reversed hazard rate is

$$\lambda_Q(u) = e^{-2au+3au^2+1}$$

Table 3.3. $\bar{\xi}(u)$ for some important quantile models.

Distribution	Quantile function	$\xi(u)$
uniform	$a + (b-a)u$	$\log u + \log(b-a)$
half logistic	$\sigma \log(\frac{1+u}{1-u})$	$2 + \log(2\sigma) + \log u - (\frac{1+u}{u})\log(1+u) + \frac{1-u}{u}\log(1-u)$
power	$\alpha u^{1/\beta}$	$\log\frac{\alpha}{\beta} + \frac{\beta-1}{\beta} + \frac{1}{\beta}\log u$
Pareto I	$\sigma(1-u)^{-1/\alpha}$	$\log(\frac{\sigma}{\alpha}) + \frac{\alpha+1}{\alpha} + \log u + \frac{\alpha+1}{\alpha}\left(\frac{1-u}{u}\right)\log(1-u)$
generalized Pareto	$\frac{b}{a}\left[(1-u)^{-\frac{a}{a+1}} - 1\right]$	$\log\frac{b}{a+1} + \frac{\alpha}{2Q+1} + \log u + \frac{Q+1}{Q+1}\frac{1-u}{u}\log(1-u)$
log logistic	$\alpha^{-1}(\frac{u}{1-u})^{1/\beta}$	$2 - \log(\alpha\beta) + \frac{1}{\beta}\log u + \frac{\beta+1}{\beta}\frac{1-u}{u}\log(1-u)$
inverted exponential	$-\frac{\lambda}{\log u}$	$1 - \frac{1}{u}\int_0^u \log(\log p)^2\, dp + \log\lambda$
linear hazard quantile function	$(a+b)^{-1}\log\frac{a+bu}{a(1+u)}$	$2 + \log\left(\frac{b-a}{a+b}\right) + \log u - \frac{1-u}{u}\log(1+u) + \frac{a}{bu}\log a - \left(\frac{a+bu}{bu}\right)\log(a+bu)$
power-Pareto	$cu^{\lambda_1}(1-u)^{-\lambda_2}$	$\log C + \lambda_2 - \lambda_1 + 1 + \lambda_1\log u + (\lambda_2+1)\frac{1-u}{u}\log(1-u) - \frac{\lambda_1\log\lambda_1}{(\lambda_2-\lambda_1)u}$ $+ \frac{\lambda_1(1-u)+\lambda_2 u}{(\lambda_2-\lambda_1)u}\log(\lambda_1(1-u) + \lambda_2(u))$
exponential	$-\frac{1}{\lambda}\log(1-u)$	$1 - \log\lambda + \log u + \frac{1-u}{u}\log(1-u)$

which is bathtub shaped. Also

$$\bar{\xi}'(u) = a(1 - 2u)$$

so that it is UBT for $a > 0$.

EXAMPLE 3.8. *In this case*

$$\bar{\xi}(u) = \alpha u + \frac{\theta}{1 + \beta u}$$

which gives a variety of shapes for past quantile entropy function. It is decreasing for $\alpha < 0$ and $\theta\beta > 0$, increasing for $\alpha > \theta\beta$ and $\alpha > 0$ and $\theta = 0$, constant for $\alpha = \beta = 0$ and BT for $0 < \alpha < \theta\beta$ (see Figure 3.7). The distribution is represented by

$$q(u) = u^{-1} \exp\left[2\alpha u + \frac{\theta}{(1 + \beta u)^2}(1 + \beta u - u) - 1\right].$$

Further

$$\lambda_Q(u) = \exp\left[-2\alpha u + \frac{\theta u}{(1 + \beta u)^2} - \frac{\theta}{(1 + \beta u)} + 1\right]$$

which is increasing for $\theta > \alpha$ and UBT for $\theta < \alpha$ (see Figure 3.8).

REMARK 3.7. *In the case of past quantile entropy, the BT, UBT definitions are similar to those in the distribution function approach. The past quantile entropy function $\bar{\xi}(u)$ is BT (UBT) according as $\bar{\xi}(u)$ is strictly decreasing (increasing) in $(0, u_0)$, constant at $u = u_0$ and strictly increasing (decreasing) in $(u_0, 1)$.*

A second advantage of the quantile formulation is that it permits to forge a relationship between $\xi(u)$ and $\bar{\xi}(u)$. From equations (2.34) and (3.16),

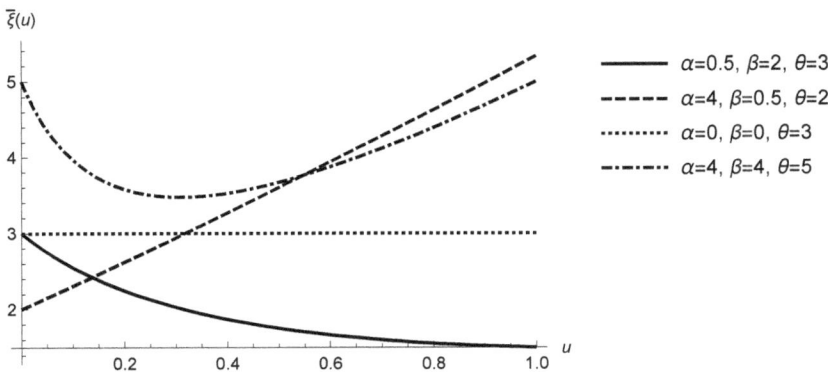

Figure 3.7. Plot of $\bar{\xi}(u)$ for different values of α, θ and β.

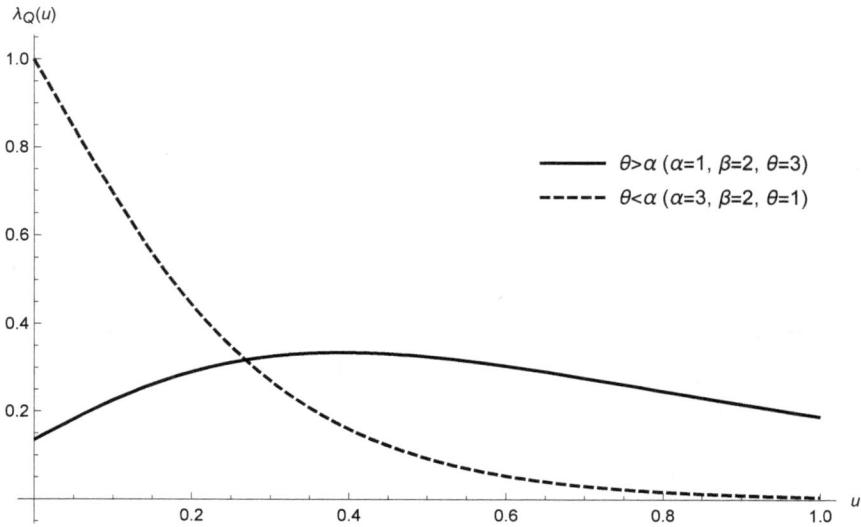

Figure 3.8. Plot of $\lambda_Q(u)$ for $\theta > \alpha$ and $\theta < \alpha$.

equating expressions of $q(u)$,

$$u\bar{\xi}'(u) + \bar{\xi}(u) - \log u - 1 = \xi(u) - (1-u)\xi'(u) - \log(1-u) - 1.$$

This simplifies to

$$\frac{d}{du}\left(u\bar{\xi}(u) + (1-u)\xi(u)\right) = \log\frac{u}{1-u}.$$

Integrating,

$$u\bar{\xi}(u) + (1-u)\xi(u) = u\log u + (1-u)(\log(1-u) - 1) + K.$$

Also $u \to 0$, $K = 1 + \xi(0)$ and as $u \to 1$, $K = \bar{\xi}(1)$. Based on these we conclude our discussions in the next theorem.

THEOREM 3.11. *The residual quantile entropy $\xi(u)$ and the past quantile entropy $\bar{\xi}(u)$ satisfy the following identities*

$$\xi(u) = \frac{u}{1-u}(\log u - \bar{\xi}(u)) + \log(1-u) + \frac{\bar{\xi}(1)}{1-u} - 1 \qquad (3.23)$$

$$\bar{\xi}(u) = \frac{1-u}{u}(\log(1-u) - \xi(u)) + \log u + \frac{\xi(0)}{u}. \qquad (3.24)$$

The primary utility of Theorem 3.11 is that for distributions in which either $\xi(u)$ or $\bar{\xi}(u)$ is known the above identity can be used to find the other, without a fresh calculation directly from the distribution.

EXAMPLE 3.9. *For the exponential distribution,* $\xi(u) = 1 - \log \lambda$ *(see Table 2.4). Using Theorem 3.11*

$$\xi(u) = \frac{1-u}{u}[\log(1-u) - (1 - \log \lambda)] + \log u + \frac{1 - \log \lambda}{u}$$

$$= \frac{1-u}{u}\log(1-u) + \log u + 1 - \log \lambda$$

as seen in Example 3.5.

Some properties of $\bar{\xi}(u)$ are listed below.

    (i) X is said to be increasing past quantile entropy (IPQE) if $\bar{\xi}_Q(u)$ is increasing for all u in $(0, 1)$. If $\lambda_Q(u)$ is decreasing for all $u > 0$, then $\bar{\xi}_Q(u)$ is IPQE. Since u is increasing whenever t is increasing, both IPQE and IUPL are equivalent. Hence the original name IUPL will be used in the sequel irrespective of whether $F(x)$ or $Q(u)$ is used.

EXAMPLE 3.10. *Consider the reciprocal exponential distribution*

$$F(x) = \exp\left(-\frac{\lambda}{x}\right), \ x > 0, \ \lambda > 0. \tag{3.25}$$

Since $\lambda(x) = \frac{\lambda}{x^2}$ *is decreasing* X *is IUPL. On the other hand if its quantile function* $Q(u) = -\frac{\lambda}{\log u}$ *is used* $\lambda_Q(u) = \frac{(\log u)^2}{\lambda}$ *and* $\lambda'_Q(u) = \frac{2\log u}{u} < 0$ *so that* $\lambda_Q(u)$ *is decreasing and* X *is IPQE.*

    (ii) If X is IUPL and $\phi(\cdot)$ is non-negative, increasing and convex, then $\phi(X)$ is also IUPL.

EXAMPLE 3.11. *Let* X *be distributed as (3.25) and* $\phi(x) = x^\alpha$, $\alpha > 0$ *has inverse Weibull distribution*

$$F(x) = \exp\left(-\frac{\lambda}{x^\alpha}\right), \ \alpha > 0, \ x > 0.$$

Also

$$\phi''(x) = \alpha(\alpha - 1)x^{\alpha-2} > 0 \text{ so that } \phi(x) \text{ is increasing}$$

and convex for $\alpha > 1$. *Hence the inverse Weibull distribution is also IUPL. However, the condition of convexity of* $\phi(x)$ *is not necessary. For example, in this case* $\lambda(x) = \frac{\lambda}{x^{\alpha+1}}$ *which is decreasing for* $\alpha > 0$ *and therefore, the distribution is IUPL. At the same time in* $0 < \alpha < 1$, $\phi(x)$ *is concave.*

    (iii) The quantile function of the equilibrium distribution of X is (Section 1.12)

$$F_E(Q_X(u)) = \mu^{-1}\int_0^u (1-p)q(p)dp, \ \mu = E(X)$$

and hence
$$f_E(Q_X(u)) = \mu^{-1}(1-u).$$
Thus the density at $Q_X(u)$ represents an exponential distribution.
(iv) In terms of the increasing mean inactivity time (IMIT)

$$\text{IMIT} \Rightarrow \text{IUPL}.$$

The proof of this result rests on the fact that exponential distributions have maximum entropy among all continuous distributions with a given mean. Also, the IUPL class is prescribed under the formation of a series system but DUPL is not. The first three results are based on Sunoj et al. (2013) and (iv) is from Qiu (2019).

3.4. Order statistics

Let $X_1, \ldots, X_n$ be independent and identically distributed random variables and $X_{i:n}$ be the corresponding order statistics. Thapliyal and Taneja (2013b) have presented several results on the past entropy of $X_{i:n}$. Recall that the Shannon's entropy of order statistics $X_{i:n}$ is

$$H_{X_{i:n}} = -\int_0^\infty f_{i:n}(x) \log f_{i:n}(x) dx$$

where

$$f_{i:n}(x) = \frac{1}{B(i, n-i+1)}[F(x)]^{i-1}[1-F(x)]^{n-1}f(x). \qquad (3.26)$$

Converting this to the quantile form by setting $U = F(X)$,

$$H_{X_{i:n}} = H_n(W_i) - E_{g_i}[\log(f F^{-1}(W_i))],$$

where

$$H_n(W_i) = \log B(i, n-i+1)$$
$$\qquad - (i-1)(\psi(i) - \psi(n+1) - (n-i)[\psi(n-i+1) - \psi(n+1)])$$

is the entropy of the ith order statistic from uniform $(0,1)$ and

$$g_i(w) = \frac{1}{B(i, n-i+1)} w^{i-1}(1-w)^{n-1}, \ 0 < w < 1$$

is its probability density function. Thus the past entropy of ith order statistics

$$\bar{H}_{X_{i:n}}(t) = 1 - \frac{1}{F_{i:n}(t)} \int_0^t f_{i:n}(x) \log \lambda_{i:n}(x) dx \qquad (3.27)$$

where $\lambda_{i:n}$ is the reversed hazard rate of $X_{i:n}$. For $i = n$, $f_{n:n}(x) = nF^{n-1}(x)f(x)$ and $F_{n:n}(x) = F^n(x)$. Substituting in (3.27) with $i : n$ and simplifying

$$\bar{H}_{X_{n:n}}(t) = n \log F(t) - \frac{1}{F^n(t)} \int_0^t nu^{n-1} \log \left(nu^{n-1} f(Q(u)) \right) du.$$

When X is exponential,

$$\lim_{t \to \infty} \bar{H}_{X_{n:n}}(t) = 1 - \log n\theta + r + \psi(n),$$

where $r = -\psi(1) = 0.5772$. Similarly for the Pareto distribution, $f(Q(u)) = \frac{\theta}{\beta}(1-u)^{\frac{\theta}{1+\theta}}$

$$\lim_{t \to \infty} \bar{H}_{X_{n:n}}(t) = 1 - \frac{1}{n} - \log n - \log \left(\frac{\theta}{\beta} \right) - \frac{\theta+1}{\theta} \left(-r - \psi(n) - \frac{1}{n} \right).$$

Thapliyal and Taneja (2013b) also proved the characterization theorem on past entropy of order statistics as given below.

THEOREM 3.12. *If $X_1, X_2, \ldots, X_n$ and $Y_1, Y_2, \ldots, Y_n$ are independent and identically distributed random variables with distribution functions $F(x)$ and $G(y)$ respectively such that $F(t_0) = G(t_0)$ for some constant t_0 and $\bar{H}_{X_{i:n}}(t_0) = \bar{H}_{Y_{i:n}}(t_0)$, $1 \leq i \leq n$, then X and Y have identical distributions.*

Some results on measure of entropy of past lifetime and k-record statistics are available in Goel et al. (2018).

3.4.1. Stochastic orders. Similar to the residual entropy, order relations and their properties can be defined for the past entropy as well (Nanda and Paul (2006a)).

DEFINITION 3.3. *If X and Y are lifetimes of two devices, X is said to be greater than Y in past entropy written as $X \geq_{PE} Y$ if*

$$\bar{H}_X(t) \leq \bar{H}_Y(t) \text{ for all } t \geq 0.$$

EXAMPLE 3.12. *Let X and Y be exponentially distributed random variables with means λ and 2λ respectively. Using the expression in Table 3.1, $X \geq_{PE} Y$.*

The stochastic order, $\geq_{PE}$ enjoys several properties.

(a) Let $\phi(\cdot)$ be a strictly increasing function with $\phi(0) = 0$ and $\phi'(0) \geq 1$. Then $\phi(X) \geq_{PE} \phi(Y)$. In this result the convexity condition need not be necessary.

(b) If $Z = aX + b$, $a > 0$, $b \geq 0$ and X is absolutely continuous, then

$$\bar{H}_z(t) = \bar{H} \left(X; \frac{t-b}{a} \right) + \log a.$$

Further if $Z_i = a_i X_i + b_i$, $i = 1, 2$, $a_i > 0$, $b_i \geq 0$ satisfying $X_1 \geq_{\mathrm{PE}} X_2$, $a_1 \geq a_2$, $b_1 \geq b_2$ then $Z_1 \geq_{\mathrm{PE}} Z_2$ provided $\bar{H}_{X_1}(t)$ or $\bar{H}_{X_2}(t)$ is increasing in $t > b_1$.

(c) Defining $X_1 = aX + b$, $Y_1 = aY + b$, where X and Y are absolutely continuous such that $X \geq_{\mathrm{PE}} Y$, then $X_1 \geq_{\mathrm{PE}} Y_1$ for $t > b$.

Sunoj et al. (2013) addressed the ordering problem within the framework of past quantile properties.

DEFINITION 3.4. *The random variable X is said to have less past quantile entropy than another random variable Y if $\bar{\xi}_X(u) \leq \bar{\xi}_Y(u)$ for all $u > 0$ and denote it as $X \leq_{PQE} Y$.*

Relating to the $\leq_{\mathrm{PQE}}$ order, we further have the following.

(i) Let $Z_i = a_i X_i + b_i$, $a_i > 0$, $b_i \geq 0$, $i = 1, 2$. Then if $a_1 \leq a_2$,

$$X_1 \leq_{\mathrm{PQE}} X_2 \Rightarrow Z_1 \leq_{\mathrm{PQE}} Z_2,$$

in other words, the PQE order is closed under linear transformations.

(ii) $X \leq_{\mathrm{RHQ}} Y \Rightarrow X \leq_{\mathrm{PQE}} Y$, that is, the reversed hazard quantile function order is stronger than the LPQE order.

(iii) If $X \leq_{\mathrm{RHQ}} Y$ and $\phi(\cdot)$ is non-negative, increasing and convex, then $\phi(X) \geq_{\mathrm{PQE}} \phi(Y)$.

Kang (2015a) in a further study of the PQE order observed that

(a)

$$X \leq_{\mathrm{PQE}} Y \Leftrightarrow \int_0^t f_X(x) \log \left\{ \frac{f_X(x)}{g_Y[G_Y^{-1} F_X(x)]} \right\} dx \geq 0$$

where G and g stands for the distribution and density functions of Y.

(b) $X \leq_{\mathrm{disp}} Y \Rightarrow X \leq_{\mathrm{PQE}} Y$

(c) $X \leq_C Y$ and $f_X(0) \geq g_Y(0) \Rightarrow X \leq_{\mathrm{PQE}} Y$.

(d) Given $X \leq_{\mathrm{PQE}} Y$, $X \leq_{\mathrm{PQE}} aY$ if $a \geq 1$ and $aX \leq_{\mathrm{PQE}} Y$ if $0 < a < 1$

(e) $X \leq_{\mathrm{PQE}} Y$ and $X \geq_{\mathrm{PQE}} Y$ hold simultaneously if $X =_{\mathrm{disp}} Y$ or equivalently $X =_{\mathrm{st}} Y + k$

(f)

$$X \leq_{\mathrm{PQE}} Y \Rightarrow X_{1:n} \leq Y_{1:n}$$
$$X_{n:n} \leq_{\mathrm{PQE}} Y_{n:n} \Rightarrow X \leq_{\mathrm{PQE}} Y.$$

The first property says that the PQE order is preserved under formation of series systems and the second has the reversed preservation property

(g) Denote by $X_{1:N} = \min_{1 \leq i \leq N} X_i$ and $X_{N:N} = \max_{1 \leq i \leq N} X_i$ and similarly $Y_{1:N}$ and $Y_{N:N}$. Then

$$X \leq_{\mathrm{PQE}} Y \Rightarrow X_{1:N} \leq_{\mathrm{PQE}} Y_{1:N}$$
$$X_{N:N} \leq_{\mathrm{PQE}} Y_{N:N} \Rightarrow X \leq_{\mathrm{PQE}} Y.$$

Here N stands for a non-negative integer valued random variable.

(h) If

$$X \geq_{\text{st}} Y \text{ and } X \leq_{\text{PQE}} Y \Rightarrow \phi(X) \leq_{\text{PQE}} Y_{1:N}$$

$$X \leq_{\text{st}} Y \text{ and } \phi(X) \leq_{\text{PQE}} \phi(Y) \Rightarrow X \leq_{\text{PQE}} Y,$$

where $\phi(\cdot)$ is a non-negative increasing concave function with $\phi(0) = 0$.

(i) Let $X(\theta)$ and $Y(\theta)$, $\theta \geq 0$ be random variables with survival functions $(\bar{F}(x))^\theta$ and $(\bar{G}(x))^\theta$ or in other words $X(\theta)$ and $Y(\theta)$ are random variables corresponding to proportional hazards models of X and Y. Then

$$X \leq_{\text{PQE}} Y \Rightarrow X(\theta) \leq_{\text{PQE}} Y(\theta), \ 0 \leq \theta \leq 1$$

and

$$X(\theta) \leq_{\text{PQE}} Y(\theta) \Rightarrow X \leq_{\text{PQE}} Y, \ \theta > 1.$$

(j) If $X \leq_{\text{su}} (\leq_*, \leq_c) Y$ and $f(0) \geq g(0)$ then $X \leq_{\text{PQE}} Y$.

(k) $X \leq_{\text{PQE}} \phi(X)$, where $\phi(\cdot)$ is a non-negative increasing function with $\phi'(x) > 1$ for all $x \geq \theta$ and if $\phi(\cdot)$ is an increasing function with $\phi'(x) \geq 1$ for all x, then $\phi(X) \leq_{\text{PQE}} X$.

(l) In the accelerated life model $F(x) = G(W(x))$ where $W(x) = \int_0^x w(t)dt$ so that $W(0) = 0$ and $W(\infty) = \infty$ and $W(x)$ is strictly increasing

$$W(x) - x \text{ is increasing in } x \Rightarrow X \leq_{\text{PQE}} Y.$$

The results in (j) through (l) are due to Qiu (2019). He also provides some results on PQE order with respect to generalized order statistics.

From the above review we have two results

$$X \leq_{\text{RHQ}} Y \Rightarrow X \leq_{\text{PQE}} Y \text{ and } X \leq_{\text{HQ}} Y \Rightarrow X \leq_{\text{RQE}} Y.$$

Since $X \leq (\geq)_{\text{RQE}} Y$ is equivalent to $X \geq (\leq)_{\text{HQ}} Y$ we have

$$X \leq_{\text{RHQ}} Y \Rightarrow X \leq_{\text{PQE}} Y \text{ and } X \geq_{\text{RHQ}} Y \Rightarrow X \leq_{\text{RQE}} Y$$

and also

$$X \geq_{\text{HQ}} Y \Rightarrow X \leq_{\text{PQE}} Y \text{ and } X \leq_{\text{HQ}} Y \Rightarrow X \leq_{\text{RQE}} Y.$$

It is learned from (ii) above that if $X \leq_{\text{RHQ}} Y$ then $X \leq_{\text{PQE}} Y$. In the next example we show that the reverse implication need not hold.

EXAMPLE 3.13. *Consider the distribution (2.36) with $a = 0$. The past entropy function is*

$$\bar{\xi}(u) = (1 - u)(1 - b) + u \log u.$$

Let X be distributed with $b = \frac{1}{4}$ and Y with $b = \frac{1}{2}$. Then $\xi_X(u) \leq \xi_Y(u)$ for all u. The reversed hazard function of X is $\lambda_X(u) = u^{-1} e^{-1/4(2u-1)}$ and that of Y is $\lambda_Y(u) = u^{-1} e^{-1/2(2u-1)}$. It is seen that when $u = \frac{1}{4}$,

$\lambda_X(x) < \lambda_X(x)$ *and when* $u = \frac{5}{8}$, $\lambda_X(x) > \lambda_Y(x)$. *Thus X and Y are not ordered with respect to reversed hazard quantile function.*

Consequent to the above example, we investigate the additional conditions required for PQE order to imply the RHQ order.

THEOREM 3.13. *If $X \geq (\leq)_{PQE}Y$ and $\frac{1-\bar{\xi}_X(u)}{1-\bar{\xi}_Y(u)}$ is non-increasing (non-decreasing) then $X \geq (\leq)_{RHQ}Y$.*

PROOF. Since $X \geq_{PQE} Y$, $\bar{\xi}_X(u) \geq \bar{\xi}_Y(u)$ and so $\frac{1-\bar{\xi}_X(u)}{1-\bar{\xi}_Y(u)} \leq 1$. Also

$$\frac{1-\bar{\xi}_X(u)}{1-\bar{\xi}_Y(u)} \text{ is non-increasing in } u$$

$$\Leftrightarrow \frac{d}{du}\left(\frac{u(1-\bar{\xi}_X(u))}{u(1-\bar{\xi}_Y(u))}\right) \leq 0$$

$$\Rightarrow \frac{\frac{d}{du}\left(u(1-\bar{\xi}_X(u))\right)}{\frac{d}{du}\left(u(1-\bar{\xi}_Y(u))\right)} \leq \frac{1-\bar{\xi}_X(u)}{1-\bar{\xi}_Y(u)} \leq 1$$

$$\Rightarrow \frac{d}{du}\left(u(1-\bar{\xi}_X(u))\right) \leq \frac{d}{du}\left(u(1-\bar{\xi}_Y(u))\right)$$

$$\Rightarrow \log \lambda_X(u) \leq \log \lambda_Y(u) \Rightarrow \lambda_X(u) \leq \lambda_Y(u).$$

The proof $X \leq_{RHQ} Y$ is similar. $\qquad\square$

3.5. Weighted past entropy

Following the concept and definition of weighted distribution and weighted entropy in Section 2.7, the weighted past entropy at time t of a random lifetime X is (Di Cresenzo and Longobardi (2006)).

$$\bar{H}_L(t) = -\int_0^t x\frac{f(x)}{F(t)}\log\frac{f(x)}{F(t)}dx, \qquad (3.28)$$

when the weight function $w(x) = x$. This is the length-biased version. Equation (3.28) can be written as

$$\bar{H}_w(t) = -\frac{1}{F(t)}\int_0^t xf(x)\log f(x)dx + \frac{\log F(t)}{F(t)}\int_0^x xf(x)dx$$

or as

$$\bar{H}_w(t) = -\int_0^t\int_0^x \frac{f(x)}{F(t)}\frac{f(x)}{F(t)}\log\frac{f(x)}{F(t)}dydx$$

$$= t\bar{H}_F(t) - \frac{1}{F(t)}\int_0^t F(y)\left[\bar{H}_F(y) + \log\frac{F(t)}{F(y)}\right]dy.$$

The expression of $\bar{H}_w(t)$ for the exponential distribution is

$$\bar{H}_w(t) = \frac{1}{1 - e^{-\lambda t}} \left[\frac{2}{\lambda} - \frac{2}{\lambda} e^{-\lambda t} - 2te^{-\lambda t} - \lambda t^2 e^{-\lambda t} \right.$$
$$\left. + (\frac{1}{\lambda} - \frac{1}{\lambda} e^{-\lambda t} - te^{-\lambda t}) \log \frac{1 - e^{-\lambda t}}{\lambda} \right]$$

Di Cresenzo and Longobardi (2006) established the following results.

(i) $\bar{H}_w(t) \leq \mu^*(t) \log \frac{t^2}{\lambda(t)}$, $\mu^*(t) = E(X|X < x)$

(ii) If $\lambda(t)$ is decreasing $\bar{H}_w(t) \leq \int_0^t x\lambda(x)dx - \mu(t)[1 + \log \lambda(t)]$.

(iii) X has decreasing (increasing) weighted uncertainty past life, DWUPL (IWUPL) if $\bar{H}_w(t)$ is decreasing (increasing) in it is DWUPL for $0 < a < \frac{1}{e}$.

(iv) Let $Y = \phi(X)$, with ϕ strictly monotonic, continuous and differentiable. Then

$$\bar{H}_Y(t) = \bar{H}_w(\phi^{-1}(t)) + E[\phi(X) \log \phi'(X)|X \leq \phi^{-1}(t)]$$

if ϕ is strictly increasing.

A brief introduction to the entropy of the equilibrium distribution and its residual counterpart was made in Section 2.7. In the distribution function approach it was found to be

$$H_E(t) = 1 + \frac{1}{m(t)\bar{F}(t)} \int_t^\infty (\log m(x))\bar{F}(x)dx \qquad (3.29)$$

where all functions mentioned on the right refer to the original random variable X. The quantile-based counterpart of (3.29) is deduced by setting $x = Q(u)$ and noting that the mean residual quantile function of X is

$$m(Q(u))M(u) = \frac{1}{1 - u} \int_u^1 (1 - p)q(p)dp.$$

Thus the quantile version of (3.29) is

$$\xi_E(u) = 1 + \frac{1}{(1 - u)M(u)} \int_u^1 (\log M(p))(1 - p)q(p)dp \qquad (3.30)$$

or alternatively in a weighted format

$$\xi_E(u) = 1 + \frac{\int_u^1 (1 - p)q(p) \log M(p)}{\int_u^1 (1 - p)q(p)dp} M(p)dp.$$

Two interesting advantages of the quantile approach are noted below.

THEOREM 3.14. *The residual quantile entropy of the equilibrium distribution is uniquely determined by the mean residual quantile function of the baseline random variable.*

PROOF. We have

$$\int_u^1 \log M(p)(1-p)q(p)dp = \int_u^1 \log M(p)\left(-\frac{d}{dp}(1-p)M(p)\right)dp$$

$$= (1-u)M(u)(\log M(u)) + \int_u^1 (1-p)M'(p)dp$$

$$= (1-u)M(u)(\log M(u)-1) + \int_u^1 M(p)dp$$

and so

$$\xi_E(u) = 1 + \frac{1}{(1-u)M(u)}\left[(1-u)M(u)(\log M(u)-1) + \int_u^1 M(p)dp\right]$$

$$= \log M(u) + \frac{\int_u^1 M(p)dp}{(1-u)M(u)}. \tag{3.31}$$

$\square$

THEOREM 3.15. *The function $\xi_E(u)$ determines the distribution of X uniquely through the quantile density function*

$$q(u) = \frac{M(u)\xi_E'(u)}{\xi_E(u) - \log M(u) - 1}.$$

PROOF. Differentiate (3.30) and use the identity $(1-u)q(u) = M(u) - (1-u)M'(u)$. $\square$

EXAMPLE 3.14. *Consider the Pareto I distribution $\bar{F}(x) = x^{-\alpha}$, $x > 1$, $\alpha > 0$. It's quantile function is $Q(u) = u^{1/\alpha}$. The equilibrium distribution of X is obtained with $\mu = \alpha - 1$ as*

$$\bar{F}_E(x) = x^{-\alpha+1}, \ \alpha > 1$$

and

$$Q_E(u) = (1-u)^{1/(1-\alpha)}.$$

We can directly calculate the residual entropy of $Q_E(u)$ using formula (2.28) as

$$\xi_E(u) = 1 + \frac{1}{\alpha - 1} - \log(\alpha - 1) - \frac{\log(1-u)}{\alpha}.$$

Since the distribution has a simple expression for $M(u)$ as $M(u) = \frac{(1-u)^{-1/\alpha}}{\alpha-1}$ it is easier to use (3.31) to obtain

$$\xi_E(u) = -\frac{1}{\alpha}\log(1-u) - \log\alpha - 1 + \frac{\alpha}{\alpha - 1}$$

which is the same as the value obtained by direct computation.

It appears that there is no simple relationship between $\xi(u)$ and $\xi_E(u)$, it being

$$\exp\left[-\frac{d}{dx}(1-u)(1-\xi(u))\right] = \frac{(1-u)M(u)\xi'_E(u)}{\xi_E(u) - \log M(u) - 1}.$$

However there can be simple identities connecting $H_E(t)$ and $H_F(t)$. For example, the generalized Pareto distribution satisfies $H_E(t) - H_F(t) = C$, where C is a constant. In particular for the exponential law, $H_E(t) = H_F(t)$.

The distribution function of the equilibrium model is

$$F_E(x) = \frac{1}{\mu}\int_0^x \bar{F}(x)dx = \frac{1}{\mu}A(x)F(x), \tag{3.32}$$

$A(x)$ being the mean time to failure in a renewal replacement policy. Accordingly the reversed hazard rate of X_E is

$$\lambda_E(x) = \frac{f_E(x)}{f_E(x)} = \frac{\bar{F}(x)}{A(x)F(x)}. \tag{3.33}$$

Equations (3.32) and (3.33) enable us to write the past entropy of $F_E(x)$ as

$$\bar{H}_E(t) = 1 - \frac{1}{F_E(t)}\int_0^t \log \lambda_E(x)f_E(x)dx$$

$$= 1 - \frac{1}{A(t)F(t)}\int_0^t \log \frac{\bar{F}(x)}{A(x)F(x)}\bar{F}(x)dx$$

$$= 1 - \frac{1}{A(t)F(t)}\left[\int_0^t \bar{F}(x)\log\bar{F}(x) + \int_0^t (\log A(x)F(x))\right]\bar{F}(x)dx.$$

The quantile version is

$$\bar{\xi}_E(u) = \bar{H}_E(Q_X(u))$$

$$= 1 - \frac{1}{uA_Q(u)}\int_0^u \left(\log\frac{1-p}{pA_Q(p)}\right)(1-p)q(p)dp,$$

where $A_Q(u) = A(Q_X(u))$.

3.6. Interval entropy

Sometimes the information about the lifetime of a device is that it has failed in between two time points t_1 and t_2. In this case the probability density function of failure time is

$$f(x|t_1 < x < t_2) = \frac{f(x)}{F(t_2) - F(t_1)}.$$

Thus the entropy is

$$IH(t_1, t_2) = - \int_{t_1}^{t_2} \frac{f(x)}{F(t_2) - F(t_1)} \log \frac{f(x)}{F(t_2) - F(t_1)} dx,$$

which is referred to as the interval entropy in (t_1, t_2). Misagh (2012) have given some properties of $IH(t_1, t_2)$ showing that it simplifies to

$$IH(t_1, t_2) = 1 - \frac{1}{F(t_2) - F(t_1)} \int_{t_1}^{t_2} (\log h(x)) f(x) dx$$
$$+ \frac{1}{F(t_2) - F(t_1)} \{ \bar{F}(t_2) \log \bar{F}(t_2) - F_1(t_1) \log F_1(t_1) \} + F(t_2)$$
$$- F(t_1) \log (F(t_2) - F(t_1)).$$

Infact $IH(t_1, t_2)$ is a generalization of both $H_F(t)$ and $\bar{H}_F(t)$ in the sense that $IH(t_1, \infty) = H_p(t_1)$ and $IH(0, t_2) = \bar{H}_F(t_2)$.

EXAMPLE 3.15. *If X has power distribution $F(x) = x^\alpha$, $0 \le x \le 1$, $\alpha > 0$ so that*

$$f(x|t_1 < X < t_2) = \frac{\alpha x^{\alpha-1}}{t_2^\alpha - t_1^\alpha},$$

and

$$IH(t_1, t_2) = - \frac{1}{t_2^\alpha - t_1^\alpha} \int_{t_1}^{t_2} \log \frac{\alpha x^{\alpha-1}}{t_2^\alpha - t_1^\alpha} \frac{d}{dx} x^\alpha dx$$
$$= \frac{1}{t_2^\alpha - t_1^\alpha} \left[t_2^\alpha \log \frac{\alpha t_2^{\alpha-1}}{t_2^\alpha - t_1^\alpha} - t_1^\alpha \log \frac{\alpha t_1^{\alpha-1}}{t_2^\alpha - t_1^\alpha} \right] + \frac{\alpha - 1}{\alpha}.$$

The properties satisfied by $IH(t_1, t_2)$ are

(i) If X is non-negative and $Y = \phi(X)$, ϕ strictly increasing,

$$IH_Y(t_1, t_2) = IH_X(t_1, t_2) + [F(\phi^{-1}(t_2)) - F(\phi^{-1}(t_1))]^{-1}$$
$$[E(\log \phi'(X) - F(\phi^{-1}(t_1))) E(\log \phi'(X)|X < \phi^{-1}(t_1))$$
$$- F(\phi^{-1}(t_2) E(\log \phi'(X)|X > \phi^{-1}(t_2)))];$$

(ii) $IH_{X+\theta}(t_1, t_2) = IH_X(t_1 - \theta, t_2 - \theta)$,

(iii) $IH_{aX}(t_1, t_2) = IH_X(\frac{t_1}{a}, \frac{t_2}{a})$,

(iv) for the uniform distribution, $IH(t_1, t_2) = \log(t_2 - t_1)$ and hence $IH(t_1, t_2) \le \log(t_1 - t_2)$ for any X.

3.7. Past extropy

The entropy of past life $\{X|X \le t\}$ is defined as

$$\bar{J}_F(t) = -\frac{1}{2} \int_0^t \left(\frac{f(x)}{F(t)} \right)^2 dx. \tag{3.34}$$

By differentiation with respect to t, (3.34) becomes

$$\lambda^2(t) + 4\bar{J}_F(t) + 2\bar{J}_F'(t) = 0 \tag{3.35}$$

as a relationship between reversed hazard rate $\lambda(t)$ and past extropy. Some examples of $\lambda(t)$ and $\bar{J}_F(t)$ are given in Table 3.4.

Since $\bar{J}$ lies in $(-\infty, 0)$, the analysis of (3.35) depends on whether $\bar{J}_F'(t)$ is zero, positive or negative. When $\bar{J}_F'(t) > 0$ there can be two roots and in the other cases there is only one positive root for (3.35). The roots are given by

$$\lambda(t) = -2\bar{J}_F(t) \pm \sqrt{4\bar{J}^2 - 2\bar{J}'}.$$

In the case of the uniform distribution $\bar{J}_F'(t) = \frac{1}{2t^2} > 0$, but the roots are identical. Hence $\bar{J}_F(t) = -\frac{1}{2t}$ characterizes the uniform law. All distributions with decreasing past entropy are characterized by the form of $\bar{J}_F(t)$. However, when $\bar{J}_F(t)$ is increasing the solutions of (3.35) have to be examined to see whether they meet conditions for the reversed hazard rate.

EXAMPLE 3.16. *In Table 3.4, $\bar{J}_F(t) = -\frac{\beta^2}{(4\beta-2)t}$, $\beta > \frac{1}{2}$ for the power distribution.*

Substituting in (3.3) and solving

$$\lambda(t) = \frac{2\beta^2}{(4\beta - 2)t} \pm \frac{2\beta(\beta - 1)}{(4\beta - 2)t}.$$

Thus the two solutions are $\lambda_1(t) = \frac{\beta}{t}$ giving the original power distribution and $\lambda_2(t) = \frac{2\beta}{(4\beta-2)t}$ giving another power distribution $F(t) = \left(\frac{t}{\beta}\right)^{\frac{2\beta}{4\beta-2}}$, $\beta > \frac{1}{2}$. Thus $\bar{J}_F(t)$ does not uniquely determine F.

Further, equation (3.35) reveals that $\bar{J}_F(t)$ is strictly increasing (decreasing) in t according as $\bar{J}_F(t) < (>) - \frac{\lambda(t)}{4}$. Examples of increasing (decreasing) past entropy, IPE (DPE) and some characterizations by relationships and reversed hazard rates are seen in the next two theorems.

THEOREM 3.16. *Let X be a non-negative random variable with support $[0, b]$. Then the property $\bar{J}_F(t) = -K\lambda(t)$ where K is some non-negative constant satisfied by all $t \in (0, b]$ if and only if X follows reversed generalized Pareto with*

$$F(x) = \left(\frac{ax + c}{ab + c} \right)^{\frac{a+1}{a}}, \ a > -1, \ c > 0. \tag{3.36}$$

Table 3.4. Extropy of past life.

Distribution	$F(x)$	$\lambda(t)$	$\bar{J}_F(t)$
uniform	$x, 0 \leq x \leq 1$	$\frac{1}{t}$	$-\frac{1}{2t}$
power	$\left(\frac{x}{\alpha}\right)^{\beta}, 0 \leq x \leq \alpha; \beta > 0$	$\frac{\beta}{t}$	$-\frac{\beta^2}{(4\beta-2)t}, \beta > \frac{1}{2}$
reciprocal exponential	$\exp\left(-\frac{\lambda}{x}\right)$	$\frac{\lambda}{t^2}$	$-\frac{(t^2+2\lambda t+2\lambda^2)}{8\lambda t^2}$
generalized exponential	$\left(1 - e^{-\lambda x}\right)^{\theta}, x > 0; \theta, \lambda > 0$	$\frac{\theta\lambda}{e^{\lambda t}-1}$	$-\frac{\lambda\theta^2}{2(2\theta-1)}\left(\frac{1}{e^{\lambda t}-1} - \frac{1}{2\theta}\right)$
Lindley	$\frac{\beta^2}{1+\beta}(1+x)e^{-\beta x}, x > 0; \beta > 0$	$\frac{\beta^2(1+t)}{1+\beta+\beta t}$	$\frac{\beta^3}{2(1+\beta+\beta t)^2}\left[\frac{1+\beta}{2\beta} - \frac{(1+t)}{\beta} - (1+t)^2\right]$
reversed exponential	$e^{a(x-b)}, 0 \leq x \leq b; a, b > 0$	a	$-\frac{1}{4a}$
reversed generalized Pareto	$\frac{(ax+c)^{(a+1)/a}}{ab+c}, a > -1, c > 0, 0 \leq x \leq b$	$\frac{a+1}{c+at}$	$\frac{-(a+1)^2}{(2a+4)(at+c)}, 0 < x \leq b$

PROOF. The necessary part follows from Table 3.4, where $K = \frac{a+1}{4a+2} > 0$. From the given property

$$\frac{\lambda'(t)}{\lambda^2(t)} = \frac{1 - 4K}{1 - 2K}$$

and on integration $\lambda(t) = (p + qt)^{-1}$ with $p > 0$ and $q = \frac{4K-1}{1-2K}$. Thus

$$F(x) = \exp\left[-\int_x^\infty \lambda(t)dt\right] = \left(\frac{p + qx}{p + qb}\right)^{1/q}, \quad 0 < x \leq b.$$

Finally $P(0 \leq x \leq b) = f(0) + \int_t^b f(x)dx$ leads to (3.36), by setting $K = \frac{a+1}{4a+2}$. □

REMARK 3.8. *As $a \to 0$ in (3.36) we have $F(x) = \exp[C^{-1}(x - b)]$, the reversed exponential distribution. In this case $K = \frac{1}{4}$ and $\bar{J}_F(t) = \frac{1}{4C}$. Two other members of the family (3.36) are of the form*

$$F(x) = \left(\frac{1 + \frac{ax}{c}}{1 + \frac{ab}{c}}\right)^{\frac{a+1}{a}}, \quad a, c > 0$$

and

$$F(x) = \left(\frac{1 - \frac{px}{c}}{1 - \frac{pb}{c}}\right)^{\frac{a+1}{a}}, \quad p, c > 0$$

obtained when $K < \frac{1}{4}$ and $K > \frac{1}{4}$ respectively.

REMARK 3.9. *X is IPE when $a > 0$, DPE when $-1 < a < 0$ and X has constant $\bar{J}_F(t)$ when $a = 0$.*

THEOREM 3.17. *If $F(x)$ is absolutely continuous satisfying $F(0) = 0$ then $\beta\bar{J}_F(t) = -K\lambda(t)$ for $K > 0$ and all t if and only if $F(x) = (\frac{x}{\alpha})^\beta$, $0 \leq x \leq \alpha$, $\beta > \frac{1}{2}$.*

PROOF. The 'if' part follows from Table 3.4. Further to prove the converse, proceeding as in the previous theorem we have

$$\bar{F}(x) = \left(\frac{qx + p}{q\alpha + p}\right)^{1/q}, \quad 0 \leq x \leq \alpha.$$

Since $F(0) = 0$ we have $p = 0$ and taking $q = \beta^{-1}$, we have the required $F(x)$. □

Much more general and powerful results can be found by considering the past quantile extropy

$$\bar{J}_Q(u) = \bar{J}_F(Q(u)) = -\frac{1}{2u^2}\int_0^u \frac{1}{q(p)}dp. \quad (3.37)$$

The major benefit of (3.37) is that $\bar{J}_Q(u)$ determines the distribution of X through the inversion formula

$$q(u) = [-2(2u\bar{J}_Q(u) + u^2 \bar{J}_Q^{-1}(u))]^{-1}. \tag{3.38}$$

This enables constriction of distributions based on past quantile entropy functions. We demonstrate the utility of (3.38) in model building.

Consider the rational function $J_Q(u) = -(a - bu + \frac{c}{u})$ where $a^2 + 4bc < 0$. By (3.38) we can write

$$q(u) = (2c + 4au - 6bu^2)^{-1}, \quad a^2 \le \min(-4bc, 3bc). \tag{3.39}$$

Equation (3.39) is a distribution with hazard quantile function

$$h_Q(u) = \frac{2c + 4au - 6bu^2}{1 - u}. \tag{3.40}$$

When $2a - 3b + c = 0$, $h_Q(u) = 6bu - 4a + 6b$, we get a linear hazard quantile distribution for X, when $c \ne 0$. When $c = 0$, the condition for the same distribution reduces to $2a = 3b$ and $J_Q(u) = bu - a$, a linear function of u satisfying $a > b$ for J to be negative. Also when $b = 0$, we have $J_Q(u)$ as a homographic function with $a > 0$, $c > 0$. Finally $b = c = 0$ leads to the exponential model. Thus (3.38) comprises of a class of distributions with exponential, half-logistic and exponential geometric as special cases.

The residual and past extropies can be related to one another in the quantile framework. Thus the knowledge of one of these is enough to determine the other which saves the computational work involved in finding it from the first principles.

THEOREM 3.18.

$$J_Q(u) = (1 - u)^{-2}[\bar{J}_Q(1) - u^2 \bar{J}_Q(u)]$$
$$\bar{J}_Q(u) = u^{-2}[J_Q(0) - (1 - u)^2 \bar{J}_Q(u)]$$
$$\bar{J}_Q(1) = J_Q(0) = J_X.$$

PROOF. We have

$$q(u) = \left[2\frac{d}{du}(1 - u)^2 J_Q(u)\right]^{-1}$$

and

$$q(u) = \left[-2\frac{d}{du}u^2 \bar{J}_Q(u)\right]^{-1}.$$

Equating these expressions and integrating

$$(1 - u)^2 J_Q(u) = -u^2 \bar{J}_Q(u) + K.$$

As $u \to 0$, $K = J(0)$ and as $u \to 1$, $K = \bar{J}(1)$. The identities in the Theorem now follow.

With reference to reversed hazard quantile function, we find

$$\lambda_Q(u) = -(4\bar{J}_Q(u) + 2u\bar{J}'_Q(u)). \tag{3.41}$$

Hence, whenever $J_Q(u)$ is increasing and convex, $\lambda(u)$ is decreasing assuming X to be absolutely continuous. As seen earlier $\lambda(u)$ can also be increasing or non-monotone when X has support $[0, b]$, $b < \infty$ and zero is a point of discontinuity. It can also be constituted that $\bar{J}_Q(u)$ determines $\lambda(u)$ uniquely. Equation (3.41) being a differential equation in $\bar{J}(u)$, can be solved to find

$$\bar{J}_Q(u) = u^{-2} \left[\int \frac{-u\lambda_Q(u)}{2} + K \right].$$

where K is determined such that $\lim_{u \to 1} \bar{J}_Q(u) = \bar{J}_X$, the extropy of X. The last equation shows that $\bar{J}_Q(u)$ is completely specified by $\lambda_Q(u)$.   □

Comparison of past extropies of two non-negative and absolutely continuous random variables X and Y can be accomplished in terms of stochastic orderings.

DEFINITION 3.5. *(i) The random variable X is less than Y in past extropy denoted by $X \leq_{PEX} Y$ if $\bar{J}_X(t) \leq \bar{J}_Y(t)$ for all $t > 0$.*
(ii) We say that X is less than Y in past quantile entropy, $X \leq_{PQEX} Y$ if $\bar{J}_{Q_X}(u) \leq \bar{J}_{Q_Y}(u)$ for all $u \in (0, 1)$.

The following are some important properties of the two orderings $\leq_{PEX}$ and $\leq_{PQEX}$.

(i) The stochastic order $\leq_{PEX}$ does not imply $\leq_{PQEX}$.

EXAMPLE 3.17. *Let $F_X(x) = x$, $0 \leq x \leq 1$ and $F_Y(x) = x^2$, $0 \leq x \leq 1$, then $J_X(t) = -\frac{1}{2t}$ and $J_Y(t) = -\frac{2}{3t}$ so that $X \geq_{PEX} Y$. However, $Q_X(u) = u$ and $Q_Y(u) = u^{1/2}$ leads to $\bar{J}_{Q_X}(u) = -\frac{1}{2u}$ and $\bar{J}_{Q_Y}(u) = -\frac{2}{3u^{1/2}}$. Now $\bar{J}_{Q_X}(u)$ and $\bar{J}_{Q_Y}(u)$ cross each other at $u = \frac{9}{16}$ and hence not ordered.*

(ii) Also $\leq_{PQEX}$ does not imply $\leq_{PEX}$.

EXAMPLE 3.18. *Let $F(x; \lambda) = e^{-x/\lambda}$, $x > 0$, $\lambda > 0$. Then*

$$J_X(t; \lambda) = -\frac{t^2 + 2\lambda t + 2\lambda^2}{8\lambda t^2},$$

and it is easy to see that $J_X(t; 1)$ and $J_X(t; 2)$ cross at $t = 2$. Hence $J_X(t; 1)$ and $J_X(t; 2)$ are not ordered by $\leq_{PEX}$. On the other hand,

$Q_X(u) = -\lambda(\log u)^{-1}$ *gives*

$$J_Q(u;\lambda) = -\frac{1}{\lambda u^2} \int_0^u p(\log p)^2 dp.$$

Hence $J_Q(u;1) \leq_{PQEX} J_Q(u;2)$.

(iii) $X \leq_{RHQ} Y \Leftrightarrow X \geq_{HQ} Y \Rightarrow X \leq_{PQEX} Y$ conversely if $\frac{J_Y(u)}{J_X(u)}$ is increasing in u, then

$$X \leq_{PQEX} Y \Rightarrow X \leq_{RHQ} Y \Rightarrow X \geq_{HQ} Y.$$

The proof is similar to that of ordering of residual quantile extropy in Theorem 2.26.

(iv) Although $\leq_{PEX}$ and $\leq_{PQEX}$ need not be mutually implied, they can do so under certain conditions. $\bar{J}_Y(t)$ be increasing and $X \leq_{hr} Y$, then $X \leq_{PEX} Y \Rightarrow X \leq_{PQEX} Y$.

PROOF.

$$X \leq_{PEX} Y \Leftrightarrow \bar{J}_X(t) \leq \bar{J}_Y(t)$$
$$\Leftrightarrow \bar{J}_X(Q_X(u)) \leq \bar{J}_Y(Q_X(u)) \leq \bar{J}_Y(Q_Y(u)).$$

Since $\bar{J}$ is increasing and $X \leq_{hr} Y \Rightarrow Q_X(u) \leq Q_Y(u)$. The last inequality is $J_{Q_X}(u) \leq J_{Q_Y}(u)$ and the result follows. $\square$

REMARK 3.10. *In the above result $X \leq_{hr} Y$ can be replaced by $X \leq_{rhr} Y$. Also one can use $X \leq_{HQ} Y$ or $X \leq_{RHQ} Y$.*

There are occasions when one has to compare the reliabilities of two devices, e.g., those with the same specifications and use but produced by different manufacturing processes. In such cases, we identity those with less residual uncertainty as more reliable. Let X and Y be non-negative random variables with zero as the left extremity of the support.

DEFINITION 3.6. *The random variable X is smaller than Y in decreasing residual extropy, denoted by $X \leq_{DRE} Y$, if $\frac{J_{Q_Y}(u)}{J_{Q_X}(u)}$ is increasing in u in $[0,1]$. This is equivalent to saying that*

$$X \leq_{DRE} Y \Leftrightarrow \frac{\int_u^1 \frac{dp}{q_Y(p)}}{\int_u^1 \frac{dp}{q_X(p)}}$$

is increasing in u for all u in $[0,1]$.

Similarly X is smaller than Y in decreasing past extropy if $\frac{\bar{J}_{Q_Y}(u)}{\bar{J}_{Q_X}(u)}$ is increasing in u and is written as $X \leq_{DPE} Y$. Obviously $X \leq_{DPE} Y$ is equivalent to

$$\frac{\int_0^u \frac{dp}{q_Y(p)}}{\int_0^u \frac{dp}{q_X(p)}}$$

increasing in u. Thus

$$X \leq_{\text{DPE}} Y \Leftrightarrow \frac{\int_0^u \frac{dp}{q_X(p)}}{\int_0^u \frac{dp}{q_Y(p)}} \geq \frac{q_Y}{q_X} \tag{3.42}$$

$$\Leftrightarrow \frac{\bar{J}_{Q_X}(u)}{\bar{J}_{Q_Y}(u)} \geq \frac{h_{Q_X}(u)}{h_{Q_Y}(u)}.$$

In cases where it is difficult to establish the monotonicity of $\frac{\bar{J}_{Q_X}(u)}{\bar{J}_{Q_Y}(u)}$, inequality (3.41) gives a graphical procedure. When the graph of the ratio $\frac{\bar{J}_{Q_X}(u)}{\bar{J}_{Q_Y}(u)}$ lies below that of $\frac{h_{Q_X}(u)}{h_{Q_Y}(u)}$, we conclude that $X \leq_{\text{DPE}} Y$. Likewise we have the bound for $X \leq_{\text{DRE}} Y$ as

$$X \leq_{\text{DRE}} Y \Leftrightarrow \frac{\bar{J}_{Q_X}(u)}{\bar{J}_{Q_Y}(u)} \leq \frac{h_{Q_X}(u)}{h_{Q_Y}(u)}. \tag{3.43}$$

Inequalities (3.41) and (3.43) are necessary and sufficient conditions. The two orders $\leq_{\text{DRE}}$ and $\leq_{\text{DPE}}$ are also useful in defining the DREX and DPEX classes. Infact we have

$$X \leq_{\text{DPEX}} E \Leftrightarrow X \text{ is DREX}$$

and

$$X \leq_{\text{DPEX}} E^* \Leftrightarrow X \text{ is DPEX}$$

where E is the exponential random variable and E^* is the reversed exponential random variable.

From Theorem 3.18 we see that for two non-negative random variables X and Y

$$[J_{Q_X}(u) - J_{Q_Y}(u)] = (1-u)^{-2}[\bar{J}_{Q_X}(1) - \bar{J}_{Q_Y}(1) - u^2(\bar{J}_{Q_X}(u) - \bar{J}_{Q_Y}(u))].$$

In lifetime models for which $\bar{J}_{Q_X}(1) = \bar{J}_{Q_Y}(1)$, that is X and Y have the same extropy measure $J_X = J_Y$, it is easy to see that

$$J_{Q_X}(u) \geq J_{Q_Y}(u) \Leftrightarrow \bar{J}_{Q_X}(u) \leq \bar{J}_{Q_Y}(u)$$

or $X \geq (\leq)_{\text{DREX}}$ is equivalent to $X \leq (\geq)_{\text{DPEX}}$. The condition $J_X = J_Y$ is not a trivial case as can be seen in the next example.

EXAMPLE 3.19. *Assume that* $F_X(x) = x^2$, $0 \le x \le 1$ *and* $F_Y(x) = 1 - (1-x)^2$, $0 \le x \le 1$. *Then* $J_X = J_Y = -\frac{2}{3}$. *Notice that* $Q_X(u) = u^{1/2}$ *and* $Q_Y(u) = 1 - (1-u)^{1/2}$. *We have*

$$J_{Q_X}(u) = -\frac{2(1 - u^{3/2})}{3(1-u)^2} \text{ and } J_{Q_Y}(u) = -\frac{2(1 - u^{-3/2})}{3(1-u)^2}$$

so that $\bar{J}_{Q_X}(u) \ge \bar{J}_{Q_Y}(u)$.

The order statistics of past life distributions are also of interest. Various results in this connection are similar to those of residual life and therefore not pursued.

CHAPTER 4

Generalized Entropies

4.1. Introduction

There have been several attempts to generalize Shannon's entropy based on the modification of underlying axioms, introduction of additional parameters, extra physical conditions, new functional forms, among others. See the observations made in Section 2.1 and names of such entropies given therein. Simultaneously efforts were on to find a general mathematical expression that can include various such generalizations.

Let (X, B_X, P_0) be a probability space in which B_X is the sigma field generalized by the subsets of X and P_θ is a probability measure where θ belongs to an open subset of $\mathbb{R}^m$. Assume that P_θ possesses a probability density function $f(x; \theta)$ with respect to a σ-finite measure μ. Then we can define a functional (Salicru et al. (1993))

$$H_\phi^h(\theta) = h\left[\int_{\mathfrak{X}} \phi(f(x; \theta))d\mu\right] \tag{4.1}$$

called the h-ϕ entropy associated with f. Here $\phi : [0, \infty] \to \mathbb{R}$ is either concave and $h : \mathbb{R} - \mathbb{R}$ is increasing or ϕ is convex and h is decreasing. The expression (4.1) contains as special cases the Shannon, Renyi, Varma, Havrda & Charvat, Armito, Sharma & Mittal, Taneja, Sharma & Taneja & Ferrari entropies in Table 4.1. A further generalization of (4.1) offered through the $H_{h,v}^{\phi_1,\phi_2}$ functional (Estaban and Morales (1995))

$$H_{h,v}^{\phi_1,\phi_2}(\theta) = h\left(\frac{\int_{\mathfrak{X}} \nu(x)\phi_1\left(f(x; \theta)\right)d\mu}{\int_{\mathfrak{X}} \nu(x)\phi_2\left(f(x; \theta)\right)d\mu}\right) \tag{4.2}$$

in which $\nu : \mathfrak{X} \to [0, \infty)$ is a weight function, $\phi_1 : [0, \infty) \to \mathbb{R}$, $\phi_2 : [0, \infty) \to \mathbb{R}$ and $h : \mathbb{R} \to \mathbb{R}$. Measures 1 to 23 in Table 4.1 belonging to the class (4.2) have also shown that

$$n^{\frac{1}{2}}\left[H_{h,v}^{\phi_1,\phi_2}(\hat{\theta}) - H_{h,v}^{\phi_1,\phi_2}(\theta)\right]$$

converges in law to $N(0, \sigma^2)$ as $n, m \to \infty$. They have also proposed tests for a predicted value of entropy with $H_0 : H_{h,v}^{\phi_1,\phi_2}(\theta) = D_0$ against the alternative $H_1 : H_{h,v}^{\phi_1,\phi_2}(\theta) \neq D_0$, equality of r populations entropies $H_0 : H_{h,v}^{\phi_1,\phi_2}(\theta_1) = \cdots = H_0 : H_{h,v}^{\phi_1,\phi_2}(\theta_r)$ and also for a common value D_0 for r such entropies. For a discrete version of (4.2) and some interesting properties and results thereof we refer to Estaban and Morales (1995). The individual members of (4.2) satisfy many special properties which makes the study of each one of them worthwhile. However, among the plethora of entropies found in literature, those of interest in reliability modelling will only be discussed in the present work. It may be noted that although Table 4.1 contains a large number of measures, it is not exhaustive. The present Chapter is confined to the important developments concerning the extensions of Shannon's work in the context of reliability.

4.2. Renyi entropy

Among various generalizations mentioned above Renyis's (Renyi (1961)) work has captured most attention. Renyi's entropy is based on an alternative axiomatic setting using an incomplete probability function $P = (p_1, p_2, \ldots, p_n)$ satisfying $W(P) = \sum_{i=1}^{n} p_i \leq 1$.

The Shannon entropy written as

$$H(P) = \sum_{i=1}^{n} p_i g(I_i), \quad I_i = -\log p_i$$

is uniquely determined by the axioms of symmetry, continuity, normalization, additivity (see Chapter 2) and the mean value condition

$$H(P \cup Q) = \frac{W(P)H(P) + W(Q)H(Q)}{W(P) + W(Q)} \qquad (4.3)$$

in which $W(P) + W(Q) \leq 1$ and $P \cup Q = (p_1, p_2, \ldots, p_n, q_1, q_2, \ldots, q_n)$. For any given function g possessing an inverse g^{-1}, the generalized mean functions can be defined as $g^{-1}(\sum p_i g(I_i))$. Extending the linear mean in (4.3) to the generalized mean yields

$$H^*(P \cup Q) = g^{-1} \frac{(W(P)g(H^*(P)) + W(Q)g(H^*(P)))}{W(P) + W(Q)}. \qquad (4.4)$$

Taking $g(\cdot)$ to be continuous and monotonic Renyi (1961) established that there are only two solutions for g that satisfy (4.3), namely $g_1(t) = at + b$ yielding Shannon's entropy and $g_2(t) = 2^{(\alpha-1)t}$ leading to the generalized version

$$H_\alpha = \frac{1}{1-\alpha} \log_2 \frac{\sum p_i^\alpha}{\sum p_i}.$$

When P is a complete distribution,

$$H_\alpha = (1-\alpha)^{-1} \log \sum p_i^\alpha, \alpha \neq 0, \alpha > 0$$

is generally called the Renyi entropy. The differential version is

$$H_\alpha = (1 - \alpha)^{-1} \log \int f^\alpha(x) dx. \qquad (4.5)$$

Like the Shannon measure, H_α in the continuous case can take negative values and also the differential version is not a limiting case of the discrete counterpart. As $\alpha \to 1$ in (4.5) we have the Shannon's entropy.

4.2.1. Residual Renyi entropy. The residual life has the entropy function (Asadi et al. (2005b))

$$\begin{aligned}
H_\alpha &= (1 - \alpha)^{-1} \log \int_t^\infty \left(\frac{f(x)}{\bar{F}(t)} \right)^\alpha dx \\
&= \log \bar{F}(t) + \frac{1}{1 - \alpha} \log \frac{1}{\bar{F}(t)} \int_t^\infty f^\alpha(x) dx.
\end{aligned} \qquad (4.6)$$

They have established the following characterization theorems.

THEOREM 4.1. *If F is differentiable with a continuous density function $f(x)$ in (t, ∞), then*
(a) $H_\alpha(t)$, $\alpha > 1$ uniquely determines F if f is strictly decreasing in (t, ∞),
(b) $H_\alpha(t)$, $0 < \alpha < 1$ uniquely determines F if f is strictly increasing over a bound support (t, b), $b < \infty$.

THEOREM 4.2. *The identity*

$$H_\alpha(t) = a - \log h(t)$$

where $h(t)$ is the hazard rate function and a is a constant, holds if and only if F is generalized Pareto with

$$f(x) = \frac{\beta + 1}{\theta} \left(1 + \frac{\beta}{\theta} x \right)^{-1/(\beta-2)}, \quad \theta > 0, \ \beta > -1.$$

REMARK 4.1. *For the uniform distribution $a = 0$, for $\beta = 1$ the limiting distribution is exponential and $H_\alpha(t)$ is a constant if and only if X is exponential.*

THEOREM 4.3. *The relationship*

$$H_\alpha(t) = b + \log m(t)$$

holds for all $t > 0$ if and only if X is distributed as generalized Pareto given in Theorem 4.2.

A comparison of residual Renyi entropies can be facilitated through various stochastic orderings. In this connection Nanda and Paul (2006b) provided the definition and some properties.

Table 4.1. Generalized entropies.

Author	Measure
1. Shannon (1948)	$-\int f(x)\log f(x)dx$
2. Renyi (1961)	$(1-r)^{-1}\log\int f^r(x)dx, r=1, r>0$
3. Aczel and Daroczy (1963)	$-\dfrac{\int f^r(x)\log f(x)dx}{\int f^r(x)dx}, r>0$
4. Aczel and Daroczy (1963)	$(s-r)^{-1}\log\dfrac{\int f^r(x)dx}{\int f^s(x)dx}, r=s, r>0, s>0$
5. Aczel and Daroczy (1963)	$\dfrac{\frac{d}{s}\arctan\int f^r(x)\sin(s\log f(x))}{\int f^r(x)\cos(s\log f(x))}, s=1, s, r>0$
6. Varma (1966)	$(m-r)^{-1}\log\int[f(x)]^{r-m+1}dx, m-1<r<m, m\geq 1$
7. Varma (1966)	$(m(m-r))^{-1}\log\int[f(x)]^{r/m}dx, 0<r<m, m\geq 1$
8. Kapur (1967)	$(1-r)^{-1}\log\left(\dfrac{\int[f(x)]^{r+s-1}}{\int[f(x)]^s}\right),$ $r=1, r>0, s\geq 1$
9. Havrda and Charvát (1967)	$(1-s)^{-1}[\int f^s(x)dx - 1], s=1, s>0$
10. Arimoto (1971)	$(r-1)^{-1}[\int[f^{\frac{1}{r}}(x)]^r - 1], r=1, r>0$
11. Sharma and Mittal (1975)	$(1-s)^{-1}[\exp\{(s-1)\int f(x)\log f(x)dx - 1\}], s=1, s>0$
12. Sharma and Mittal (1975)	$(1-s)^{-1}(\int f^r(x)dx)^{\frac{s-1}{r-1}} - 1), r, s=1, r, s>0$
13. Sharma and Taneja (1977)	$-r^{-1}\int f^r(x)\log x dx, r>0$
14. Sharma and Taneja (1975)	$(s-r)^{-1}[f^r(x) - f^s(x)]dx, r, s>0, r=s$
15. Sharma and Taneja (1975)	$-(\sin s)^{-1}\int f^r(x)\sin(s\log f(x))dx, s=k\pi, k=0,1,2,\ldots$
16. Ferreri (1980)	$(1+\lambda)\log(1+\lambda) - \dfrac{1}{\lambda}[\int\{1+\lambda f(x)\}\log\{1+\lambda f(x)\}dx], \lambda>0$
17. Santanna and Taneja (1985)	$-\int f(x)\log\left(\dfrac{\sin sf(x)}{2\sin\frac{s}{2}}\right)dx, 0<s<\pi$
18. Santanna and Taneja (1985)	$-\int\dfrac{\sin sf(x)}{2\sin s/2}\log\sin\left(\dfrac{sf(x)}{2\sin s/2}\right)dx, 0<s<\pi$

Table 4.1. Continued

Author	Measure
19. Belis and Guiasu (1968)	$-\int f(x)w(x)\log f(x)dx,\; w(x) > 0$
20. Picard (1979)	$-\dfrac{\int v(x)\log f(x)dx}{\int v(x)dx}$
21. Picard (1979)	$(1-r)^{-1}\log\left(\dfrac{\int v(x)f^{r-1}(x)dx}{\int v(x)dx}\right),\; r > 0,\, r\neq 1$
22. Picard (1979)	$(1-r)^{-1}\exp\left[(r-1)\dfrac{\int v(x)f(x)dx}{\int v(x)dx}\right],\; r\neq 1,\, r > 0$
23. Picard (1979)	$(1-s)^{-1}\left(\left[\dfrac{\int f^{r-1}(x)v(x)dx}{\int v(x)dx}\right]^{\frac{s-1}{r-1}} - 1\right),\; r\neq 1,\, s\neq 1,\, r,s > 0$
24. Kapur (1988)	$-\int \log\Gamma(1+f(x))dx$
25. Santanna and Taneja (1985)	$\int \dfrac{\sin s f(x)}{2\sin(s/2)}dx,\; 0 < s < \pi$
26. Picard (1979)	$-\left(\int v(x)\log f(x)dx\right)/\int v(x)dx$
27. Mathai and Haubold (2006)	$\dfrac{1}{\alpha-1}\int (f(x))^{2-\alpha}dx - 1,\; \alpha\neq 1,\, -\infty < d < 2$
28. Rathie (1970)	$(1-r)^{-1}\log\left(\int f^{r+s(x)-1}(x)dx/\int f(x)s(x)dx\right),\; \alpha\neq 1$
29. Aczel and Daroczy (1963)	$-\left[\int f^r(x)\log f(x)dx/\int f^r(x)dx\right]$
30. Behara and Chawla (1974)	$\dfrac{1-\int f^{1/r}(x)dx}{1-e^{r-1}},\; r > 0,\, r\neq 1$
31. Landsberg and Vedral (1998)	$\dfrac{1}{q-1}\left(\dfrac{1}{\int f^{q-1}(x)dx} - 1\right)$
32. Abe (1997)	$-\int \left(f^q(x) - f^{q^{-1}}(x)\right)dx$
33. Kaniadakis (2002)	$-\dfrac{1}{2k}\int \left[f^{1+k}(x) - f^{1-k}(x)\right]dx$
34. Koski and Persson (1992)	$\exp\left[\dfrac{1}{\beta-\alpha}\log\left(\dfrac{\int f^\alpha(x)dx}{\int f^\beta(x)dx}\right)\right]$
35. Sharma and Mittal (1975)	$-\dfrac{1}{K}\int f^r(x)(f^k(x) - f^{-k}(x))dx$
36. Khinchin (1957)	$-\dfrac{1}{K}\int f^r(x)(f^k(x) - f^{-k}(x))dx$
37. Mathai and Haubold (2006)	(a) $\dfrac{1}{\alpha-1}\left[\int_{-\infty}^\infty f^{2-\alpha}(x)dx - 1\right],\; \alpha\neq 1$ (b) $\dfrac{1}{\alpha-1}\log\int_{-\infty}^\infty f^{2-\alpha}(x)dx$

DEFINITION 4.1. *Let X and Y be non-negative random variables with probability density functions $f_X(x)$ and $f_Y(x)$. Then X is less than Y in residual Renyi entropy denoted by $X \leq_{RRE} Y$ if $H_{\alpha,X}(t) \leq H_{\alpha,Y}(t)$ for all $t > 0$.*

Some properties of the order are

(a) If $Z_1 = a_1 X + b_1$ and $Z_2 = a_2 Y + b_2$, $a_1, a_2 > 0$, $b_1, b_2 > 0$, $a_1 \leq a_2$, $b_1 \leq b_2$ then $X \leq_{RRE} Y$ if $H_{\alpha,X}(t)$ or $H_{\alpha,Y}(t)$ is decreasing in t. If $a_1 = a_2 = a > 0$ and $b_1 = b_2 = b \geq 0$, then $Z_1 \leq_{RRE} Z_2$ if $X \leq_{RRE} Y$.
(b) If $h_X(t) \leq h_Y(t)$ for all t and either X or Y is DFR, then $X \geq_{RRE} Y$. For application of this result in characterizing DFR distributions as maximum dynamic Renyi entropy models, see Asadi et al. (2005a,b).
(c) Let $(S(n), n = 1, 2, \ldots)$ be an increasing sequence of non-negative random variables $X(n) = S_n - S(n-1)$ and $N(t) = \sup(n|S_n \leq t)$ be the number of failures in $(0, t)$. Then the point process $\{N(t), t \geq 0\}$ is increasing (decreasing) in residual Renyi entropy if $Y_i \leq (\geq)_{RRE} Y_j$ for all $t \geq 0$ and $1 \leq i \leq j \leq n$, where $Y_t = [X(t)|s_{(1)} = s_{(1)}, \ldots, s_{(k-1)} = s_{k-1}]$, $k = 1, 2, \ldots$.

Non-parametric classes of life distributions can be defined in terms of the monotonic behaviour of $H_\alpha(t)$ discussed in Abraham and Sankaran (2005).

DEFINITION 4.2. *A non-negative random variable X is decreasing (increasing) uncertainty in residual Renyi entropy, DURL-α (IURL-α) if $H_\alpha(t)$ is decreasing (increasing) in $t \geq 0$.*

As a consequence of Definition 4.2, the exponential distribution is DURL-α as well as IURL-α. Also DURL (IURL) $\subset$ DURL-α (IURL-α). These non-parametric classes are closed under increasing linear transformations. Differentiating (4.6) and simplifying we have a relation connecting $H_\alpha(t)$ and the hazard rate $h(t)$ of X in the form

$$(1 - \alpha)H'_\alpha(t) = [\alpha h(t) - h^\alpha(t) \exp[-(1 - \alpha)H_\alpha(t)]]. \qquad (4.7)$$

Li and Zhang (2011) have more results on the above classes. Their main results are

(i) X is DURL-α if and only if $H_{\alpha,X_t}(t) \leq H_{\alpha,X}(t)$ for all $s, t \geq 0$.
(ii) If $X \leq_{lr} Y$ and be increasing in $t \geq 0$ then Y is DURL-α if X is so.
(iii) If X is DURL-α then $X_{n:n}$ is also DURL-α. Thus DURL-α property is preserved under the formation of parallel systems. However, it is not preserved for series systems. Further if $X_{r:n}$ is DURL-α, then $X_{r+1:n}, X_{r:n-1}$ and $X_{r+1:n+1}$ are also DURL-α.

Mahmoudi and Asadi (2010) have considered two non-negative and absolutely continuous random variables X and Y with their hazard rates satisfying the condition $h_Y(t) = \theta h_X(t)$ for a non-negative increasing θ defined on $[0, 1]$. If $\frac{\bar{F}_Y(t)}{\bar{F}_X(t)} < \infty$, Y is DURL-α if X is DURL-α. The result holds for proportional hazards models when $\theta(t) = \theta$, a constant. It also has applications in proportional odd families. Define a family of distributions

$$\frac{F(x|\beta)}{\bar{F}(x|\beta)} = \frac{1}{\eta} \frac{F(x)}{\bar{F}(x)}.$$

Then $F(x|\beta)$ is called the proportional odds family. Then its hazard rate satisfies

$$h(x|\beta) = \theta(x)h(x), \quad \theta(x) = \frac{1}{F(x) + \eta\bar{F}(x)}.$$

Thus for $\eta > 1$, if X is DURL-α then $F(x|\eta)$ is also DURL-α.

Renyi's residual entropy of order statistics have been studied by Zarezadeh and Asadi (2010).

THEOREM 4.4. *If $F(x)$ is absolutely continuous, the residual entropy of $X_{r:n}$ can be expressed as*

$$H_\alpha(X_{r:n}, t) = H_\alpha(U_{r:n}, t) + (1 - \alpha)^{-1} \log(f^{\alpha-1}(F^{-1}(Y_k)))$$

where $Y_k = B_{F(t)}(\alpha(r-1) + 1, \alpha(n-r+1))$ and $B_x(a,b) = \int_a^1 u^{a-1}(1-u)^{b-1}du$, $0 < x < 1$.

REMARK 4.2. *In the case of the exponential law with parameter λ,*

$$H_\alpha(X_{1:n}, t) - H_\alpha(X; t) = -\log n.$$

This means that in a series system the difference between its residual entropy and that of a component depends only on the number of components.

THEOREM 4.5. *(a) If $X_1, X_2, \ldots, X_n$ are independent and identically distributed lifetimes of a parallel (series) system with common distribution function $F(x)$ with increasing (decreasing) density function, then the residual system lifetime is decreasing in n.*
(b) If r_1 and r_2 are integers, $r_1 \leq r_2 \leq n$, then $H_\alpha(X_{r_1:n}, t) \leq H_\alpha(X_{r_2:n}, t)$.

The weighted version of the Renyi dynamic information measure can be written as

$$H_\alpha^L(t) = (1 - \alpha)^{-1} \log \int_t^\infty x \left(\frac{f(x)}{\bar{F}(t)}\right)^\alpha dx.$$

Sekeh et al. (2014) have shown that the weighted Renyi residual entropy is a constant in t for $\alpha \neq 2$ if and only if X has a Weibull distribution and for

$\alpha = 2$ if and only if X has a power distribution. Further if ϕ is a one-to-one transformation

$$H_\alpha^L(t) = \begin{cases} \log \bar{F}(\phi^{-1}(t)) + (1-\alpha)^{-1} \log E\{\phi(X)(\frac{f(X)}{\phi'(X)})^{\alpha-1}|X > \phi^{-1}(t)\} \\ \quad \text{if } \phi \text{ is increasing} \\ \log F(\phi^{-1}(t)) + (1-\alpha)^{-1} \log E\{\phi(X)(\frac{f(X)}{|\phi'(X)|})^{\alpha-1}|X > \phi^{-1}(t)\} \\ \quad \text{if } \phi \text{ is decreasing} \end{cases}$$

In an alternative formulation, instead of weighting the information measure, Nourbaksh and Yari (2017) considered the weighted distribution of X to define the weighted entropy of order α as,

$$H_\alpha^w = (1-\alpha)^{-1} \log \int_0^\infty (xf(x))^\alpha dx$$
$$= (1-\alpha)^{-1} \log E(X^\alpha f^{\alpha-1}(X)).$$

Different distributions may have identical values of H_α^W, for example uniform distribution order $[0,1]$ and exponential with parameter $\frac{1}{4}$ when $\alpha = 2$. However, it is necessary that for such distributions the weighted versions are equal. In the case of residual entropy we have

$$H_\alpha^w(t) = (1-\alpha)^{-1} \log \int_t^\infty \left(\frac{xf(x)}{\bar{F}(t)}\right)^\alpha dx, \ \alpha \neq 1, \alpha > 0 \qquad (4.8)$$
$$= (1-\alpha)^{-1} \log(t^\alpha \exp(1-\alpha)H_\alpha(t)$$
$$+ \alpha \int_t^\infty x^{\beta-1} \left(\frac{\bar{F}(x)}{\bar{F}(t)}\right)^\alpha \exp(1-\alpha)H_\alpha(x)dx.$$

The weighted entropy can be ordered as follows. We say that X is smaller than Y in weighted residual Renyi entropy written as $X \leq_{\text{WRRE}} Y$ if $H_{\alpha,X}^w(t) \leq H_{\alpha,Y}^w(t)$ for all $t > 0$. As an example if X_i has distribution

$$f(x_i) = \frac{1}{\alpha_i} \exp\left(-\frac{x_i - \mu}{\alpha_i}\right), \ x_i > \mu_i, \ i = 1, 2$$

then $X_1 \leq_{\text{WRRE}} Y$ whenever $\alpha_1 \leq \alpha_2$. Denoting by

$$H_\alpha^w(\phi(X), t) = (1-\alpha)^{-1} \log \int_t^\infty \left(\phi(x)\frac{f(x)}{\bar{F}(t)}\right)^\alpha dx$$

where $\phi(X)$ is a linear function, we have $H_\alpha^W(\phi(x), t) = H_\alpha^W(\phi(x), \phi^{-1}(t))$ when $\phi(\cdot)$ is strictly increasing. Also if $H_\alpha^w(t)$ is increasing (decreasing) in t then

$$h(t) \geq (\leq)[\alpha t^{-\alpha} \exp[(1-\alpha)H_\alpha^w(t)]]^{\frac{1}{\alpha-1}}.$$

A relationship between the weighted version and original entropy exists in the form

$$H_\alpha^w(t) \geq (\leq) H_\alpha + \frac{\alpha}{1-\alpha} \int_t^\infty \frac{f(x)}{\bar{F}(t)} \log x \, dx, \ 0 < \beta < 1 (\beta > 1).$$

In this case

$$H_\alpha^w(t) \geq (\alpha - 1)^{-1} \left(1 - x \frac{f(x)}{\bar{F}(t)}\right)^\alpha dx.$$

The above notion of weighted Renyi entropy was further extended by Singh and Kundu (2018) to two-sided truncated random variable defined by

$$H_{\alpha,W}(t_1, t_2) = \frac{1}{\alpha - 1} \int_{t_1}^{t_2} \left(\frac{x f(x)}{F(t_2) - F(t_1)}\right)^\alpha dx.$$

They have studied several properties of the measure which are extensions of the measure given in (4.8).

Maya et al. (2014) have suggested non-parametric estimates of $H_\alpha(t)$ based on complete and censored data using kernel type estimation. It is assumed that the data is assumed to be α-mixing in the sense that

$$\alpha(n) = \sup_{k \geq 1}\{|P(A \cap B) - P(A)P(B)|\}$$

where $A \in S_1^K$ and $B \in S_{k+n}^\infty$ tends to zero as $n \to \infty$, where S_i^k is the sigma field of events generated by $(X_j | i \leq j \leq k)$. The estimator proposed for $H_\alpha(t)$ is

$$\hat{H}_\alpha(t) = (1 - \alpha)^{-1} \log \left\{ \frac{1}{n} \frac{\sum_{i=1}^n f_n^{\alpha-1}(X_i) I(X_i > t)}{\bar{F}_n^\alpha(t)} \right\} \qquad (4.9)$$

where $f_n(X_i) = \frac{1}{n-1} \sum_{j \neq i} \frac{1}{b_n} K(\frac{X_i - X_j}{b_n})$ is the kernel estimator obtained from the sample without X_i and $\bar{F}_n$ is either the empirical survival function or a kernel estimator of $\bar{F}(x)$. The expression for the bias and variance are

$$\text{bias} \doteq \frac{b_n^s C_s}{s!} f^{(s)}(x) \text{ and variance } \doteq \frac{1}{n b_n} f(x) C_K,$$

where $C_s = \int_{-\infty}^\infty u^s K(u) du$. Also one can have the kernel estimator.

$$\tilde{H}_\alpha(t) = (1 - \alpha)^{-1} \log \int_t^\infty f_n^\alpha(x) dx - \alpha \log \bar{F}_n(t)$$

with

$$f_n(x) = (n b_n)^{-1} \sum_{j=1}^n K\left(\frac{x - X_j}{b_n}\right) \text{ and } \bar{F}_n(t) = \int_t^\infty f_n(x) dx.$$

When the sample observations are assumed to have a common distribution function, but not independent and X_i is censored in the right by a random variable Y_i, the distribution function $G(\cdot)$ of $Z_i = \min(X_i, Y_i)$ is

$$1 - G(t) = (1 - F(t))(1 - P(t))$$

where $P(\cdot)$ the distribution function of the Y_i's. Two estimators suggested by them are

$$H_\alpha^*(t) = (1 - \alpha)^{-1} \log \left[\frac{\frac{1}{n} \sum_{i=1}^n f_n^{\alpha-1}(z_i) I(Z_i > t)}{\bar{F}_n^*(\alpha)} \right]$$

where

$$f_n(z_i) = (n-1)^{-1} \sum_{j \neq i}^n b_n^{-1} K \left(\frac{Z_i - Z_j}{b_n} \right)$$

and $\bar{F}_n^*$ is either the kernel estimator based on f_n or the Kaplan-Meier estimator and

$$\tilde{H}_\alpha^*(t) = (1 - \alpha)^{-1} \log \int_t^\infty f_n^\alpha(x) dx - \alpha \log \bar{F}_n^*(t)$$

in which

$$f_n^*(x) = b_n^{-1} \int_t^\infty \frac{K \left(\frac{t-x}{b_n} \right)}{1 - P(x)} dG_n^*(x), \quad G_n^*(x) = \frac{1}{n} \sum_{i=1}^n I(Z_i \leq x, S_i = 1).$$

It is shown that the estimators are consistent and asymptotic normal under regularity conditions.

Another type of censoring that has been frequently discussed in reliability and life testing is type II progressive censoring. We consider a life testing experiment in which there are n independent and identically distributed failure terms $X_1, X_2, \ldots, X_n$. Following the first failure R_1 items are removed from the experiment, immediately after second failure R_2 items are removed after the mth failure R_m items are removed from the test at random. We have $n = m + \sum_{i=1}^n R_i$. If $R_1 = R_2 = \cdots = R_{m-1} = 0$, $R_m = n - m$ which is called the type II censoring and if all the R_i's are zero, we have the order statistics $(X_{(1)}, \ldots, X_{(n)})$. The joint density function of the m ordered failure times $(X_{(1)}^m, \ldots, X_{(m)}^m)$ in the progressive censored case

$$f(x_1, \ldots, x_m) = \prod_{i=1}^m C_i f(x_i)(1 - F(x_i))^{R_i}, \; 0 < x_1 < \cdots < m$$

where x_i is the realization of $X_{(i)}^m$ and $c_i = \sum_{j=i}^m (R_j + 1)$. The distribution of the rth order statistic $X_{(r)}^m$ is

$$f_r(x) = K_{r-1} f(x) \sum_{i=1}^r a_i (1 - F(x))^{c_i - 1}$$

in which

$$a_i = \begin{cases} 1, & i = 1, \\ \prod_{j=1}^r \frac{1}{c_j - c_i}, & i \neq j, r > 1 \end{cases}$$

and $A_{s-1} = \prod_{i=1}^s$. Further, the corresponding survival function becomes

$$\bar{F}_r(x) = A_{r-1} \sum_{i=1}^r \frac{a_i}{c_i} (1 - F(x))^{c_i}.$$

The residual Renyi entropy of $X_{(r)}^m$ has the form

$$H_\alpha(X_{(r)}^m; t) = H_\alpha(U_{(r)}^m, F(t)) + (1 - \alpha)^{-1} \log E(f^{\alpha-1}(F^{-1}(Y_r)))$$

with

$$f_{Y_r}(x) = \left(\frac{\sum_{i=1}^r (a_i (1-x)^{c_i-1})^\alpha}{\int_{F(t)}^1 (\sum_{i=1}^r a_i (1-u)^{C_i-1})^\alpha du} \right). \qquad (4.10)$$

From (4.10) a plug-in estimator for H_α is

$$\hat{H}_\alpha(X_{(r)}^m, t) = H\left(Y_{(r)}^m, \hat{F}_{m,n}(t) + (1-\alpha)^{-1} \left[\log \frac{\sum_{i: x_{(i)}^m > t} \hat{f}_n^{\alpha-1}(X_{(i)}^m)}{\sum I(X_{(i)}^m > t)} \right] \right)$$

where

$$\hat{f}_n(x_j) = c_j^{-1} \prod_{(1 \leq i \leq n, x_i < x_j)} \left(\frac{c_i - 1}{c_i} \right), \qquad j = 1, 2, \ldots, m.$$

These results are due to Jamboori and Yousefzadeh (2014). In their work some simulation studies have been carried out to assess the merit of the estimator also.

4.2.2. Quantile version.

Following the procedure adopted earlier, the quantile version of the Renyi residual entropy is given by

$$H_{\alpha_x Q_x}(u) = (1 - u)^{-1} \log \int_u^1 (1 - u)^{-\alpha} q(p)^{1-\alpha} dp. \qquad (4.11)$$

Differentiation and simplification yields,

$$q(u) = [\alpha - (1 - \alpha)(1 - u) H'_{\alpha, Q}(u)]^{\frac{1}{1-\alpha}} \frac{\exp[H_{\alpha, Q}(u)]}{1 - u} \qquad (4.12)$$

Table 4.2. Renyi residual quantile entropy.

Distribution	Quantile function	$H_{\alpha,Q}(u)$
exponential	$-\frac{1}{\lambda}\log(1-u)$	$-\frac{\log\alpha}{1-\alpha} - \log\lambda$
uniform	$a + (b-a)u$	$\log[(b-a)(1-u)]$
Pareto II	$\beta[(1-u)^{-\frac{1}{c}} - 1]$	$\log(\frac{\beta}{c}) + \frac{\log(\frac{c}{\alpha(c+1)-1})}{1-\alpha}$
rescaled beta	$R[1 - (1-u)^{-\frac{1}{c}}]$	$\log(\frac{R}{c}) + \frac{\log(\frac{c}{\alpha(c-1)-1})}{1-\alpha}$
power	$au^{-\frac{1}{c}}$	$\log(\frac{a}{c}) + \frac{\log(\frac{c}{\alpha(c-1)+1})}{1-c}$

showing that $H_{\alpha,Q}(u)$ determines the distribution of X uniquely. This is a stronger result than the one mentioned earlier in Asadi et al. (2005a). These results are due to Nanda et al. (2014). The expressions for $H_{\alpha,Q}(u)$ for some distributions are presented in Table 4.2.

When $0 < \alpha < 1$ and the $H_{\alpha,Q}$ is increasing we have the bound

$$\exp[(1-\alpha)H_{\alpha,Q}(u)] \geq \frac{[h_Q(u)]^{\alpha-1}}{\alpha} \qquad (4.13)$$

and if $\alpha > 1$ and $H_{\alpha,Q}$ is increasing

$$\exp[(1-\alpha)H_{\alpha,Q}(u)] \leq \frac{[h_Q(u)]^{\alpha-1}}{\alpha}. \qquad (4.14)$$

Nanda et al. (2014) have characterized the exponential, Pareto and rescaled beta distributions by the property

$$H_{\alpha,Q}(u) = A + B\log(1-u), \quad A > 0$$

when $B = 0$, $B < 0$ and $B > 0$ respectively.

DEFINITION 4.3. *The lifetime X has increasing (decreasing) Renyi residual quantile entropy, (DRRQE) if $H_{\alpha,Q}(u)$ is non-decreasing (non-increasing in u). This IRPQE is equivalent to the DURL-α (IURL-α) given in Definition 4.2.*

If $Y = \phi(X)$ where ϕ is non negative, increasing and convex (concave) then for $0 < \alpha < 1$, Y is IURL-α (DURL-α) whenever X is so and for $\alpha > 1$, Y is IURL-α (DURL-α) if X is DURL-α (IURL-α). Comparison of $H_{\alpha,Q}(u)$ for two random variables can be effected by defining an ordering among them.

DEFINITION 4.4. *We say that X is smaller than Y in Renyi residual quantile entropy order, $X \leq_{RPQ} Y$ if $H_{\alpha,Q_x}(u) \leq H_{\alpha,Q_y}(u)$ for all u in $(0,1)$.*

For any non-negative increasing convex function $\phi(\cdot)$, $\phi(X) \leq_{RRQ} \phi(Y)$. Also if $X \leq_{HQ} Y$ then $X \leq_{RRQ} Y$. Yan and Kang (2016) gave some new results in this connection. They proved that

(i)

$$X \leq_{\text{RRQ}} Y \Leftrightarrow \begin{cases} \int_t^\infty |f_X(x)|^\alpha \left[\left(\dfrac{f(x)}{g_Y[G_Y^{-1}F_X(x)]} \right)^{1-\alpha} - 1 \right] dx \geq 0, & 0 < \alpha < 1 \\[4mm] \int_t^\infty |f_X(x)|^\alpha \left[1 - \left(\dfrac{f_X(x)}{g_Y[G_Y^{-1}F_X(x)]} \right)^{1-\alpha} \right] dx \geq 0, & \alpha > 1 \end{cases}$$

(ii) $X \leq_c Y \Rightarrow X \leq_{\text{RRQ}} Y$ provided $f_X(0) \geq g_Y(0) > 0$

(iii) If $a > 1$, then $X \leq_{\text{RRQ}} aX$ and if $0 < a \leq 1$, $aX \leq_{\text{RRQ}} X$.
Also, if $a \geq 1$ then $X \leq_{\text{RRQ}} aY$ and $0 < a \leq 1$, $aX \leq_{\text{RRQ}} Y$.

(iv)

$$X \leq_{\text{RRQ}} Y \Rightarrow \begin{cases} aX + b \leq_{\text{RRQ}} cY + d, & 0 < a \leq c, \, 0 \leq b \leq d \\[2mm] aX + b \leq_{\text{RRQ}} aY + d, & a > 0, \, b \geq 0 \end{cases}$$

(v) $X \leq_{\text{RRQ}} Y \Rightarrow X_{n:n} \leq_{\text{RRQ}} Y_{n:n}$
$X_{1:n} \leq_{\text{RRQ}} Y_{1:n} \Rightarrow X \leq_{\text{RRQ}} Y$

(vi) Let $\phi(\cdot)$ be an increasing concave function such that $\phi(0) = 0$.
Given $X \leq_{\text{st}} Y$, $\phi(X) \leq_{\text{RRQ}} \phi(Y) \Rightarrow X \leq_{\text{RRQ}} Y$

(vii) If $\theta \geq 1$, $X \leq_{\text{RRQ}} Y \Rightarrow X(\theta) \leq_{\text{RRQ}} Y(\theta)$ where $X(\theta)$ and $Y(\theta)$ are proportional hazards models of X and Y respectively. Also, if $0 < \theta \leq 1$,

$$X(\theta) \leq_{\text{RRQ}} Y(\theta) \Rightarrow X \leq_{\text{RRQ}} Y.$$

In the case of reversed hazard models $X^*(\theta)$ and $Y^*(\theta)$ of X and Y

$$X \leq_{\text{RRQ}} Y \Rightarrow X^*(\theta) \leq_R Y^*(\theta), \; \theta \geq 1$$

and

$$X^*(\theta) \leq_{\text{RRQ}} Y \Rightarrow X \leq_{\text{RRQ}} Y, \; 0 < \theta \leq 1.$$

(viii) The proportional odds random variable X_p is defined by

$$\bar{F}_{X_P}(x) = \frac{\theta \bar{F}_X(x)}{1 - (1-\theta)\bar{F}_X(x)}, \quad \theta > 0.$$

If Y_p is defined for a survival function $\bar{G}_{Y_p}(x)$, then

$$X \leq_{\text{RRQ}} Y \Rightarrow X_p \leq_{\text{RRQ}} Y_p, \quad \theta \geq 1$$
$$X_p \leq_{\text{RRQ}} Y_p \Rightarrow X \leq_{\text{RRQ}} Y, \quad 0 < \theta \leq 1$$

To get further insight into the role of Renyi's residual entropy in reliability modelling, we need to relate it to the ageing properties. Towards this objective from (4.11) we find

$$\exp[(1-\alpha)H_{\alpha,Q}(u)] = \int_u^1 (1-u)^{-\alpha}[q(p)]^{1-\alpha}dp.$$

Differentiating with respect to u and simplifying,

$$\exp[(1 - \alpha)H_{\alpha,Q}(u)][\alpha - (1 - u)(1 - \alpha)H'_{\alpha,Q}(u)]$$
$$= [q(u)(1 - u)]^{1-\alpha} = h_Q^{\alpha-1}(u).$$

or

$$e^{H_{\alpha,Q}(u)}h_Q(u) = [\alpha - (1 - u)(1 - \alpha)H'_{\alpha,Q}(u)]^{\frac{1}{\alpha-1}}. \qquad (4.15)$$

Since the left side is non-negative, from (4.15),

$$H'_{\alpha,Q}(u) \geq \frac{\alpha}{(1 - \alpha)(1 - u)}.$$

Integrating from 0 to u we have a lower bound to the Renyi residual entropy

$$H_{\alpha,Q}(u) > H_\alpha + \frac{\alpha}{\alpha - 1}\log(1 - u) \qquad (4.16)$$

since at the equality, $H_{\alpha,Q}(u) = H_\alpha + \frac{\alpha}{\alpha-1}\log(1 - u)$, we have $q(u) = 0$.

The relationship (4.15), when further analyzed gives some results about the conditions under which monotone behaviour of $H_{\alpha,Q}(u)$ and $h_Q(u)$ can be ensured.

THEOREM 4.6. *The random variable X is IFR (DFA) if $H_{\alpha,Q}(u)$ is decreasing (increasing) and convex (concave) for $0 < \alpha < 1$ ($\alpha > 1$).*

PROOF. Taking logarithms in (4.15),

$$\log h_Q(u) = -H_{\alpha,Q}(u) + \frac{1}{\alpha - 1}\log[\alpha - (1 - u)(1 - \alpha)H'_{\alpha,Q}(u)].$$

Differentiating and simplifying

$$\frac{h'_Q(u)}{h_Q(u)} = \frac{[(1 - u)(1 - \alpha)(H'_{\alpha,Q}(u))^2 + (1 - u)H''_{\alpha,Q}(u) - (1 + \alpha)H'_{\alpha,Q}(u)]}{\alpha(1 - u)(1 - \alpha)H'_{\alpha,Q}(u)}.$$

The result follows by applying the conditions on $H_{\alpha,Q}(u)$ given in the Theorem.   □

THEOREM 4.7. *The Renyi residual quantile entropy is increasing (decreasing) if $h_Q(u) \leq (\geq)\alpha^{\frac{1}{\alpha-1}}e^{-H_{\alpha,Q}(u)}$ for all u in $(0, 1)$ and $0 < \alpha < 1$ ($\alpha > 1$).*

PROOF. We have

$$(1 - u)(1 - \alpha)H'_{\alpha,Q}(u)\exp[(1 - \alpha)H_{\alpha,Q}(u)]$$
$$= \alpha\exp[(1 - \alpha)H_{\alpha,Q}(u)] - h_Q^{\alpha-1}(u)$$
$$H'_{\alpha,Q}(u) = \frac{\alpha\exp[(1 - \alpha)H_{\alpha,Q}(u)] - h_Q^{\alpha-1}(u)}{(1 - u)(1 - \alpha)\exp(1 - \alpha)H_{\alpha,Q}(u)}.$$

The denominator is positive for $0 < \alpha < 1$ and negative for $\alpha > 1$. This proves the result. $\qquad\qquad\square$

EXAMPLE 4.1. *Consider the power distribution* $Q(u) = u^m$, $0 \le u \le 1$, $m > 0$. *We have*

$$H_{\alpha,Q}(u) = (1 - \alpha)^{-1} \left[\log m^{1-\alpha} (1 - u)^{-\alpha} \frac{1 - u^{(m-1)(1-\alpha)+1}}{(m - 1)(1 - \alpha) + 1} \right].$$

In the special case $m = \frac{1}{2}$, $\alpha = \frac{1}{2}$,

$$H'_{\alpha,Q}(u) = \frac{2}{1 - u} - \frac{3}{4(u^{1/4} - u)}$$

we have $H'(u) < 0$ *and it is increasing. Thus X is IFR. Infact in this case*

$$h_Q(u) = 2(1 - u)^{-1} u^{1/2} \text{ and } h'_Q(u) = u^{-\frac{1}{2}}(1 + u) > 0.$$

A different scenario results when $\alpha = 3$, $m = 2$, giving

$$H'_{\alpha,Q}(u) = \frac{1 - 3u}{2u(1 - u)}$$

so that $H_{\alpha,Q}(u)$ is increasing in $(0, \frac{1}{3})$ and decreasing in $(\frac{1}{3}.1]$. Thus H is upside down bathtub-shaped with change point $u = \frac{1}{3}$. The hazard quantile function $h_Q(u) = \frac{1}{2u(1-u)}$ is bathtub-shaped with change point $u = \frac{1}{2}$. It is obvious that the monotonicity of H in the theorem cannot be dropped.

Finally $m = 1$, provides the uniform distribution $Q(u) = u$. The Renyi residual quantile entropy is

$$H_{\alpha,Q}(u) = \log(1 - u)$$

which is decreasing and convex. Thus by Theorem 4.7, X is IFR, a well known result. However, $H_{\alpha,Q}(u)$ is independent of α and therefore, the requirement $0 < \alpha < 1$ can be dispensed with.

4.2.3. Renyi past entropy. Gupta and Nanda (2002) have defined the Renyi past entropy as

$$\bar{H}_\alpha(t) = \frac{1}{1 - \alpha} \log \int_0^t \left[\frac{f(x)}{F(t)} \right]^\alpha dx \qquad (4.17)$$

and Nanda and Paul (2006a) have studied its properties in detail. On ordering $\bar{H}_\alpha$, they have proposed that the random variable X is greater than Y in Renyi past entropy if $\bar{H}_{\alpha,X}(t) \le \bar{H}_{\alpha,Y}(t)$ for all t and denoted by $X \ge_{\text{RPE}} Y$. In general $X \ge_{\text{RPE}} Y$ need not imply $X \ge_{\text{PE}} Y$. Under a linear transformation $z = aX + b$

$$\bar{H}_{\alpha,Z}(t) = \log a + \bar{H}_{\alpha,X}\left(\frac{t - b}{a}\right).$$

If we look at two such transformations $Z_1 = aX + b$, $Z_2 = aY + b$, $a > 0$ and $b > 0$ such that $X \geq_{\text{RPE}} Y$, then $Z_1 \geq_{\text{RPE}} Z_2$ if either H_{α,z_1} and H_{α,z_2} is increasing in $t \geq b$.

If $\bar{H}_\alpha(t)$ is increasing in t, we say that X belongs to the class of increasing uncertainty in the past life of order or IUPL-α. This class is closed under linear transformations. Also it follows that DRHR $\subset$ IUPL $\subset$ IUPL-α. X has a uniform distribution over (a, b), if and only if $\bar{H}_\alpha(t) = \log(t - a)$. Li and Zhang (2011) have established that (i) if $X \geq_{\text{lr}} Y$ and $\frac{\lambda_Y(t)}{\lambda_X(t)}$ is decreasing in t, then Y is IUPL-α and (ii) if X is IUPL-α, then $X_{1:n}$ is also IUPL-α, but $X_{n:n}$ does not preserve the same property.

We can write (4.17) is

$$\bar{H}_\alpha(t) = \log F(t) + (1 - \alpha)^{-1} \log E(f^{\alpha-1}(X)|X \leq t) \qquad (4.18)$$

or, alternatively,

$$\{F(t) \exp[-\bar{H}_\alpha(t)]\}^{\alpha-1} = E[f^{\alpha-1}(X)|X \leq t]. \qquad (4.19)$$

Again,

$$F^\alpha(t) \exp[(1 - \alpha)\bar{H}_\alpha(t)] = \int_0^t f^\alpha(x)dx.$$

Differentiating with respect to t and using $\lambda(t) = \frac{f(t)}{F(t)}$, leaves an identity

$$[\alpha\lambda(t) + (1 - \alpha)\bar{H}'_\alpha(t)] \exp[(1 - \alpha)\bar{H}_\alpha(t)] = \lambda^\alpha(t) \qquad (4.20)$$

connecting $\bar{H}_\alpha$ and $\lambda(t)$. This identity has many applications, one of which is in characterization problems.

THEOREM 4.8. *An absolutely continuous random variable X with support $[0, 1]$ and $F(0) = 0$ follows power distribution $F(x) = x^p$, $0 \leq x \leq 1$ if and only if*

$$\bar{H}_\alpha(t) = K - \log \lambda(t) \qquad (4.21)$$

for all t in $[0, 1]$ and some constant K.

PROOF. For the power distribution

$$\bar{H}_\alpha(t) = \log t + \frac{1}{1 - \alpha} \left[\log \frac{p^\alpha}{\alpha(p - 1) + 1} \right]$$

and $\lambda(t) = pt^{-1}$, proves the necessary part. To prove sufficiency, substitute (4.21) in (4.20) to find

$$\frac{\lambda'(t)}{\lambda^2(t)} = K_1 \qquad (4.22)$$

or

$$\lambda(t) = -(K_1 t + K_2)^{-1}.$$

Since $F(t) = \exp[-\int_x^1 \lambda(t)dt]$,

$$F(t) = \left(\frac{K_1 x + K_2}{K_1 + K_2}\right)^{\frac{1}{K_1}}.$$

Using $F(0) = 0$, $K_2 = 0$ and the power distribution is recovered.   □

There is a similar characterization by relationship between $\bar{H}_\alpha(t)$ and the mean inactivity time $\mu(t)$.

THEOREM 4.9. *Under the conditions of Theorem 4.8*

$$\bar{H}_\alpha(t) = K + \log\mu(t) \qquad (4.23)$$

for all t and some K, if and only if X has power distribution.

PROOF. When X has a power distribution $\mu(t) = (p+1)^{-1}t$ and this proves (4.23). Invoking the relationship $\lambda(t) = \frac{1-\mu'(t)}{\mu(t)}$ in (4.20).

$$\left[\alpha\left(\frac{1-\mu'}{\mu}\right) + (1-\alpha)\frac{\mu'}{\mu}\right]\mu^{1-d}K_1 = \left(\frac{1-\mu'}{\mu}\right)^\alpha, \quad K_1 = exp[(1-\alpha)K]$$

which simplifies to

$$K_1[\alpha + (1-2\alpha)\mu'(t)] = (1-\mu'(t))^\alpha.$$

Differentiating

$$\mu''(t)[K_1(1-2\alpha) + \alpha(1-\mu'(t))^{\alpha-1}] = 0$$

resulting in either $\mu''(t) = 0$ or $\mu'(t) = C_1$, a constant. Both give $\mu(t) = C_2 t$ and hence X has a power distribution.   □

When the condition of absolute continuity of $F(x)$ is dropped in Theorem 4.3, the relationship (4.21) can give rise to a few families of distributions with a jump at $x = 0$. Such a model arises in life testing of devices in which failure can happen at the instant it begins to function with positive probability which does not follow a pattern of failures in the rest of their life. Once this failure is overcome the device functions with a continuous lifetime.

THEOREM 4.10. *If X is a non-negative random variable, then*

$$\bar{H}_\alpha(t) = K - \log\lambda(t), \quad 0 < t \le c \qquad (4.24)$$

for some K depending on α if and only if

$$F(x) = \left(\frac{a+bx}{a+bc}\right)^{\frac{1}{b}}, \quad 0 \le x \le c. \qquad (4.25)$$

PROOF. Assume the distribution in (4.25). Then

$$f(x) = \begin{cases} \left(\frac{a}{a+bc}\right)^{\frac{1}{b}}, & x = 0 \\ \frac{(a+bx)^{\frac{1}{b}-1}}{(a+bc)^{\frac{1}{b}}}, & 0 < x \le c. \end{cases}$$

This gives

$$\lambda(t) = \begin{cases} 1, & x = 0 \\ \frac{1}{a+bt}, & 0 < x \le c \end{cases}$$

and

$$\bar{H}_\alpha(t) = \begin{cases} 0, & x = 0 \\ \log(a + bt) + \frac{1}{1-\alpha} \log \frac{a^{\alpha(\frac{1}{b}-1)-1}}{[b+(1-b)\alpha]}, & 0 < x \le c, \end{cases}$$

and hence the relationship (4.24) is true.

Conversely, if (4.24) holds for the absolutely continuous part we have as in Theorem 4.8

$$F(t) = \left(\frac{K_1 x + K_2}{K_1 c + K_2} \right)^{\frac{1}{K_1}}$$

which is of form (4.25) in $(0, c]$. Also $P(0 \le X \le c) = 1$ gives $F(0) = \left(\frac{K_2}{K_1 c + K_2} \right)$ so that (4.25) is recovered.   □

REMARK 4.3. *The relationship* $\bar{H}_\alpha(t) = K + \log \mu(t)$, *for all t in* $(0, c]$ *and some K, characterizes* (4.25) *whose proof is similar to that of Theorem 4.9. However, for a reparametrization* (4.25) *is the same as the distribution in* (3.9) *and therefore, the remarks offered on the latter hold for* (4.25) *too.*

The Renyi's entropy can be thought of as introducing an additional parameter α in the Shannon entropy. Hence it is of interest to examine how α influences the extent of uncertainty by comparing the two entropies.

THEOREM 4.11. *The Renyi and Shannon entropies satisfy the relationships* $\bar{H}_\alpha(t) \ge \bar{H}(t)$ *for* $\alpha > 1$ *and* $\bar{H}_\alpha(t) \le \bar{H}(t)$ *for* $0 \le \alpha < 1$.

PROOF. In equation (4.19), $E(f^{\alpha-1}(X)|X \le t) = A_\alpha(t)$ represents the truncated arithmetic mean of $f^{\alpha-1}(X)$. If one considers the corresponding geometric mean $G_\alpha(t)$, we have

$$\log G_\alpha(t) = E(\log f^{\alpha-1}(X)|X \le t) = (\alpha - 1)E(\log f(X)|X \le t)$$

$$= \frac{(\alpha - 1)}{F(t)} \int_0^t f(x) \log f(x) dx$$

$$= (\alpha - 1) \int_0^t \left[\log \frac{f(x)}{F(t)} + \log F(t) \right] \frac{f(x)}{F(t)} dx$$

$$= \log F^{\alpha-1} - (\alpha - 1)\bar{H}(t),$$

where $\bar{H}(t)$ is the Shannon's past entropy in (3.1). This gives

$$G_\alpha(t) = F^{\alpha-1}(t)\exp[(1-\alpha)\bar{H}(t)].$$

From (4.19)

$$A_\alpha(t) = F^{\alpha-1}(t)\exp[(1-\alpha)\bar{H}(t)].$$

Thus using $A_\alpha(t) \geq G_\alpha(t)$, the last two expressions lead to $\bar{H}_\alpha(t) \geq \bar{H}(t)$ if $\alpha > 1$ and $\bar{H}_\alpha(t) \leq \bar{H}(t)$ if $0 \leq \alpha < 1$. $\qquad\square$

EXAMPLE 4.2. *Consider the power distribution given in Theorem 4.8 with $\bar{H}_\alpha(t)$ as specified in it and $\bar{H}(t) = \frac{p-1}{p} + \log\left(\frac{t}{p}\right)$, then the relationship between Renyi and Shannon entropy in the past lifetime based on Theorem 4.11 is evident from the Figures 4.1 and 4.2. Further, $\bar{H}(t)$ of power distribution is BT as given in Example 3.3, evident from Figure 4.3.*

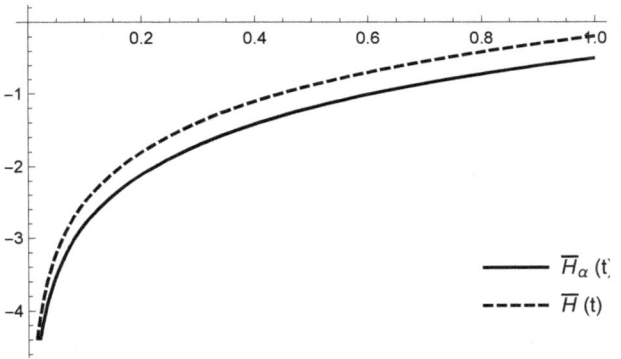

Figure 4.1. Plot of $\bar{H}_\alpha(t)$ and $\bar{H}(t)$ for $p = 2$ and $\alpha = 10$.

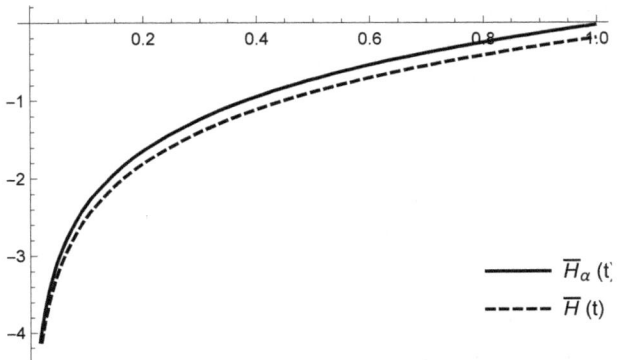

Figure 4.2. Plot of $\bar{H}_\alpha(t)$ and $\bar{H}(t)$ for $p = 2$ and $\alpha = 0.1$.

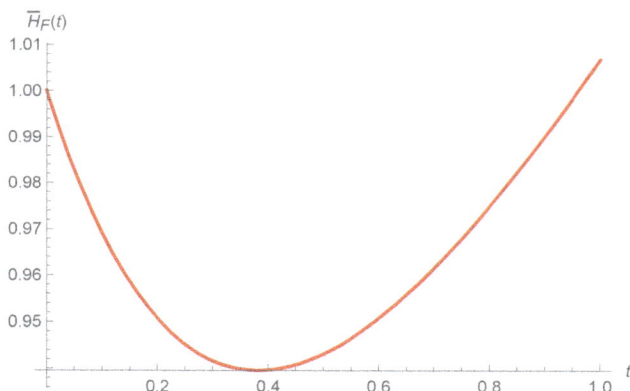

Figure 4.3. Plot of $\bar{H}(t)$.

Besides the IUPL-α (DUPL-α) classes discussed earlier, some classes defined below also exist with reference to Renyi entropy of past life.

DEFINITION 4.5. *The random variable X is said to be (i) decreasing uncertainty in Renyi past entropy (DUPL-α) if $\bar{H}_\alpha(t)$ is decreasing in $(0, b]$, $b < \infty$, and (ii) BT (UBT) if*

$$\bar{H}_\alpha(t) = \begin{cases} \bar{H}_1(t) & t \leq t_1 \\ c, \text{ a constant} & t_1 \leq t \leq t_2 \\ \bar{H}_2(t) & t \geq t_2 \end{cases}$$

where $\bar{H}_1(t)$ is strictly decreasing (increasing) and $\bar{H}_2(t)$ is strictly increasing (decreasing) for all t and α.

As illustrations of the above classes, we point out model (4.25) for DUPL-α when $b < 0$ and Example 3.3 where $\bar{H}_\alpha$ is UBT shaped.

Li and Zhang (2011) have discussed various properties of $\bar{H}_\alpha$. These include

(a) If $X \geq_{\text{lr}} Y$ and $\frac{\lambda_Y(t)}{\lambda_X(t)}$ is decreasing in $t \geq 0$, then Y is also IUPL-α.

(b) If X is IPUP-α then $X_{1:n}$ is also IUPL-α and $X_{k:n}$ is IUPL-α implies the same property for $X_{k:n-1}$, $X_{k:n+1}$ and $X_{k-1:n-1}$.

(c) When $w(x)$ is a weight function which is decreasing (increasing) and $E[w(X)|X \leq t]/w(t)$ is increasing (decreasing) then $X_w(X)$ is also IUPL-α if $X(X_w)$ is.

Much discussions do not appear to have been on the quantile version of $\bar{H}_\alpha(t)$ in the literature. We define the quantile form of Renyi's past entropy as

$$\bar{H}_{\alpha,Q}(u) = (1 - \alpha)^{-1} \log \int_0^u u^{-\alpha} (q(p))^{1-\alpha} dp. \qquad (4.26)$$

On simplifying and suppressing α,

$$q(u) = e^{\bar{H}_Q(u)}[(1-\alpha)u^\alpha H_Q(u) + \alpha u^{\alpha-1}]^{\frac{1}{1-\alpha}} \qquad (4.27)$$

which it is clear that $\bar{H}_Q(u)$ determines the distribution of X uniquely. Further from (4.27) we have two useful results

$$\bar{H}_Q(u) = \log(uq(u)) + \frac{1}{\alpha-1}\log(\alpha + (1-\alpha)u\bar{H}'_Q(u)) \qquad (4.28)$$

and

$$e^{\bar{H}_Q(u)}\lambda_Q(u) = [\alpha + (1-\alpha)u\bar{H}'_Q(u)]^{\frac{1}{\alpha-1}}. \qquad (4.29)$$

Equation (4.29) can provide several applications. First, we note that

$$\alpha + (1-\alpha)u\bar{H}'_Q(u) \geq 0$$

and integrating from u to 1,

$$\bar{H}_Q(u) \leq H_\alpha + \frac{\alpha}{1-\alpha}\log u,$$

an upper bound to the Renyi past entropy. Secondly differentiating (4.29)

$$\frac{\lambda'_Q(u)}{\lambda_Q(u)} = -\left[\bar{H}'_Q(u) + \frac{u\bar{H}''_Q(u) + \bar{H}'_Q(u)}{\alpha + (1-\alpha)u\bar{H}'_Q(u)}\right].$$

According when $\bar{H}_Q(u)$ is increasing and convex, the reversed hazard quantile function is decreasing or X is DRHR whenever $0 < \alpha < 1$. Finally, from (4.29),

$$e^{(\alpha-1)\bar{H}_Q}\lambda_Q^{\alpha-1}(u) = \alpha + (1-\alpha)u\bar{H}'_Q(u)$$

or

$$\frac{\lambda_Q^{\alpha-1}(u) - \alpha e^{(1-\alpha)\bar{H}_Q}}{(1-\alpha)ue^{(1-\alpha)\bar{H}_Q}} = \bar{H}'_Q(u).$$

When $0 < \alpha < 1$, $\bar{H}'_Q(u) > 0$ if $\lambda_Q^{\alpha-1}(u) > \alpha e^{(1-\alpha)\bar{H}_Q}$ or $\lambda_Q(u) \leq \alpha^{\frac{1}{\alpha-1}}\bar{e}^{\bar{H}_Q}$ and for $\alpha > 1$, $\lambda_Q(u) \geq \alpha^{\frac{1}{\alpha-1}}\bar{e}^{\bar{H}_Q(u)}$. Thus $\bar{H}_Q$ is increasing (decreasing) according as $0 < \alpha < 1$ ($\alpha > 1$) and $\lambda_Q(u) \leq (\geq)\alpha^{\frac{1}{\alpha-1}}\bar{e}^{\bar{H}_Q}$. The monotonicity of the Renyi past quantile entropy does not depend on the monotonicity of the reversed hazard quantile function, but its magnitude.

4.2.4. Relative Renyi residual entropy. Let X_1 and X_2 be two nonnegative absolutely continuous random variables with survival functions $\bar{F}_1, \bar{F}_2$, hazard rates h_1 and h_2 and cumulative hazard rates H_1 and H_2. Then the relative Renyi entropy of X_1 with respect to X_2 is defined as

$$H_{1:2}(t) = (1-\alpha)^{-1}\log\int_t^\infty \left[\frac{f_2(x)}{\bar{F}_2(t)}\right]^\alpha dH_1(x) \qquad (4.30)$$

for all t for which $\bar{F}_2(t) > 0$. This definition is using the same ideas in Wei (1992) wherein the author defines relative mean residual life. The interpretation here is that when X_1 follows the unit exponential distribution $H_1(x) = x$ and therefore $H_{1:2}(t)$ is the same as the residual entropy of X_2. That is in the case of no-ageing of X_1, there is no change in the residual entropy of X_2. Now, with this interpretation, $H_{1:2}(t)$ provides the relative value for a general distribution $\bar{F}_1$. Zaradasht (2020) has used the same notion to define relative cumulative residual entropy, see Chapter 8.

EXAMPLE 4.3. *Assume that X_1 is Pareto II with $\bar{F}_1(x) = \left(1 + \frac{x}{\beta}\right)^{-c}$ so that $h_1(x) = c(x + \beta)^{-1}$. Then with respect to the exponential distribution with parameter λ for X_2, we have*

$$H_{1,2}(t) = (1-\alpha)^{-1} \log \left(\frac{\lambda \bar{e}^{-\lambda x}}{\bar{e}^{-\lambda t}}\right)^{\alpha} \frac{c}{x + \beta} dx$$

$$= (1-\alpha)^{-1} \left[\alpha \log \lambda + \log C + \alpha \lambda t + \log \int_t^{\infty} \frac{\bar{e}^{\lambda \alpha x}}{x + \beta} dx\right]$$

$$= (1-\alpha)^{-1} \left[\alpha \lambda t + \lambda \alpha \beta + \log c \lambda^{\alpha} E_1(\lambda \alpha (t + \beta))\right]$$

(4.31)

where $E_1(z) = \int_z^{\infty} \frac{\bar{e}^t}{t} dt$, is the exponential integral.

Comparison of life distributions are necessitated in several practical situations in which the reliabilities of two devices are of interest. For example when the same kind of device is produced by two manufactures the choice between them rests on their relative ageing which determines which of the two devices ages faster than the other. One general procedure available for this purpose is to find stochastic orders in terms of ageing criteria like IFR, IFRA, DMRL, NBU, among others. Other measures include specific measures that quantify relative aging in the form of ageing intensity function, specific ageing factor and relative ageing factor. A good account of these approaches is available in Kochar and Wiens (1987), Kochar (1989), Sengupta and Deshpande (1994), Wei (1992), Abraham and Nair (2013), Jiang et al. (2003), Finkelstein (2006), Misra et al. (2017) and their references.

From (4.30), we can write

$$F_2^{\alpha}(t) \exp[(1-\alpha)H_{1,2}(t)] = \int_t^{\infty} f_2^{\alpha}(x) h_1(x) dx \qquad (4.32)$$

and on differentiation,

$$h_1(t) h_2^{\alpha}(t) = [\alpha h_2(t) - (1-\alpha) H_{1,2}'(t)] \exp[(1-\alpha) H_{1,2}(t)].$$

Thus we have the identity

$$H_{1,2}'(t) = (1-\alpha)^{-1}[\alpha h_2(t) - h_1(t) h_2^{\alpha}(t) \exp(-(1-\alpha) H_{1,2}(t))]. \quad (4.33)$$

Some bounds for $H_{1.2}(t)$ follow from (4.33).

THEOREM 4.12. *If $H_{1.2}(t)$ is strictly increasing (strictly decreasing) and $0 < \alpha < 1$, then*

$$H_{1.2}(t) > (<)(1-\alpha)^{-1}\log\left(\frac{h_1(t)h_2^{\alpha-1}(t)}{\alpha}\right)$$

and if $H_{1.2}(t)$ is strictly increasing (strictly decreasing) and $\alpha > 1$, then

$$H_{1.2}(t) < (>)(1-\alpha)^{-1}\log\left(\frac{h_1(t)h_2^{\alpha-1}(t)}{\alpha}\right).$$

REMARK 4.4. *As $\alpha \to 1$, the Shannon entropy is deduced from the Renyi entropy. In this case*

$$H_{1.2}(t) = -\int_t^\infty \frac{f_2(x)}{\bar{F}_2(t)}\log\left(\frac{f_2(x)}{\bar{F}_2(t)}\right)h_1(x)dx. \qquad (4.34)$$

It is difficult to obtain bounds as in Theorem 4.14, since algebraic manipulations of (4.34) yields

$$h_2(t)H''_{1.2}(t) - (h_2^2(t) + h_2'(t))H'_{1.2}(t) + H_{1.2}(t)h_2^3(t)$$
$$= h_1(t)[h_2^3(t) + h_2(t)h_2'(t) + h_1^2(t)\log h_2(t) - h_2^3(t)\log h_2(t)].$$

For complete discussions on the nature of the $H_{1.2}(t)$ function, it is necessary to investigate whether it can be a constant.

THEOREM 4.13. *The relative entropy $H_{1.2}(t)$ is a constant if and only if*

$$h_1(t) = Kh_2^{1-\alpha}(t) \qquad (4.35)$$

for some K.

PROOF. First assume that $H_{1.2} = C$, a constant. Then from (4.33),

$$(1-\alpha)^{-1}h_2(t)[\alpha - Ch_1(t)h_2^{\alpha-1}(t)] = 0$$

from which (4.35) follows. Conversely if the relationship (4.35) is true then equation (4.32) simplifies to

$$\exp[(1-\alpha)H_{1.2}(t)] = \int_t^\infty \left(\frac{f_2(x)}{\bar{F}_2(t)}\right)^\alpha Kh_2^{1-\alpha}(x)dx$$
$$= K\int_t^\infty \left(\frac{f_2(x)}{\bar{F}_2(t)}\right)^\alpha \left(\frac{f_2(x)}{\bar{F}_2(t)}\right)^{1-\alpha} dx$$
$$= \frac{K}{\bar{F}_2^\alpha(t)}\int_t^\infty F_2^{\alpha-1}(x)f_2(x)dx$$
$$= \frac{K}{\alpha},$$

a constant and this proves the assertion. □

REMARK 4.5. *The distribution of X_1 for which $H_{1.2}(t)$ is a constant is specified by*

$$\bar{F}_1(x) = \exp\left[-K \int_0^x h_2^{1-\alpha}(t)dt\right].$$

As an example if X_2 has Rayleigh distribution $\bar{F}_2(x) = \exp[-x^2]$, $x > 0$, then X_1 has Weibull form

$$\bar{F}_1(x) = \exp\left[-\frac{K2^{1-\alpha}}{2-\alpha}x^{2-\alpha}\right], \ x > 0, \ \alpha < 2, K > 0.$$

The relative Renyi residual entropy is $(1-\alpha)^{-1}\frac{K}{\alpha}$, which is the same for all choices of X_1 and X_2 satisfying (4.35).

To place the above results from the perspective of application, we note that in the spirit of Zellener (1971), Renyi's residual entropy of X is a measure of the uncertainty in the residual life distribution about the prediction of $X|X > t$. Thus an increasing $H_{1.2}(t)$ indicates that relative to X_2, X_1 has more increasing uncertainty in its residual life distribution. In other words, X_1 has more increasing uncertainty in residual life (IURL-α) than in X_2, meaning X_1 is less reliable than X_2 in predicting the residual life. Thus the notion of monotonicity of $H_{1.2}(t)$ can provide an indicator as to which of the devices with lifetimes X_1 and X_2 can be chosen.

There are many areas in reliability modelling that call for comparison of life distributions and we will illustrate the use of $H_{1.2}$ in some such cases.

1. Replacement policy

Among various schemes of replacement of items to ensure the continued operation of a system one that makes use of relevation transforms has assumed importance. Here, a component with life distribution F, upon failure is replaced by another of the same age and with the same life distribution. The distribution after $(n-1)$ replacements turns out to be (Baxter (1982))

$$\bar{G}_n(x) = \bar{F}(x)\left[\sum_{k=0}^{n-1}\frac{(-\log\bar{F})^k}{k!}\right], \ n \geq 1, x > 0.$$

For simplicity, the survival function of the first replacement from the beginning of service until failure of the unit is

$$\bar{G}_2(x) = \bar{F}(x)(1 - \log\bar{F}(x)). \tag{4.36}$$

Choosing X_2 as the lifetime if the unit is not replaced and X_1 the lifetime after the unit is replaced, the relative entropy of $\bar{G}$ relative to $\bar{F}$ is calculated when $\bar{F}_2(x) = e^{-\lambda x}$. Notice that

$$\bar{F}_1(x) = e^{-\lambda x}(1 + \lambda x)$$

with hazard rate $h_1(x) = \frac{\lambda^2 x}{1+\lambda x}$ so that

$$H_{1.2}(t) = (1-\alpha)^{-1} \log \int_t^\infty \frac{\lambda^\alpha \bar{e}^{\lambda \alpha x}}{\bar{e}^{\alpha \lambda t}} \frac{\lambda^2 x}{1+\lambda x} dx$$

$$= (1-\alpha)^{-1} \left[\alpha \lambda t + \log \lambda^\alpha + \log \left(\frac{\bar{e}^{\alpha \lambda t}}{\alpha} - e^\alpha \int_{1+\lambda t}^\infty \frac{\bar{e}^{\alpha y}}{y} dy \right) \right]$$

$$= (1-\alpha)^{-1} \left[\log \lambda^\alpha \left\{ 1 - e^{\alpha(\lambda t - 1)} E_1 \left(\frac{1+\lambda \alpha}{t} \right) \right\} \right].$$

2. Dynamic proportional hazards model

Nanda and Das (2011) have considered two non-negative absolutely continuous random variables X_1 and X_2 such that their hazard rates satisfy $h_2(x) = \theta(x) h_1(x)$ for a non-negative increasing $\theta(x)$ defined on $[0, 1]$. In this case to compare the distributions of X_1 and X_2, we use

$$H'_{1.2}(t) = (1-\alpha)^{-1} \left[\alpha \theta(t) - h_2^\alpha(t) \bar{e}^{(1-\alpha)H_{1.2}} \right] h_1(t)$$

using (4.33) and conclude that $H_{1.2}(t)$ is increasing in t if

$$0 < \alpha < 1 \text{ and } H_{1.2}(t) > \frac{1}{1-\alpha} \log \frac{h_2^\alpha(t)}{\alpha \theta(t)} \qquad (4.37)$$

and for

$$\alpha > 1 \text{ and } H_{1.2}(t) > \frac{1}{1-\alpha} \log \frac{h_2^\alpha(t)}{\alpha \theta(t)}. \qquad (4.38)$$

In the replacement policy mentioned above, in the case of a general survival function $\bar{F}_1(x)$, the hazard rate of X_2 is derived from (4.37)

$$h_2(x) = -\frac{f_1(x) \log \bar{F}_1(x)}{\bar{F}_1(x)(1 - \log \bar{F}_1(x))} = h_1(x) \frac{H_1(x)}{1 + H_1(x)}$$

where $H_1(x) = -\log \bar{F}_1(x)$, is the cumulative hazard rate of X_1. It is easy to recognize that it is a case of dynamic proportional hazard rate. We see that
$H_{1.2}(t)$ is increasing if $0 < \alpha < 1$ $(\alpha > 1)$ and

$$\frac{H_1(t)}{1 + H_1(t)} > (<) h_2^\alpha(t) e^{(1-\alpha)H_{1.2}(t)}$$

and $H_{1.2}(t)$ is decreasing if $0 < \alpha < 1$ $(\alpha > 1)$ and

$$\frac{H_1(t)}{1 + H_1(t)} < (>) h_2^\alpha(t) e^{(1-\alpha)H_{1.2}(t)}.$$

Thus we have a real situation in which the dynamic proportional hazard rate holds and also various other properties of the model given becomes valid.

3. Proportional odds family

The proportional odds family of distributions is defined in terms of an additional parameter ϕ (Marshall and Olkin (2007)) as

$$\frac{F(x|\phi)}{\bar{F}(x|\phi)} = \frac{1}{\beta}\frac{F_1(x)}{\bar{F}_1(x)}, \ \beta > 0.$$

The hazard rate of $\bar{F}(x|\phi)$ taken as $\bar{F}_2(x)$ in our notation, gives a hazard rate

$$h_2(x) = \theta(x)h_1(x). \tag{4.39}$$

where $\theta(x) = (F + \beta\bar{F})^{-1}$. Equation (4.39) is still not a dynamic proportional hazards model, since

$$\theta'(x) = \frac{(\beta - 1)f(x)}{[F(x) + \beta\bar{F}(x)]^2}$$

can be increasing or decreasing according as $\beta > 1$ or $\beta < 1$. Thus for $\beta > 1$, the conditions in (4.37) and (4.38) hold.

4. Environment effect

The reliability of a device assessed in a laboratory based on life tests may not be the same when operated in a real world environment. Factors such as temperature, humidity and rate of usage constituting environments may produce a change in the hazard rate. For example in the exponential case, with hazard rate λ, the influence of the operating conditions changes the hazard rate to $\eta\lambda$ with $\eta > 1 (< 1, = 1)$ suggesting a harsher (milder, same) condition than that in a laboratory. For details of such models we refer to Lindley and Singpurwalla (1986). They have assumed η to follow a gamma distribution with scale parameter m and shape parameter p. The survival function under the new environment is the Pareto distribution encountered in the Example and $H_{1,2}(t)$ will be as in (4.31).

It is of interest to compare the concept of relative ageing using reliability functions with relative entropy. The appropriate concept from the reliability context is the relative hazard rate of X_1 with respect to X_2 defined as $\frac{h_1(x)}{h_2(x)}$, see Wei (1992). We have

$$\frac{h_1(t)}{h_2(t)} = [\alpha h_2(t) - (1-\alpha)H'_{12}(t)]h_2^{-(\alpha+1)}\exp[(1-\alpha)H_{1.2}(t)]. \tag{4.40}$$

We say that X_1 is ageing faster than X_2 if $\frac{h_1(t)}{h_2(t)}$ is increasing in t which happens when $h_2(t)$ is decreasing and $H_{1.2}(t)$ is increasing, for any $\alpha > 0$ ($\alpha \neq 1$).

4.3. Tsallis entropy

Another useful generalized entropy was proposed by Constantive Tallis in 1998 as a basis for generalizing the standard statistical mechanics. It is identical in form with the Havrda and Charvát (1967) entropy considered in information theory in the context of cybernetics. In the continuous case the Tallis entropy is defined as

$$T_\alpha = \frac{1}{\alpha - 1} \left[1 - \int_0^\infty f^\alpha(x) dx \right], \ \alpha > 0. \tag{4.41}$$

As $\alpha \to 1$, T_α reduces to the Shannon-entropy. Also the Tsallis and Renyi entropies are related through

$$\exp[(1 - \alpha)H_R] = 1 + (1 - \alpha)H_T$$

where H_R and H_T denote the Renyi and Tsallis entropies.

In Physics entropy is generally used to quantify the missing information on the concrete state of a system. Although the form was known since Havrda and Charvát (1967), Tsallis was for the first time proposed to generalize statistical mechanics using this form. Tsallis entropy being a monotone function of Renyi entropy and so both attain the maximum at the same value, but a missing property of the latter in theoretical discussions is the concavity of T_α. Also when discrete probabilities p_i are considered T_α is more or less stable under small perturbations in p_i's which is not shared by other entropies.

4.3.1. Tsallis residual entropy. Nanda and Das (2006) have discussed the properties of the Tsallis residual entropy

$$T_\alpha(t) = \frac{1}{\alpha - 1} \left[1 - \int_t^\alpha \left(\frac{f(x)}{\bar{F}(t)} \right)^\alpha \right] dx$$

in some detail.

DEFINITION 4.6. *If X and Y are non-negative random variables, then X is less than Y in Tsallis residual entropy, $X \leq_{T_\alpha} Y$ if $T_{X,\alpha}(t) \leq T_{Y,\alpha}(t)$ for all $t > 0$. With reference to this order we have*

(i) *for $Z_1 = a_1 X + b_1$, $Z_2 = a_2 Y + b_2$, $a_1, a_2 > 0$; $b_1, b_2 \geq 0$, $Z_1 \leq_{T_\alpha} Z_2$ provided that $X \leq_{T_\alpha} Y$, $a_1 \leq a_2$, $b_1 \leq b_2$ and either of $T_{X,\alpha}(t)$ or $T_{Y,\alpha}(t)$ is decreasing in t*

(ii) *a point process $\{N(t), t \geq 0\}$ with interarrival times $\{X_n | n = 1, 2, \dots \}$ is increasing (decreasing) in Tsallis residual entropy if $T_{Y_i,\alpha}(t) \leq (\geq) T_{Y_j,\alpha}(t)$ for all $t \geq 0$ and $i \neq j \leq n$ and $Y_k = [X_k | S(k - 1) = k - 1, \dots, S_1 = s_1]$, $k = 1, 2, \dots$ and $X_n = S_n - S_{n-1}$.*

DEFINITION 4.7. *The random variable X is said to have decreasing (increasing) uncertainty in Tsallis residual entropy, DURL-T_α (IURL-T_α) if $T_\alpha(t)$ is decreasing (increasing) in $t > 0$.*

The DURL-T_α is closed under linear transformations $Z = aX + b$, $a \geq 0$, $b > 0$. Also if X is DURL-T_α (IURL-T_α) then

$$h(t) \leq (\geq)[\alpha(1 - (\alpha - 1)T_\alpha(t))]^{\frac{1}{\alpha-1}}.$$

THEOREM 4.14. *The relationship*

$$T_\alpha(t) = (\alpha - 1)^{-1}[1 - Ch^{\alpha-1}(t)],$$

is satisfied for some constant C, $\alpha > 0$ and all $t > 0$ if and only if X has
  (i) *exponential distribution for $C = \frac{1}{\alpha}$,*
 (ii) *Pareto II for $C < \frac{1}{\alpha}$, and*
(iii) *rescaled beta for $C > \frac{1}{\alpha}$.*

Some further analysis of (4.41) will bring to focus the relationship the classes DURL-T_α and IURL-T_α have with the hazard rate of X. We note that

$$[1 - (\alpha - 1)T_\alpha(t)]\bar{F}^\alpha(t) = \int_t^\infty f^\alpha(x)dx$$

giving

$$(\alpha - 1)T'_\alpha(t) = h^\alpha(t) - [1 - (\alpha - 1)T_\alpha(t)]\alpha h(t).$$

THEOREM 4.15. *The lifetime X is DURL-T_α (IURL-T_α) according as*

$$h(t) > (<)[\alpha(1 - (\alpha - 1)T_\alpha(t))]^{\frac{1}{\alpha-1}}, \ \alpha \neq 1$$

and X is exponential if $h(t) = [\alpha(1 - (1 - \alpha)T_\alpha(t))]^{\frac{1}{\alpha-1}}.$

Kumar and Taneja (2011a) have further extended the results of monotone T_α in the following manner.

THEOREM 4.16. *If X is IURL-T_α (DURL-T_α) and $\phi(\cdot)$ is non-negative increasing and convex $\phi(X)$ is DURL-T_α (IURL-T_α).*

Baratpour and Khammar (2015) have studied various properties of the Tsallis entropy of order statistics and recorded values. The Tsallis entropy associated with the ith order statistic is

$$T_\alpha(X_{i:n}) = \frac{1}{\alpha - 1}\left[1 - \int_{-\infty}^\infty f_{i:n}^\alpha(x)dx\right] \qquad (4.42)$$

where $f_{i:n}$ has the expression given in (4.41). For a random sample $(X_1, X_2, \ldots, X_n)$ from a continuous $F(x)$, (4.42) simplifies to

$$T_\alpha(X_{i:n}) = \frac{1}{\alpha - 1}[1 - C_i E(f^{\alpha-1}(F(Z_i)))]$$

where $c_i = B(\alpha(i-1)+1, \alpha(n-i)+1)/B(i, n-i+1)$ and Z_i follows beta distribution $(\alpha(i-1)+1, \alpha(n-i)+1)$. Further if $X = Y + C$, then $T_\alpha(X_{m:n}) = T_\alpha(Y_{m:n})$, $1 \le m \le n$ and for a symmetric $F(x)$, $T_\alpha(X_{m:n}) = T_\alpha(Y_{m:n})$, $1 \le m \le n$ and for a symmetric $F(x)$, $T_\alpha(X_{m:n}) = T_\alpha(X_{n-m+1};n)$.

4.3.2. Tsallis past entropy. The Tsallis entropy of past lifetime does not appear to have received much attention in literature. It is defined by

$$\bar{T}_\alpha(t) = \frac{1}{\alpha - 1}\left[1 - \int_0^t \left(\frac{f(x)}{F(t)}\right)^\alpha dx\right]. \qquad (4.43)$$

Differentiating (4.43) and simplifying

$$(\alpha - 1)\bar{T}_\alpha'(t) = \lambda^\alpha(t) - \alpha(1 - (\alpha - 1)\bar{T}_\alpha(t))\lambda(t), \qquad (4.44)$$

giving the relationship between $\bar{T}_\alpha$ and the reversed hazard rate. This relation provides some elegant identities when specialized to certain life distributions.

THEOREM 4.17. *The random variable X has power distribution $F(x) = x^\beta$, $0 \le x \le 1$, $\beta > 0$ if and only if*

$$\bar{T}_\alpha(t) = \frac{1}{\alpha - 1}[1 - c\lambda^{\alpha-1}(t)] \qquad (4.45)$$

for all $t > 0$.

PROOF. First we show that (4.45) holds if X has a power distribution. This follows from the fact that for the given power distribution

$$\bar{T}_\alpha(t) = \frac{1}{\alpha - 1}\left[1 - \frac{\beta^\alpha}{1 + \alpha(\beta - 1)}t^{1-\alpha}\right]$$

and $\lambda(t) = \frac{\beta}{t}$. Note that $C = \frac{\beta}{1+\alpha(\beta-1)}$. Conversely assuming (4.45) and making use of (4.44) lead to

$$\frac{\lambda'(t)}{\lambda^2(t)} = -K, \quad K = \frac{c(\alpha - 1)}{1 - \alpha C} > 0.$$

Solving $\lambda(t) = \frac{K}{t}$, the reversed hazard rate of the assumed model. $\square$

THEOREM 4.18. *The identity*

$$\bar{T}_\alpha(t) = \frac{1}{\alpha - 1}[1 - K\mu^{1-\alpha}(t)]$$

is satisfied for all t and some positive constant K if and only if X has a power distribution $F(x) = x^\beta$, $0 \le x \le 1$, $\beta > 0$.

The proof of this result is exactly along the lines of Theorem 4.9 and is therefore not given. Theorem 4.10 also has a parallel result in the case of Tsallis entropy.

Another aspect of Tsallis entropy is that it enhances the flexibility of Shannon's entropy by giving various values α. In the context of reliability the change in the extent of uncertainty, in the past or residual life have a discrete impact on the predictable lifetime of the device, when we utilize the two entropies. This is explained in the next theorem.

THEOREM 4.19. *The relationship between Tsallis entropy and Shannon entropy for the past lifetimes is given by*

$$\bar{T}_\alpha(t) \geq \frac{e^{(1-\alpha)\bar{H}(t)} - 1}{1 - \alpha}.$$

PROOF. From Theorem 4.11, the truncated geometric mean $G_\alpha(t)$ is given by

$$G_\alpha(t) = F^{\alpha-1}(t) \exp[(1 - \alpha)\bar{H}(t)]. \qquad (4.46)$$

On the other hand

$$1 - (\alpha - 1)\bar{T}_\alpha(t) = \frac{1}{F^\alpha(t)} \int_0^t f^\alpha(x)dx$$

or

$$\frac{1}{F^\alpha(t)} \int_0^t f^{\alpha-1}(x)f(x)dx = \frac{1}{F^{\alpha-1}(t)}E(f^{\alpha-1}(X)|X \leq t) = \frac{A_\alpha(t)}{F^{\alpha-1}(t)}$$

where $A_\alpha(t)$ is the arithmetic mean of $[f^{\alpha-1}(X)|X \leq t]$. Thus

$$A_\alpha(t) = F^{\alpha-1}(t)[1 - (\alpha - 1)\bar{T}_\alpha(t)]. \qquad (4.47)$$

Since $A_\alpha(t) \geq G_\alpha(t)$ we have the required result. Obviously when $\alpha < 1$, $\bar{T}_\alpha(t) \geq \bar{H}(t)$ and the reverse inequality holds when $\alpha > 1$. □

The discussions on classes based on $\bar{T}_\alpha(t)$ and order statistics being very similar to those of $T_\alpha(t)$, we do not proceed further on these topics.

q-distributions.

The q-distributions or Tsallis distributions as they are sometimes called, are derived by maximizing the Tsallis entropy subject to certain constrains. A continuous real parameter q adds to the flexibility of these distributions and creates models with heavy tails. Defining q-logarithms

$$\log_q(x) = \begin{cases} \log x, & x > 0,\ q = 1 \\ \frac{x^{1-q}-1}{1-q}, & x > 0,\ q \neq 1 \\ \text{undefined}, & x \leq 0 \end{cases}$$

gives on inversion, the q-exponential function

$$e_q(x) = \begin{cases} [1 + (1-q)x]^{\frac{1}{1-q}}, & q \neq 1, \ 1 + (1-q)x > 0 \\ e^x, & q = 1 \end{cases}.$$

Originally proposed by Box and Cox (1964) known as the reverse Box-Cox transformation, the q-exponential distribution has probability density function

$$f_q(x) = (2-q)\lambda[1 + (q-1)\lambda x]^{\frac{1}{1-q}};$$

$$x > 0, \ \lambda > 0, \ 1 < q < 2, \ 0 < x < \frac{1}{\lambda(1-q)}, q < 1.$$
$$(4.48)$$

It has survival function

$$\bar{F}(x) = [1 + \lambda(1-q)x]^{\frac{1}{1-q}+1}.$$
$$(4.49)$$

In the context of entropy (4.48) arises the solution for the maximum Tsallis entropy

$$H_q = \frac{1}{1-q}\left(\int_0^\infty f^q(x)dx - 1\right)$$

subject to the conditions $\int_0^\infty xf^q(x)dx = \theta$ and $\int_0^\infty f(x)dx = 1$ (Bercher and Vignat (2008)). The same solution is reached if Tsallis entropy is replaced by Renyi entropy. In the limiting case $q \to 1$, Tsallis entropy becomes Shannon entropy and the well known solution results as the exponential distribution. From the point of view of the distribution theory, the q-exponential law can also be derived as a mixture distribution

$$f_q(x) = \int_0^\infty f_x(x|\lambda)f_n(\lambda)d\lambda$$

where $f_x(x|\lambda) = \lambda e^{-\lambda x}$, the usual exponential and $f_n(\lambda) = \frac{e^{-\lambda}\lambda^{\frac{n}{2}-1}}{\Gamma(\frac{n}{2})}$ is a gamma distribution. It is interesting to note that the generalized Pareto

$$f\langle x; \sigma, \alpha\rangle = \frac{1}{\alpha}\left(1 + \frac{\alpha x}{\sigma}\right)^{-\frac{1}{\alpha}-1}$$

coincides with the q-exponential when $\alpha = -\frac{1-q}{2-q}$, $\sigma = \frac{1}{\lambda(2-q)}$. The q-exponential characteristics are

$$\text{mean } \mu = \frac{1}{\lambda(3-2q)}, \quad q < \frac{3}{2}$$

mode zero,

$$\text{variance} = \frac{q-2}{(2q-3)^2(3q-4)\lambda^2}, \quad q < \frac{3}{4}.$$

The reliability aspects of the distribution are the same as those of the generalized Pareto model with the corresponding values of q.

Assis et al. (2013) have studied the q-Weibull distribution with density function

$$f_q(t) = (2-q)\frac{\beta}{\eta}\left(\frac{t}{\eta}\right)^{\beta-1} e_q\left(-\left(\frac{t}{\eta}\right)^\beta\right), \ t \geq 0$$

$$= (2-q)\frac{\beta}{\eta}\left(\frac{t}{\eta}\right)^{\beta-1} [1-(1-q)\left(\frac{t}{\eta}\right)^\beta]^{\frac{1}{1-q}}, \ t > 0, 1 \leq q < 2 \cdot$$

$$\text{and } t \in [0, \frac{\eta}{(1-q)^{\frac{1}{\beta}}}], \ q < 1$$

$$(4.50)$$

The survival function is

$$R_q(t) = \left[1-(1-q)\left(\frac{t}{\eta}\right)^\beta\right]^{\frac{2-q}{1-q}}.$$

Special cases of (4.50) are the Weibull distribution as $q \to 1$, exponential as $q \to 1$, $\beta = 1$, q-exponential as $\beta \to 1$. It can be considered as an extension of the usual Weibull law in the range $t \geq 0$ and also in the finite range $[0, \frac{\eta}{(1-q)} y_\beta]$. The hazard rate is

$$h_q(t) = \frac{(2-q)\frac{\beta}{\eta\beta}t^{\beta-1}}{1-(1-q)\left(\frac{t}{\eta}\right)^\beta}$$

which is more flexible than the usual Weibull hazard rate, offering DFR in $1 < q < 2, 0 < \beta < 1$; IFR in $q < 1, \beta > 1$; unimodal in $1 < q < 2, \beta > 1$ and bathtub shaped in $q < 1, 0 < \beta < 1$. For more details we refer to Assis et al. (2013).

4.3.3. Quantile form of Tsallis entropy.
By virtue of the transformation $x = Q(u)$ in (4.43), the Tsallis entropy in the quantile form can be written as

$$T_{Q,\alpha} = \frac{1}{1-\alpha}\left[\int_0^1 (q(p))^{1-\alpha}dp - 1\right], \ \alpha > 1, \ \alpha \neq 1,$$

and its residual life version as

$$T_Q(u) = \frac{1}{1-\alpha}\int_u^1 \frac{q^{1-\alpha}(p)}{(1-u)^\alpha}dp - 1.$$

Kumar and Rani (2017) have studied the properties of the quantile-based Tsallis entropy in residual and inactivity time. Their main results are

(i) the distribution of X is uniquely determined by $H_Q(u)$ through the inversion formula

$$q(u) = (1-u)^{-1}[\alpha + (1-\alpha)H_\alpha(u) - (1-\alpha)(1-u)H'_\alpha(u)]^{\frac{1}{1-\alpha}}$$

(ii) if $X_1, \ldots, X_n$ are independent and identically distributed random variables with exponential distribution function, the Tallis quantile entropy of $Z = \min(X_1, X_2, \ldots, X_n)$ is independent of α.

(iii) $T_Q(u) \geq (\leq) \frac{h_Q^{\alpha-1}(u) - \alpha}{\alpha(1-\alpha)}$.

The quantile version of the past lifetime is

$$\bar{T}_Q(u) = \frac{1}{1-\alpha} \left(\int_0^u \frac{q^{1-\alpha}(p)dp}{u^\alpha} - 1 \right).$$

When $T_Q(u)$ is increasing (decreasing) in u,

$$\bar{T}_Q(u) \geq (\leq) \frac{\lambda^{\alpha-1}(u) - \alpha}{\alpha(1-\alpha)}.$$

A characterization of the power distribution can be offered in terms of the identity

$$\bar{T}_Q(u) = \frac{1}{1-\alpha}[c\lambda^{\alpha-1}(u) - 1], \ c > 0.$$

Khammar and Jahanshahi (2018b) point out that

$$T_Q(u) = (\alpha - 1)^{-1} \left[1 - \frac{\int_u^1 (1-p)}{(1-u)^\alpha} h_Q^{\alpha-1}(p)dp \right]$$

and

$$T_Q(u) \leq (\geq)(\alpha - 1)^{-1} \left[1 - \frac{(1-u)^\alpha e^{(1-\alpha)}}{(-\log u)^\alpha} H_{\alpha,Q(u)} \right]$$

when $\alpha > 1$ $(0 < \alpha < 1)$. They define classes of life distributions based on increasing (decreasing) $T_Q(u)$. However these are identical with the classes defined in Section 4.2.1. We say that X is smaller than Y in residual quantile Tallis entropy order, $X \leq_{\text{RQTE}} Y$ if $T_{Q_X}(u) \leq T_{Q_Y}(u)$ for all u in $(0,1)$. With reference to this order, the following properties hold

(i) $X \leq_{\text{HQ}} Y \Rightarrow Z_Q^X \leq (\geq)_{\text{RQTE}} Y$ according as $\alpha > 1$, $(0 < \alpha < 1)$

(ii) If $\phi(\cdot)$ is non-negative increasing convex function such that $g(0) = 0$, then $X \leq_{\text{RQTE}} Y \Rightarrow \phi(X) \leq_{\text{RQTE}} \phi(Y)$

(iii) $X \leq (\geq)_{\text{RQTE}} Y \Rightarrow X_{n:n} \leq (\geq)_{\text{RQTE}} Y_{n:n}$ for $\alpha > 1$ $(0 < \alpha < 1)$
$X_{1:n} \leq (\geq)_{\text{RQTE}} Y_{1:n} \Rightarrow X \leq (\geq)_{\text{RQTE}} Y$ for $\alpha > 1$ $(0 < \alpha < 1)$.

Khammar and Jahanshahi (2018b) further show that if X is non-negative,

$$(\alpha - 1)T_Q(u) = 1 - A(1-u)^B, \ A > 0, \ \alpha \neq 1$$

holds for all in $(0,1)$ if and only X is uniform if $B = 1 - \alpha$, exponential if $B = 0$, generalized Pareto if $B = \frac{\alpha(\alpha-3)-2}{\alpha+1}$ and rescaled beta if $B = \frac{(1-\alpha)(1-2h)}{b}$.

The Tsallis entropy of the ith order statistics $X_{i:n}$ is given by (Kumar and Rekha (2018)).

$$T_{\alpha, X_{i:n}} = \frac{1}{(1-\alpha)} \left[\int_0^1 g_i^\alpha(u) q^{1-\alpha}(u) du - 1 \right]$$

where $g_i(u)$ is the density function of the beta distribution $(i, n-1+1)$. It is seen that

$$T_{\alpha, X_{i:n}} = -T_\alpha(g_i)$$

and in the case of the residual life

$$T_{\alpha, X_{i:n}}(u) = \frac{1}{(1-\alpha)(\bar{\beta}_u(i, n-1+1))} \int_u^1 p^{\alpha(i-1)}(1-p)^{\alpha(n-i)} q^{1-\alpha}(p) dp - 1$$

in which

$$\bar{\beta}_x(a, b) = \int_x^1 t^{a-1}(1-t)^{b-1} dt.$$

Further

$$T_{\alpha, X_{1:n}}(u) = (1-\alpha)^{-1} \{ Ch_{Q, X_{1:n}}^{\alpha-1}(u) - 1 \}, \ \alpha > 1$$

if and only if X has exponential distribution $(C = \frac{1}{\alpha})$ or Pareto II distribution $(C < \frac{1}{\alpha})$ on rescaled beta $(C > \frac{1}{\alpha})$. It is to be noted that there is no general closed form expression available for the entropy of the order statistics.

4.4. Varma entropy

The Varma entropy (Varma (1966)) is an extension of the Renyi entropy arrived at by the introduction of an additional parameter in the former and is expressed as

$$V_{\alpha,\beta} = \frac{1}{\beta - \alpha} \log \int f^{\alpha+\beta-1}(x) dx, \ \beta - 1 < \alpha < \beta, \ \beta \geq 1. \quad (4.51)$$

Notice the structured similarity between (4.5) and (4.51), so that when $\beta \to 1$, we have the Renyi entropy and when α also tends to unity the Shannon entropy results. Accordingly most of the results of Varma entropy and their proofs by and large resemble those of Renyi entropy. The Varma entropy of residual life is

$$V_{\alpha,\beta}(t) = \frac{1}{\beta - \alpha} \log \frac{\int_t^\infty f^{\alpha+\beta-1}(x) dx}{\bar{F}^{\alpha+\beta-1}(t)} \quad (4.52)$$

or

$$(\beta - \alpha)V_{\alpha,\beta}(t) = \log \int_t^\infty f^{\alpha+\beta-1} - (\alpha + \beta - 1)\log \bar{F}(t).$$

Baig and Dar (2008) have shown that the relationship

$$V_{\alpha,\beta}(t) = (\beta - \alpha)^{-1}[\log K - (2 - \alpha - \beta)\log h(t)]$$

is satisfied for all $t > 0$ if and only if X is exponential ($K = (\alpha + \beta - 1)^{-1}$) or Pareto II ($K < (\alpha + \beta - 1)^{-1}$) or rescaled beta ($K > (\alpha + \beta - 1)^{-1}$). They have classified the life distribution as increasing (decreasing) Varma entropy of residual life abbreviated as IVERL (DVERL) whenever $V_{\alpha,\beta}(t)$ is increasing (decreasing) in t. Differentiating (4.52) we have

$$(\alpha + \beta - 1)h(t) - h^{\alpha+\beta-1}(t)e^{-(\alpha-\beta)V(t)} = (\beta - \alpha)V'(t).$$

It follows that X is IVERL (VERL) if

$$h(t) \leq (\geq)(\alpha + \beta - 1)^{\frac{1}{\alpha+\beta-2}} \exp\left[-\frac{\alpha - \beta}{\alpha + \beta - 2}V(t)\right] \qquad (4.53)$$

whenever $\alpha + \beta > 2$ and IVERL (DVERL) if the expression

$$h(t) \geq (\leq)(\alpha + \beta - 1)^{\frac{1}{\alpha+\beta-2}} \exp\left[-\frac{\alpha - \beta}{\alpha + \beta - 2}V(t)\right]$$

holds for $\alpha + \beta < 2$. Thapliyal and Taneja (2012) have worked out the entropy of the order statistic $X_{i:n}$ as

$$V_{X_{i:n}} = V(Z_i) - \frac{1}{\alpha - \beta}\log E\left[f^{\alpha+\beta-2}(F^{-1}(Y_i))\right]$$

where $V(Z_i)$ is the Varma entropy of a beta distribution $(i, n - i + 1)$ and $Y_i - g_i$ is beta $((\alpha + \beta - 1)(i - 1) + 1, (\alpha + \beta - 1)(n - i) + 1)$, and for the residual life

$$V_{X_{i:n}}(t) = V(Z_i, F(t)) + \frac{1}{\beta - \alpha}\log E[f^{\alpha+\beta-2}(F^{-1}(W_i))]$$

where $W_i = \bar{B}_{F(t)}((\alpha + \beta - 1)(i - 1) + 1, (\alpha + \beta - 1)(n - i + 1))$ and $\bar{B}_t(p, q) = \int_t^1 x^{p-1}(1 - x)^{q-1}dx$. Kayal (2014) has some further results on the entropy of order statistics.

There have been some discussions in the quantile form of the Varma entropy which is obtained under the transformation $x = Q(u)$ in (4.51) and (4.52). We thus have

$$V_{\alpha,\beta} = \frac{1}{\beta - \alpha}\int_0^1 q^{2-\alpha-\beta}(p)dp \qquad (4.54)$$

and for the residual life

$$V_Q(u) = \frac{\int_u^1 q^{2-\alpha-\beta}(p)dp}{(\beta - \alpha)(1 - u)^{\alpha+\beta-1}}. \tag{4.55}$$

From $V_Q(u)$ by differentiation it is found that

$$q(u) = \frac{\exp[\frac{\beta-\alpha}{2-\beta-\alpha}]}{(1 - u)}[(\alpha + \beta - 1) + (\alpha - \beta)(1 - u)V'(u)]^{\frac{1}{2-\alpha-\beta}}$$

which shows that the Varma residual entropy determines the life distribution uniquely. Besides this, Kumar and Rani (2018) have the following results.

(a) An upper bound to the entropy is

$$V_Q(u) \le \frac{\alpha + \beta - 2}{\beta - \alpha} \log h_Q(u) - \frac{\log \alpha + \beta - 1}{\beta - 1}.$$

(b) If $Y = \phi(\cdot)$ is strictly increasing continuous and differentiable,

$$V_Y(u) = \frac{(\beta - \alpha)^{-1}}{(1 - u)^{\alpha+\beta-1}} \log \int_u^1 (q(p)\phi'(Q(p)))^{2-\alpha-\beta}dp$$

and in particular when $Y = aX + b, \ a, b > 0$

$$V_Y(u) = V_X(u) + \frac{2 - \alpha - \beta}{\beta - \alpha} \log a$$

(c) If $Z = \psi(\cdot)$ is non-negative increasing and convex (concave) for $0 < \alpha+\beta < 2$, $V_Z(u)$ is increasing (decreasing) whenever $V_X(u)$ is increasing (decreasing) and for $\alpha + \beta > 2$, $V_Z(u)$ is decreasing (increasing) whenever $V_X(u)$ is increasing (decreasing)

The Varma entropy for past lifetime is

$$\bar{V}_Q(u) = (\beta - \alpha)^{-1} \log \left(\frac{\int_0^u q^{2-\alpha-\beta}(p)}{u^{\alpha+\beta-1}} dp \right) \tag{4.56}$$

which determine the distribution of X through

$$q(u) = \exp\left[\frac{\beta - \alpha}{2 - \beta - \alpha}\bar{V}(u)\right] u^{-1}\{(\alpha + \beta - 1) + (\beta - \alpha)u\bar{V}(u)\}.$$

The power distribution is characterized by the property

$$\bar{V}_Q(u) = \{\log C + (\alpha + \beta - 2) \log \lambda_Q(u)\}$$

and the entropy satisfies the bound

$$\bar{V}(u) \le \frac{\alpha + \beta - 2}{\beta - \alpha} \log \lambda_Q(u) - \log \frac{\alpha + \beta - 1}{\beta - 1}.$$

The interval Varma entropy is considered in Kumar and Singh (2018) and the weighted version is discussed in Kayal (2015). For a given weight function $w(\cdot)$, if $E[w(X)|X > t]$ or $w(t)$ is decreasing (increasing) and X or X_w is DFR then

$$V_X(t) \leq (\geq)V_{X_w}(t) \text{ for } \alpha + \beta > (<)2$$

and

$$V_X(t) \geq (\leq)V_{X_w}(t) \text{ for } \alpha + \beta > (< 2).$$

As a Corollary in the case of the length-biased random variable X_L of X,

$$V_X(t) \geq (\leq)V_{X_L}(t) \text{ whenever } \alpha + \beta > (<)2$$

and for the equilibrium variable X_E,

$$V_X(t) \geq (\leq)V_{X_E}(t) \text{ whenever } \alpha + \beta > (<)2.$$

4.5. Other entropies

There have been some limited attempts to study other entropies in the context of residual life. Among these is the form of Mathai and Haubold (2006).

$$M_\alpha = \frac{1}{\alpha - 1} \int_{-\infty}^{\infty} [f^{2-\alpha}(x) - 1],$$

whose residual life version is

$$M_\alpha(t) = \frac{1}{\alpha - 1} \int_t^{\infty} \left[\frac{f^{2-\alpha}(x)dx}{\bar{F}^{2-\alpha}(t)} dx - 1 \right], \quad \alpha \neq 1. \tag{4.57}$$

Dar and Al-Zahrani (2013) state that if $M_\alpha(t)$ is decreasing in t, the corresponding distribution is uniquely determined. This restriction on $M_\alpha(t)$ is not required to characterize the distribution of X. To see this the quantile formulation of (4.57) is

$$M_Q(u) = (\alpha - 1)^{-1} \int_u^1 \frac{q^{\alpha-1}}{(1-u)^{2-\alpha}} du$$

which on differentiation leads to

$$q(u) = [(\alpha - 1)\{(2 - \alpha)(1 - u)^{1-\alpha}M(u) - (1 - u)^{2-\alpha}M'(u)\}]^{\frac{1}{\alpha-1}}.$$

The last equation shows that given $M(u)$, $q(u)$ can be written in terms of $M(u)$. Dar and Al-Zahrani (2013) have characterized exponential distribution by the constancy of $M_\alpha(t)$ and further they show that

$$M'_\alpha(t) = Ch^{2-\alpha}(t)$$

if and only if X is exponential ($C = 0$) or Pareto II ($c > 0$) or rescaled beta ($c < 0$). In terms of the mean residual life function $r(t)$,

$$M_\alpha(t) + r(t) = M_\alpha + r(0),$$

only if X is exponential. They have also considered classes of life distributions using the monotonicity of $M_\alpha(t)$ as in the case of other entropies. In a similar manner Nanda and Das (2006) have reported a study of the residual form of R-norm entropy

$$B_R = \frac{R}{R-1} \left[1 - \left(\int_0^\infty f^R(x) dx \right)^{1/R} \right], \ R > 0, R \neq 1$$

of Boekee and van der Lubbe (1980) viz.

$$B_R(t) = \frac{R}{R-1} \left[1 - \left(\int_t^\infty \frac{f^R(x) d}{\bar{F}^R(t)} \right)^{1/R} \right]$$

It can be seen that only a few of the generalized entropies given in Table 4.1 have been considered for modelling reliability data. Discussions relating to others capable of generating more flexible models of uncertainty in residual and past lifetimes remain open problems. It may also be observed that some misgivings about these measures have been voiced as to their role in ascertaining uncertainty. Some of these entropies, e.g., Renyi, are scale dependent when applied to continuous distributions which makes their numerical values less meaning full and therefore except in comparative or differential processes care should be taken to use their absolute values for interpretation. More over many of the generalized forms have been derived mathematically by making changes in the original axioms of Shannon with the newly provided axioms not fully reflecting the basic understanding one should generally associate with the notion of uncertainty.

CHAPTER 5

Divergence Measures

5.1. Introduction

The notions of divergence and distance are used interchangeably in many problems of discrimination of probability distributions, although they are slightly different. If x and y are elements of a space empowered with a function $d(x, y)$ satisfying (i) $d(x, y) \geq 0$ with equality sign holding if and only if $x = y$ (ii) $d(x, y) = d(y, x)$ and (iii) $d(x, y) + d(y, z) \geq d(x, z)$ for all x, y, z belonging to the space is called a distance. On the other hand if S is a function defined over all probability distributions with the same support and D is measured such that $D(P, Q) \geq 0$ with equality sign holding if and only if $P = Q$ for all $P, Q \in S$, then D is a divergence measure. It appears that the concept of distance between two distributions was introduced by Mahalanobis (1936) in connection with inference problems involving multivariate distributions. A little later Bhattacharya (1943), Bhattacharya (1946) discussed a measure of a divergence of two statistical populations through their distributions and certain analogues to the amount of information in estimation theory. The present day divergence measures in information theory and statistics surfaced in literature with the seminal work of Kullback and Leibler (1951) which was essentially viewed as a means of discriminatiing two distributions. Since then, a large number of distance as well as divergence measures have been proposed, some of them arising out of the practical exigencies and others as mere mathematical extensions. A list of some of these measures can be found in Table 5.1. In the present chapter we discuss a few important divergence measures in so far as they are concerned with reliability modelling. It can be noted from the deliberations in this chapter that the most of the divergence functions in Table 4.1 have not yet appeared in the context of lifetimes.

5.2. Kullback-Leibler divergence

Let P and Q be two probability measures defined on a sample space $\mathcal{X}$ and Q absolutely continuous with respect to P. Then the Kullback-Leibler divergence from Q to P is defined as

$$D_{F,G} = - \int_{\mathcal{X}} \log\left(\frac{dQ}{dP}\right) dP.$$

When P and Q have densities f and g with respect to Lebesgue measure we get after suppressing the suffixes F and G,

$$D = \int f(x) \log\left(\frac{f(x)}{g(x)}\right) dx \qquad (5.1)$$

and the integral is taken over the support of the random variable X representing $f(x)$. The measure D in (5.1) is variously called, directed divergence, cross entropy, relative entropy or discrimination measure. One may write (5.1) as

$$D = E\left[\log\left(\frac{f(X)}{g(X)}\right)\right]$$

and as such it is a geometric mean. Being the expected value of a likelihood ratio in favour of a true model $f(x)$ of X against another possible distribution specified by $g(x)$, D is sometimes interpreted as the evidence data for $f(x)$ against $g(x)$. In the Bayesian context, it is the gain in information when the belief on the basis of the prior $g(x)$ is revised to the posterior $f(x)$ or equivalently, the amount lost when $g(x)$ is used to approximate $f(x)$. From the point of view of discrimination between f and g, we obtain the mean information per observation x from f for the discrimination of f in favour of g. The divergence D can also be related to the probability of error in the test of a hypothesis.

Table 5.1. Discrimination measures.

Title	function
Cramer-von Mises	$\int_{-\infty}^{\infty} (F - G)^2 dF$
Renyi divergence of order 2	$\log \int_{-\infty}^{\infty} f^2(x)(g(x))^{-1} dx$
Hellinger (1999)	$\frac{1}{2} \int_{-\infty}^{\infty} (\sqrt{f(x)} - \sqrt{g(x)})^2 dx$
Bhattacharya	$-\log \int \sqrt{f(x)g(x)} dx$
Renyi information divergence of order α	$\frac{1}{\alpha-1} \log \int f_t^\alpha(x) g^{1-\alpha}(x) d\mu(x)$
Tsallis divergence	$\frac{1}{\alpha-1} \int [f^\alpha(x) g^{1-\alpha}(x) - 1] d\mu(x)$
Kullback and Leibler (1951)	$\int \log(\frac{f(x)}{g(x)}) f(x) dx$
Jeffreys (Jeffreys (1946))	$\int f(x) \log \frac{f(x)}{g(x)} d\mu + \int g(x) \log \frac{g(x)}{f(x)} d\mu$
Csiszar ϕ divergence (Csiszar (1967))	$\int g(x)\phi(\frac{f(x)}{g(x)}) d\mu,\ \phi$ convex

Continued on next page

Title	function
Chi-square	$\int \frac{(f(x)-g(x))^2}{g(x)} dx$
Relative J divergence (Dragomir et al. (2001))	$\int (f(x) - g(x)) \log \frac{f(x)+g(x)}{2g(x)} dx$
Jain and Chhabra (2014)	$\int \frac{(f^2(x)-g^2(x))^{2m}}{(f(x)g(x))^{\frac{2m-1}{2}}} dx$
Jain and Sarawath (2012)	$\int \frac{(f(x)-g(x))^{2m}}{(f(x)+g(x))^{2m-1}} \exp\left[\frac{(f(x)-g(x))^2}{(f(x)+g(x))^2}\right] dx$
Kumar and Hunter (2004)	$\int \left[\frac{(f(x)-g(x))}{f(x)+g(x)} \frac{\log f(x)g(x)}{2\sqrt{f(x)g(x)}}\right] dx$
Kumar and Johnson (2005)	$\int \frac{(f^2(x)-g^2(x))^2}{2(f(x)g(x))^{3/2}} dx$
Osterreicher et al. (1999)	$\int \left[\frac{(f(x)-g(x))^{2m}}{(f(x)+g(x))^{2m-1}} dx\right], m=1,2,3,\ldots$
Burbea and Rao (1982)	$\frac{1}{2}\int \left[f(x) \log \frac{2f(x)}{f(x)+g(x)} + g(x) \log \frac{2g(x)}{f(x)+g(x)}\right] dx$
power divergence	$\frac{1}{a(a+1)}\int_0^\infty g(x)\left[\frac{f^{a+1}(x)}{g^{a+1}(x)} - \frac{f(x)}{g(x)} - a(\frac{f(x)}{g(x)} - 1)\right] dx, a>0$
Matusita (1967)	$\int_0^\infty g(x)\left[1 - (\frac{f(x)}{g(x)})^{\frac{1}{2}}\right] dx$
Kumar and Hunter (2004)	$\int_0^\infty \left[1 - (1+\frac{1}{a})\frac{f(x)}{g(x)} + \frac{1}{a}(\frac{f(x)}{g(x)})^{1+a}\right] g(x) dx, a>0$
Mattheou et al. (2009)	$\int \left[(\frac{f(x)}{g(x)})^{1+a} - (1+\frac{1}{a})(\frac{f(x)}{g(x)})^a + \frac{1}{a}\right] g(x) dx$
root mean square	$\int \frac{f^2(x)+g^2(x)}{2} dx$
harmonic mean	$\int \frac{2f(x)g(x)}{f(x)+g(x)} dx$
arithmetic mean	$\int \frac{f(x)+g(x)}{2} dx$
square root mean	$\int (\frac{\sqrt{f(x)}+\sqrt{g(x)}}{2})^2 dx$
logarithmic mean	$\int \frac{f(x)-g(x)}{\log f(x)-\log g(x)} dx, f \neq g$
geometric mean	$\int \sqrt{f(x)g(x)} dx$
square root arithmetic mean	$\int \sqrt{\frac{f^2(x)+g^2(x)}{2}} - 1$
square root geometric mean	$\int \sqrt{\frac{f^2(x)+g^2(x)}{2}} - \sqrt{f(x)g(x)}$
square root harmonic mean	$\int \sqrt{\frac{f^2(x)+g^2(x)}{2}} - \frac{2f(x)g(x)}{f(x)+g(x)}$
Chernoff	$\frac{4}{1-\alpha^2}(1 - \int \frac{1-\alpha}{f^2(x)} \frac{1+\alpha}{g^2(x)} dx)$
exponential	$\int (\log g(x) - \log f(x))^2 f(x) dx$
Kagan	$\frac{1}{2}\int \frac{(f(x)-g(x))^2}{f(x)} dx$
α, β product divergence	$\frac{2}{(1-\alpha)(1-\beta)} \int \left[(1-(\frac{f(x)}{g(x)})^{\frac{1-\alpha}{2}}) (1-(\frac{g(x)}{f(x)})^{\frac{1-\beta}{2}})\right] f(x) dx$
Jain and Chhabra (2014)	$\int \frac{(f^2(x)-g^2(x))^{2m}}{(f(x)g(x))^{\frac{2m-1}{2}} g^{2m}(x)} \exp\left[\frac{(f^2(x)-g^2(x))^2}{f(x)g^3(x)}\right], m=1,2,3$
relative arithmetic geometric divergence (Taneja (2011a))	$\int (\frac{f(x)-g(x)}{2}) \log(\frac{f(x)+g(x)}{2}) dx$
arithmetric-geometric mean divergence (Taneja (2011a))	$\int (\frac{f(x)+g(x)}{2}) \log(\frac{f(x)+g(x)}{2}) dx$
d-divergence (Taneja (2001))	$1 - \int (\frac{\sqrt{f(x)}+\sqrt{g(x)}}{2}) \sqrt{\frac{g(x)-f(x)}{2}} dx$
relative Jensen-Shannon divergence (Sibson (1969))	$\int f(x) \log(\frac{2f(x)}{f(x)+g(x)}) dx$
triangular	$\int \frac{(f(x)-g(x))^2}{f(x)+g(x)} dx$ $\frac{1}{S(S-1)}\left[\int f^s(x)g^{1-s}(x) dx - 1\right]$

EXAMPLE 5.1. *Let F and G be Stacy distributions having density functions of the form*

$$f(x; \alpha, \beta, \sigma) = \frac{x^{\beta-1}}{\Gamma(\beta/\alpha)} \exp\left[-\left(\frac{x}{\sigma}\right)^{\alpha}\right], x > 0; \ \alpha, \beta, \gamma > 0, \tag{5.2}$$

$$x > 0; \ \alpha, \beta, \gamma > 0$$

with parameters $(\alpha_1, \beta_1, \sigma_1)$ and $(\alpha_2, \beta_2, \sigma_2)$ respectively. Then

$$D = \log\left[\frac{\alpha_1 \sigma_2^{\beta_2} \Gamma(\frac{\beta_2}{\alpha_2})}{\alpha_2 \sigma_1^{\beta_1} \Gamma(\frac{\beta_1}{\alpha_1})}\right]$$

$$+ (\beta_1 - \beta_2)\left[\frac{\psi(\frac{\beta_1}{\alpha_1})}{\alpha_1} + \log \sigma_1\right] + \frac{\Gamma(\frac{\beta_1 + \alpha_2}{\alpha_1})}{\Gamma(\frac{\beta_1}{\alpha_1})}(\frac{\sigma_1}{\sigma_2})^{\alpha_2} - \frac{\beta_1}{\alpha_1} \tag{5.3}$$

where $\psi(p) = \frac{d \log \Gamma(p)}{dp}$. The model (5.2) contains as special cases, the gamma distribution when $\alpha = 1$, the Weibull when $\alpha = \beta$, Rayleigh when $\alpha = \beta$, $\alpha = 2$, the exponential when $\alpha = 1$, $\beta = 1$ and hence (5.3) provides the divergence in all these special cases and their combinations.

Some special properties enjoyed by D are (a) $D \geq 0$ with equality sign holding when $F = G$, (b) D is neither symmetric (*i.e.*, $D(F, G) \neq D(G, F)$) nor satisfies the triangular inequality for distances, (c) D is convex in pairs (f, g), (d) When X_1 and X_2 are independent observations their joint measure is

$$D(F(x_1, x_2), G(x_1, x_2)) = D(F_1(x_1), G_1(x_1)) + D(F_2(x_2), G_2(x_2))$$

where F_i and G_i are the distribution functions of X_i, $i = 1, 2$. (e) Statistical data processing will not alter D. Intuitively the Shannon's entropy is related to how much $f(x)$ diverges from the uniform distribution on the support of X.

5.2.1. Residual divergence. As considered in Chapter 2, we examine the structure and properties of the Kullback-Leibler divergence of the residual life $X_t = (X - t | X > t)$, which will be referred to as the residual divergence. It is given by

$$D(t) = \int_t^\infty \frac{f(x)}{\bar{F}(t)} \log \frac{f(x)/\bar{F}(t)}{g(x)/\bar{G}(t)} dx. \tag{5.4}$$

Ebrahimi and Kirmani (1996b) have studied various properties of the function $D(t)$, and they refer to $D(t)$ as the disparity between two systems at age t. Their main results are

(1) $D(t) = \log \bar{G}(t) + H_F(t) - \int_t^\infty \frac{f(x)}{\bar{F}(x)} \log g(x) dx$, where $H_F(t)$ is the Shannon entropy in (2.7)

(2) For all increasing functions $\phi(\cdot)$,

$$D(X, Y, \phi^{-1}(t)) = D(\phi(X), \phi(Y), t)$$

(3) If $\frac{h_F(t)}{h_G(t)}$ is increasing (decreasing) and both F and G are NBU (NWU) then $D(t) \geq (\leq)D$

(4) If $\frac{h_F(t)}{h_G(t)}$ is increasing (decreasing) in t and both F and G are IFR, then $D(t)$ is increasing (decreasing) in t

(5) If X_1 and X_2 and Y are lifetimes with density functions f_1, f_2 and g, $\frac{f_1}{g}$ is increasing in t and $h_{F_2}(t) \leq h_{F_1}(t)$, then $D_{F_1,G}(t) \leq D_{F_2,G}(t)$

(6) If $h_F(t) \leq \lambda$ and $m(t)$ is increasing and $G = 1 - e^{-\lambda t}$, then $D(t)$ is increasing in t.

(7) The distribution G for which $D_{F,G}$ is minimum subject to conditions $\int_0^\infty f(x)dx = 1$ and $w_r = \int_0^\infty T_r(x)f(x)dx$, $r = 1, 2, \ldots, n$, for some specified functions $T_r(x)$ and specified quantities w_r, is called the minimum discrimination information (MDI) distribution and is denoted by $F^*(x)$. Minimization of $D_{F,G}$ with constraints $\int_0^\infty f(x)dx = 1$ and $\theta(t) = -\int_t^\infty (\log \frac{\bar{G}(x)}{\bar{G}(t)})f(x)dx$ leads to $F^*(x)$ that satisfies

$$\frac{\bar{F}^*(x)}{\bar{F}^*(t)} = \left(\frac{\bar{G}(x)}{\bar{G}(t)} \right)^{\frac{1}{\theta(t)}}$$

showing that the residual life distributions relating to $\bar{G}(x)$ are the proportional residual life distributions of $F^*(x)$ with proportionality $\theta(t)$.

(8) A characterization of the usual proportional hazards model given in Ebrahimi and Kirmani (1996a) states that $D(t)$ is independent of t if and only if $\bar{G}(t) = \bar{F}^\theta(t)$

(9) A test of hypothesis that the distribution F is uniform employing $D(t)$ is discussed in Ebrahimi (2001).

Now, we show that a direct relationship exists between $D(t)$ and the hazard rates of F and G. Equation (4.3) gives

$$D(t)\bar{F}(t) = \int_t^\infty f(x) \log \frac{f(x)}{g(x)} dx + \bar{F}(t) \log \left(\frac{\bar{G}(t)}{\bar{F}(t)} \right). \tag{5.5}$$

Differentiating (5.5) and dividing by $\bar{F}(t)$,

$$D'(t) - D(t)h_F(t) = -h_F(t)\log f(t) + h_F(t)\log g(t) - h_G(t) + h_F(t)$$
$$- h_F(t)\log \bar{G}(t) + h_F(t)\log \bar{F}(t)$$
$$= h_F(t)\left[\log \frac{h_G(t)}{h_F(t)} + 1 - \frac{h_G(t)}{h_F(t)}\right] \qquad (5.6)$$

gives,

$$D(t) = \frac{1}{\bar{F}(t)}\int_t^\infty \left[\frac{h_G(x)}{h_F(x)} - 1 - \log \frac{h_G(x)}{h_F(x)}\right] f(x)dx. \qquad (5.7)$$

As mentioned in Section 4.2, $\frac{h_G(t)}{h_F(t)} = \theta(t)$, is the relative hazard rate of G with respect to F and $h_G(t) = \theta(t)h_F(t)$ is the dynamic proportional hazards model. An equivalent representation (5.6) is

$$D'(t) = h_F(t)D(t) + E[\theta(X) - 1 - \log \theta(X)|X > t] \qquad (5.8)$$

Several new properties of $D(t)$ can be derived from the above identities.

THEOREM 5.1. *The residual divergence $D(t)$ is increasing (decreasing) if either $\theta(x)$ is increasing (decreasing) and $\theta(x) > 1$ or when $\theta(x)$ is decreasing (increasing) and $\theta(x) < 1$.*

PROOF. Equation (5.8) can be written as

$$D'(t) = [D(t) + \log \theta(t) + 1 - \theta(t)]h_F(t) \qquad (5.9)$$

and by virtue of (5.7),

$$D'(t) = \frac{1}{\bar{F}(t)}h_F(t)[\int_t^\infty [\theta(x) - 1 - \log \theta(x) - (\theta(t) - 1 - \log \theta(t))]f(x)dx].$$

When $\theta(x) - 1 - \log \theta(x)$ is an increasing (decreasing) function $D'(t) \geq (\leq)0$ which happens when $\theta'(x)\left(1 - \frac{1}{\theta(x)}\right) > (<)0$. Thus $D(t)$ is increasing if either $\theta(x)$ is decreasing and $\theta(x) < 1$. Similarly $D(t)$ is increasing if either $\theta(x)$ is decreasing and $\theta(x) > 1$ or when $\theta(x)$ is increasing and $\theta(x) < 1$. □

REMARK 5.1. *The relative hazard rate $\theta(t)$ is increasing (decreasing) when Y is more IFR (less IFR) than X and $\theta(x) > (<)1$ if $h_Y(x) < h_X(x)$. Thus when Y is more IFR or DFR than X, the divergence becomes larger and smaller divergence is the result of Y having lesser IFR or DFR than X.*

REMARK 5.2. *Theorem gives different sets of conditions for monotonic $D(t)$ than given in Ebrahimi and Kirmani (1996b) when the monotonicity of the hazard rates is an essential requirement. It easily follows form (4.9)* *that when $\theta(t) = $ a constant representing proportional hazards model, if and only if $D(t)$ is also a constant.*

EXAMPLE 5.2. *In the proportional hazards model $D(t) = \theta - \log\theta - 1$, where θ is the ratio of the hazard rates. When X and Y are exponential with parameters λ_1 and λ_2, we have directly*

$$D(t) = \log\frac{\lambda_1}{\lambda_2} - \frac{\lambda_1 - \lambda_2}{\lambda_1}$$

without the aid of the formula (5.5).

EXAMPLE 5.3. *Assume that X has generalized Pareto distribution*

$$\bar{F}(x) = \left(1 + \frac{ax}{b}\right)^{-\frac{a+1}{a}}, \quad x > 0; \ b > 0, \ a > -1. \tag{5.10}$$

The hazard rate of X is $h_F(x) = \frac{a+1}{ax+b}$. We can write the distribution of Y in the case of proportional hazards model using the hazard rate

$$h_G(x) = \theta h_F(x) = \frac{\theta(a+1)}{ax+b}$$

as

$$\bar{G}(x) = \left(1 + \frac{ax}{b}\right)^{-\frac{(a+1)\theta}{a}}.$$

There are many problems in reliability analysis in which the distribution of Y has special forms and the distributions of X and Y need a comparative study. We examine how the residual divergence is of assistance in such situations.

Equilibrium distribution. The equilibrium distribution of X has density function $g(x) = \frac{\bar{F}(x)}{\mu}$, $\mu = E(X)$ representing the asymptotic distribution of residual life in a renewal replacement process in which a failed unit is replaced by a new unit with the same life distribution. Many papers discuss the comparison of reliability aspects of $\bar{F}(x)$ and $\bar{G}(x)$. A review of the associated results are available in Gupta (2007) and his references. The divergence between $\bar{F}$ and $\bar{G}$

$$
\begin{aligned}
D_E(t) &= \frac{1}{\bar{F}(t)} \int_t^\infty \log\left[\frac{\mu f(x)}{\bar{F}(x)} \frac{\int_t^\infty \bar{F}(x)dx}{\mu\bar{F}(t)}\right] f(x)dx \\
&= \frac{1}{\bar{F}(t)} \int_t^\infty \log(h_F(x)m_F(t))f(x)dx.
\end{aligned}
$$

As in (5.7)

$$D_E(t) = \frac{1}{\bar{F}(t)} \int_t^\infty ((m_F(x)h_F(t))^{-1} - 1 - \log(m_F(x)h_F(x))^{-1}) f(x) dx$$

(5.11)

and

$$D'(t) = h_F(t)[D(t) + \log(h_F m_F)^{-1} + 1 - (h_F m_F)^{-1}]$$ (5.12)

Equations (5.7) and (5.9) give scope for some interesting results on the divergence of equilibrium distribution from the parent distribution. The first of these is a characterization.

THEOREM 5.2. *A necessary and sufficient condition that $D_E(t)$ is a constant is that X follows the generalized Pareto distribution in (5.10).*

PROOF. When X follows generalized Pareto distribution $m_F(x) = ax + b$ and $h_p(x) = \frac{a+1}{ax+b}$ so that $m_F(x)h_F(x) = a + 1$. Hence from (5.11)

$$D_E(t) = \log(a+1) - \frac{a}{a+1},$$

which is a constant. On the other hand when $D_E = C$, a constant

$$C\bar{F}(t) = \int_t^\infty (m_F(x)h_F(x))^{-1} - 1 - \log(m_F(x)h_F(x))^{-1} f(x) dx.$$

Differentiating

$$C = \theta(t) - 1 - \log \theta(t), \quad \theta(t) = m_F(t)h_F(t)$$

giving

$$\theta'(t) \left(1 - \frac{1}{\theta(t)} \right) = 0.$$

Thus $\theta(t) = $ a constant or $\theta(t) = 1$. The theorem follows from the fact that $\theta(t) = m_F(t)h_F(t)$ is a constant if and only if X follows the generalized Pareto law with $\theta = 1$ specializing to the exponential distribution, a member of (5.10). □

THEOREM 5.3. *The residual divergence function $D_E(t)$ is increasing (decreasing) if the mean residual function is concave (convex) and decreasing or when it is convex (concave) and increasing.*

PROOF. From Theorem the condition for increasing $D_E(t)$ is an increasing $\theta(x)$ greater than unity. However,

$$\theta(x) = \frac{1}{h_F(x)m_F(x)} = (1 + m'(x))^{-1}$$

so that $\theta(x)$ increases when $1 + m'(x)$ is decreasing or $m_F(x)$ is concave. Also $\theta(x) > 1$ is equivalent to $(1 + m'(x))^{-1} > 1$ showing that $m_F(x)$ is decreasing. The proof for decreasing $D_E(t)$ is similar. □

REMARK 5.3. *One may wonder while the constancy of divergence is limited to the generalized Pareto law in the equilibrium case while it is true for all proportional hazards models. It is justified by the fact that among all proportional hazards models, the generalized Pareto model is the only one for which the equilibrium distribution admits constant hazard rate ratios.*

Relevation transform. It is sometimes difficult to replace a failed item by making it as good as new through repair, but easier to repair it to nearly the same state as just before failure. Hence another repair replacement strategy is to replace the failed unit by another item of the same age x. In general if the survival function of the replaced item is $\bar{T}(x)$,

$$\bar{G}(x) = \bar{F}(x) - \bar{T}(x) \int_0^x \frac{dF(t)}{\bar{T}(t)}$$

is called the relevation transform of $\bar{F}(x)$ by $\bar{T}(x)$. When $\bar{T}(x) = \bar{F}(x)$, we have the auto relevation transform

$$\bar{G}(x) = \bar{F}(x)[1 - \log \bar{F}(x)]. \tag{5.13}$$

For a discussion of the properties and application of relevation transforms in reliability theory we refer to Baxter (1982).

The density function of $\bar{G}(x)$ or Y is

$$g(x) = -f(x) \log \bar{F}(x)$$

leading to

$$h_G(x) = -\frac{\log \bar{F}(x)}{1 - \log \bar{F}(x)} h_F(x)$$

$$= \frac{H_F(x)}{1 + H_F(x)} h_F(x) \tag{5.14}$$

where $H_F(x) = -\log \bar{F}(x)$ is the cumulative hazard rate of X. By virtue of (5.3) the general formula (5.7) reduces to the residual diverging between $\bar{F}$ and $\bar{G}$

$$D_R(t) = \frac{1}{\bar{F}(t)} \int_t^\infty \left[\left(\frac{H_F(x)}{1 + H_F(x)} \right) - 1 - \log \left(\frac{H_F(x)}{1 + H_F(x)} \right) \right] f(x) dx.$$

In this case $\theta(t) = \frac{H(t)}{1+H(t)}$ and accordingly $\theta'(t) = \frac{h(t)}{(1+H(t))^2} > 0$ and $\theta(t) < 1$ for all t. Hence by Theorem $D_R(t)$ is decreasing indicating that the reliability improves by this replacement strategy. Further it is easy to see that there is no proper life distribution for which $D_R(t) = C$, a constant.

When replacement is made in the above manner, often called replacement after minimal repair, it is of interest to know the advantage of the policy. This can be accomplished by comparing h_F and h_G or equivalently $\theta(t)$. Since $\theta(t) < 1$, $h_G \leq h_F$ or $X \leq_{\mathrm{hr}} Y$.

EXAMPLE 5.4. *When X is exponential, $\bar{F}(x) = e^{-\lambda x}$, $x > 0$, $\lambda > 0$,*

$$\bar{G}(x) = (1 + \lambda x)e^{-\lambda x},$$

$h_F(t) = \lambda$, $h_G(t) = \frac{\lambda^2 t}{1+\lambda t}$, $H_F(t) = \lambda(t)$ *and* $\theta(t) = \frac{\lambda t}{1+\lambda t}$. *Hence*

$$D_R(t) = \log \frac{1 + \lambda t}{\lambda t} - e^{\lambda t} E_1(\lambda t).$$

where $E_1(z) = \int_z^\infty \frac{e^{-t}}{t} dt$.

Series and parallel systems. We consider a parallel system of n components with independent and identically distributed lifetimes. The system life is represented by the largest order statistics $X_{n:n}$ with survival function,

$$\bar{F}_n(x) = 1 - F^n(x).$$

Similarly, we have another parallel system with component life distribution $G(x)$ and system life specified by

$$\bar{G}_n(x) = 1 - G^n(x).$$

Our interest is to compare the performances of the two systems through their residual divergences,

$$D_p(t) = \frac{1}{1 - F^n(x)} \int_t^\infty \log\left(\frac{f_n(x)}{g_n(x)} \frac{\bar{G}_n(t)}{\bar{F}_n(t)}\right) f_n(x) dx. \qquad (5.15)$$

Using the general formula (4.7) with

$$h_{G_n}(x) = \frac{nG^n(x)h_G(x)}{1 - G^n(x)}$$

and

$$h_{G_n}(x) = \frac{nG^n(x)h_G(x)}{1 - G^n(x)}$$

equation (5.15) simplifies to

$$D_P(t) = \frac{1}{1 - F^n(x)} \int_t^\infty [\theta_n(x) - 1 - \log \theta_n(x)] f_n(x) dx$$

in which

$$\theta_n(x) = \frac{h_{G_n}(x)}{h_{F_n}(x)} = \frac{G^n(x)(1 - F^n(x))}{F^n(x)(1 - G^n(x))} \theta(x).$$

When a series system of n components with independent and identically distributed lifetimes with survival function $\bar{F}(x)$ is considered, we have the system life as $X_{1:n}$ whose distribution is obtained as $\bar{F}_1(x) = \bar{F}^n(x)$.

For a second series system with component distribution $\bar{G}(x)$, we have $\bar{G}_1(x) = \bar{G}^n(x)$. Accordingly the hazard rates of the two systems are $h_{F_1}(x) = n h_F(x)$ and $h_{G_1}(x) = n h_G(x)$ giving the relative rates of the two systems as

$$\theta_1(x) = \frac{h_{G_1}(x)}{h_{F_1}(x)} = \frac{h_G(x)}{h_F(x)} = \theta(x).$$

Thus the divergence function of the systems are

$$D_s(t) = \frac{1}{\bar{F}^n(x)} \int_t^\infty [\theta(x) - 1 - \log\theta(x)] n \bar{F}^{n-1}(x) f(x) dx. \qquad (5.16)$$

Irrespective of the differences in the survival functions in (5.7) and (5.14), the monotonicity of $D_s(t)$ has the same conditions as in Theorem 5.1. That is $D_s(t)$ is increasing (decreasing) if either $\theta(x)$ is increasing (decreasing) and $\theta(x) > 1$ or when $\theta(x)$ is decreasing (increasing) and $\theta(x) < 1$.

There is a relationship between $D_s(t)$ and $D(t)$, the divergence of the component distributions F and G. Equation (5.7) is

$$\frac{d}{dt}\bar{F}(t)D(t) = [1 - \theta(t) + \log\theta(t)]f(t)$$

and (5.16) is

$$\frac{d}{dt}\bar{F}^n(t)D_s(t) = [1 - \theta(t) + \log\theta(t)]n\bar{F}^{n-1}(t)f(t).$$

Thus

$$\frac{d}{dt}(\bar{F}^n(t)D_s(t)) = n\bar{F}^{n-1}(t)\frac{d}{dt}(\bar{F}(t)D(t)).$$

Further simplification yields

$$D_s'(t) - nh_F(t)D_S(t) = n(D'(t) - h_F(t)D(t))$$

or

$$h_F(t) = \frac{D_s'(t) - nD'(t)}{n[D_s(t) - D(t)]}.$$

EXAMPLE 5.5. *Assuming that $\bar{F}(x) = (1+x)^{-\alpha}$ and $\bar{G}(x) = e^{-\lambda x}$, we have $\bar{F}_1(x) = (1+x)^{-n\alpha}$ and $\bar{G}_1(x) = e^{-n\lambda x}$ so that*

$$\theta_1(x) = \frac{h_{G_1}(x)}{h_{F_1}(x)} = \frac{\lambda(1+x)}{\alpha}.$$

This yields

$$D_s(t) = (1+t)^{\alpha n} \int_t^\infty \left[\frac{\lambda(1+x)}{\alpha} - 1 - \log\frac{\lambda(1+x)}{\alpha}\right] \alpha n (1+x)^{-n\alpha-1}$$

$$= -\frac{1}{\alpha n} - 1 + \frac{\lambda n(1+t)}{\alpha n - 1} - \log\frac{\lambda(1+t)}{\alpha},$$

and

$$D'_s(t) = \frac{\lambda n}{\alpha n - 1} - \frac{\alpha}{1 + t},$$

showing that $D_s(t)$ is an increasing function for $0 < \alpha < \frac{1}{n}$ and inverted bathtub shaped for $\alpha > \lambda + \frac{1}{n}$ with change point at $t = \frac{n(\alpha - \lambda) - 1}{n\lambda}$. Thus $D_s(t)$ need not always be monotonic.

Asadi et al. (2005a) have considered minimum dynamic discrimination information models (MDDI) relative to G, in which $D_{F,G}$ is minimized subject to certain constraints. The MDDI model among the class of distributions $\Omega_F = \{F\}$ is F^* satisfying $D_{F^*,G}(t) \le D_{F,G}(t)$ for all F and $t \ge 0$. Taking $\Omega_F = \{F | m_F(t) \le q(t)\}$ and $m_F^*(t) = q(t)$, if $\log \frac{f^*(x)}{g(x)}$ is decreasing and concave (increasing and convex) then F^* is the MDDI distribution. As an example if $q(t) = \mu + \alpha t$ then F^* is generalized Pareto which is increasing and convex when $\lambda \ge \frac{2\alpha + 1}{\beta}$ and hence is the MDDI model relative to the exponential distribution with parameter λ satisfying $\lambda \ge \frac{2\alpha + 1}{\beta}$. Here, the generalized Pareto has the form

$$f^*(x) = \frac{\alpha + 1}{\beta} \left(1 + \frac{\alpha}{\beta} x\right)^{-\frac{1}{\alpha} - 2}.$$

Arising from this result the condition on $\log \frac{f^*(x)}{g(x)}$ can be relaxed to give the following results that accommodate more distributions

(a) Let $\psi_F = \{F | m_F(t) \le (\ge) g(t); \frac{m'_F(t)}{m_F(t)} \ge (\le) \frac{q'(t)}{q(t)}\}$. If F^* is such that $m_F^*(t) = \frac{q(t)f^*}{g}$ decreasing (increasing) then F^* is the MDDI model relative to G.

(b) Let $\psi_F = \{F | h_F(t) \le (\ge) p(t)\}$ and F^* satisfies $h_{F^*}(t) = p(t)$. Then if $\frac{f^*}{g}$ is decreasing (increasing), then F^* is the MDDI distribution. The result also holds if $\frac{h'_F(x)}{h_F(x)} \ge (\le) \frac{p'(x)}{p(x)}$, where $p(x)$ is a probability density function.

5.2.2. Residual quantile divergence. Sankaran et al. (2016) have given a formula for the Kullback-Leibler divergence when the distribution of X is specified by a quantile function. This is arrived at by setting $x = Q_X(u)$ in (4.1) to write

$$\begin{aligned} D_{Q_X} &= \int_0^1 \log \left\{ \frac{f(Q_X(p))}{g(Q_x(p))} \right\} f(Q_X(p)) dQ_x(p) \\ &= -\int_0^1 \log \frac{d}{dp} G(Q_X(p)) dp. \end{aligned} \tag{5.17}$$

When G has density g,

$$D_{Qx} = - \int_0^1 \log g(Q_X(p))q_X(p)dp.$$

For the residual life, a similar method yields the quantile version of the residual divergence measure as

$$D_{Q_X}(u) = \log \left(\frac{\bar{G}(Q_X(u))}{1-u} \right) - \frac{1}{1-u} \int_u^1 \log \frac{d}{dp} G(Q_X(p))dp$$

$$= \log \left(\frac{\bar{G}(Q_X(u))}{1-u} \right) - (1-u)^{-1} \int_u^1 \log g(Q_X(p))q_X(p)dp.$$

$$(5.18)$$

EXAMPLE 5.6. *Let X follow the linear hazard quantile function distribution*

$$Q_x(u) = (a+b)^{-1} \log \frac{a+bu}{a(1-u)}$$

and Y follow exponential law $\bar{G} = e^{-\lambda x}$. Then

$$\bar{G}(Q_X(u)) = \exp \left[-\lambda(a+b)^{-1} \log \frac{a+bu}{a(1+u)} \right]$$

$$= \left(\frac{a+bu}{a(1+u)} \right)^{-\frac{\lambda}{a+b}}, \quad 0 \le u \le 1.$$

Substituting in (5.17),

$$D_Q = \log \frac{a(a+b)}{\lambda(b-a)} - \left(\frac{\lambda}{a+b} - 1 \right)(\log 4 - 1)$$

$$- \left(\frac{\lambda}{a+b} + 1 \right) \left(\frac{a+b}{b} \log \frac{a}{a+b} + 1 \right).$$

EXAMPLE 5.7. *When $Q_X(u) = -(a+b)\log(1-u) - 2bu$, $a > 0$, $a + b > 0$, $F(x)$ cannot be obtained in algebraic form and the definition (5.1) is difficult to use. Taking $Q_Y(u) = -\frac{1}{\lambda}\log(1-u)$,*

$$\frac{h_Y(Q_X(u))}{h_X(Q_X(u))} = \frac{\lambda - 1}{(a-b+2bu)^{-1}} = A + Bu \qquad (5.19)$$

where $A = (a-b)\lambda$ and $B = 2b\lambda$.

Differentiating (5.18),

$$\frac{d}{du}(1-u)D_{Q_X}(u) = 1 + \log(1-u) - \frac{h_Y(Q_X(u))}{h_X(Q_X(u))} + \log\frac{g\left(Q_X(u)\right)q_X(u)}{\bar{G}(Q_X(u))}$$

$$= 1 - \frac{h_Y(Q_X(u))}{h_X(Q_X(u))} + \log\left\{\frac{g(Q_X(u))}{\bar{G}(Q_X(u))}(1-u)q_X(u)\right\}$$

$$= 1 - \frac{h_Y(Q_X(u))}{h_X(Q_X(u))} + \log\frac{h_Y(Q_X(u))}{h_X(Q_X(u))}.$$

$$(5.20)$$

Substituting (5.19),

$$D_{Q_X}(u) = A + \frac{B}{2}(1+u) - 1 + \frac{1}{B}((A+B)\log(A+B) - (A+Bu)\log(A+Bu)).$$

REMARK 5.4. *Apart from providing an alternative method for determining $D_{Q_x}(u)$, (4.18) also gives an interesting result which does not follow in the distribution function approach.*

THEOREM 5.4. *The residual divergence quantile function $D_{Q_x}(u)$ is uniquely determined by the relative hazard quantile function $\phi(u) = \frac{h_Y(Q_X(u))}{h_X(Q_X(u))}$ as*

$$D_{Q_x}(u) = \frac{1}{1-u}\int_u^1 [\phi(p) - 1 - \log\phi(p)]dp \qquad (5.21)$$

Several consequences of (5.18) and (5.20) are discussed in detail in Nair et al. (2021c) When the hazards ratio can be postulated from physical conditions or from empirical evidence, formula (5.20) provides an exact method to arrive at the divergence function. For example let $\phi(u)$ be in the from

$$\phi(u) = (a + bu)^{-1}.$$

Then

$$D_Q(u) = \frac{a+b+1}{1-u}\log(a+b-1) - \frac{a+bu+1}{1-u}\log(a+bu) - 2b.$$

Such a relative hazards quantile function can arise from a choice of

$$Q_X(u) = (A+B)^{-1}\log\frac{A+Bu}{A(1-u)}$$

considered earlier with $h_{Q_X}(Q_X(u)) = (A+Bu)^{-1}$ and $Q_Y(u)$, the exponential with $h_Y(Q_X(u)) = \lambda$ and a reparametrization $a = \frac{A}{\lambda}$ and $b = \frac{B}{\lambda}$.

5.2.3. Divergence of past lives. The divergence measure between past lives $X|X \leq t$ and $Y|Y \leq t$ is deduced from (4.1) as Di Cresenzo and Longobardi (2004)

$$
\begin{aligned}
\bar{D}(t) &= \int_0^t \frac{f(x)}{F(t)} \log \frac{f(x)/F(t)}{g(x)/G(t)} dx \\
&= \log \frac{G(t)}{F(t)} + \frac{1}{F(t)} \int_0^x f(x) \log \frac{f(x)}{g(x)} dx \qquad (5.22) \\
&= -\int_0^t \frac{f(x)}{F(t)} \log \frac{g(x)}{\bar{G}(t)} - \bar{H}_X(t),
\end{aligned}
$$

where $\bar{H}_X(t) = -\int_0^t \frac{f(x)}{F(t)} \log \frac{f(x)}{F(t)} dx$. Di Cresenzo and Longobardi (2004) established that

(i) if $\frac{f(t)}{g(t)}$ is increasing (decreasing) then

$$
\bar{D}(t) \leq (\geq) \log \frac{\lambda_X(t)}{\lambda_Y(t)}
$$

(ii) if $g(t)$ is decreasing then $\bar{D}(t) \leq -\log \lambda_Y(t) - \bar{H}_X(t)$

(iii) if X, Y have support $(0, \infty)$ and $\phi(\cdot) : (0, \infty) \to (0, \infty)$ is a bijective transformation, then if ϕ is strictly increasing

$$
\bar{D}_{\phi(X),\phi(Y)}(t) = \bar{D}_{X,Y}(\phi^{-1}(t))
$$

and if ϕ is strictly decreasing

$$
\bar{D}_{\phi(X),\phi(Y)}(t) = \theta_X(\phi^{-1}(t)) \left[\bar{D}_{X,Y}(\phi^{-1}(t)) + \log \frac{\theta_X(\phi^{-1}(t))}{\theta_Y(\phi^{-1}(t))} \right]
$$

where $\theta_z(t) = \frac{\bar{F}_2(t)}{F_z(t)}$, the odds ratio

(iv) when X_1, X_2, Y are non-negative random variables with density functions f_1, f_2 and g, with $\frac{f_1((x)}{g(x)}$ decreasing and $\lambda_{X_1}(t) \leq \lambda_{X_2}(t)$ then $\bar{D}_{X_1,Y}(t) \leq \bar{D}_{X_2,Y}(t)$. Also for random variables X, Y_1, Y_2 with densities f, g_1 and g_2 satisfying $\frac{g_1(x)}{g_2(x)}$ increasing,

$$
\bar{D}_{X,Y_1}(t) \leq \bar{D}_{X,Y_2}(t) - \log \frac{\lambda_{X_1}(t)}{\lambda_{X_2}(t)}.
$$

(v) when $G(x)$ is the proportional reversed hazards model of $F(x)$ that $G(x) = F^\theta(x)$, $D(t)$ will be constant $\theta - 1 - \log \theta$ and conversely.

Apart from the above we have more results on the properties of $\bar{D}(t)$ in terms of the reversed hazard rates λ_F and λ_G.

THEOREM 5.5. *The divergence function $\bar{D}(t)$ can be expressed as*

$$\bar{D}(t) = \frac{1}{F(t)} \int_0^t \left(\frac{\lambda_G(x)}{\lambda_F(x)} - 1 - \log \frac{\lambda_G}{\lambda_F} \right) f(x)dx. \qquad (5.23)$$

PROOF. From (5.22),

$$\bar{D}(t)F(t) = \int_0^t f(x)[\log \frac{f(x)}{g(x)} + \log G(t) - \log F(t)]dx$$

$$= \int_0^t f(x) \log \frac{f(x)}{g(x)} + F(t) \log G(t) - F(t) \log F(t).$$

Hence

$$\bar{D}'(t)F(t) + \bar{D}(t)F(t) = f(t) \log f(t) - f(t) \log G(t) + F(t)\lambda_G(t)$$
$$+ f(t) \log G(t) - F(t)\lambda_F(t) - f(t) \log F(t)$$

$$\bar{D}'(t) + \lambda_F(t)D(t) = \lambda_F(t) \log \lambda_F(t) - \lambda_G(t) + \lambda_G(t) - \lambda_F(t)$$

$$= \lambda_F(t) \left[\frac{\lambda_G(t)}{\lambda_F(t)} - 1 - \log \frac{\lambda_G(t)}{\lambda_F(t)} \right]$$

or

$$\bar{D}'(t)F(t) + f(t)D(t) = f(t) \left[\frac{\lambda_G(t)}{\lambda_F(t)} - 1 - \log \frac{\lambda_G(t)}{\lambda_F(t)} \right]$$

$$\frac{d}{dt}F(t)D(t) = \left[\frac{\lambda_G(t)}{\lambda_F(t)} - 1 - \log \frac{\lambda_G(t)}{\lambda_F(t)} \right] f(t) \qquad (5.24)$$

Integrating from 0 to t, we have (5.23). $\qquad\qquad\qquad \square$

REMARK 5.5. *The ratio $\bar{\theta}(t) = \frac{\lambda_G(t)}{\lambda_F(t)}$ is called the relative reversed hazard rate of G with respect to F. Various properties of $\bar{\theta}(t)$ are discussed in Nair et al. (2022).*

REMARK 5.6. *Using the quantile function $Q_X(u)$ of K, (5.22) takes the form*

$$\bar{D}(Q_X(u)) = \bar{D}_{Q_X}(u) = \frac{1}{u} \int_0^u [\bar{\phi}(p) - 1 - \log \bar{\phi}(p)]dp, \ \bar{\phi}(u) = \bar{\theta}(Q_X(u))$$
$$(5.25)$$

From (5.25) we have the following theorem.

THEOREM 5.6. *The divergence quantile function $\bar{D}_{Q_X}(u)$ is uniquely determined by the relative reversed hazard quantile function at the $100(1 - u)\%$ point of the distribution of X.*

The identity $\lambda_G(x) = \bar{\theta}(x)\lambda_F(x)$ defines the dynamic proportional reversed hazard model of Nanda and Das (2011). Its quantile version is

$\lambda_G(u) = \bar{\phi}(u)\lambda_Q(u)$. Equation (4.22) states that $\bar{\phi}(u)$ uniquely determines the past divergence. There are several special interesting cases when $\bar{\phi}(u)$ has specific functional forms. The simplest among them is when $\bar{\phi}(u)$ is a constant or equivalently $\bar{\theta}(x) = $ a constant, say $\bar{\theta}$. In this case $\bar{D}(t) = \bar{\theta} - 1 - \log \bar{\theta}$ and conversely if $\bar{D}(t) = K$, a constant then $\bar{\theta}(t)$ is also constant, a result proved in Di Cresenzo and Longobardi (2004) by a different method.

THEOREM 5.7. *The divergence of past life $\bar{D}(t)$ is increasing (decreasing) if $\theta(t)$ is increasing (decreasing) and $\theta(t) > 1$ or when $\bar{\theta}(t)$ is decreasing (increasing) and $\theta(t) < 1$.*

The proof of the theorem follows the same lines as that of Theorem 5.1. We give some examples on the application of the above results.

EXAMPLE 5.8. *The reversed relevation transform of the random variable X with respect to a distribution function $A(x)$ is*

$$G(x) = A(x) + F(x) \int_x^\infty \frac{a(t)}{F(t)} dx_r$$

and the corresponding auto transform is obtained when $A(x) = F(x)$. Then

$$G(x) = F(x) + F(x) \int_x^\infty \frac{f(t)}{F(t)} dt. \qquad (5.26)$$

Using $-\log F(x) = \int_x^\infty \lambda(t)dt$, equation (5.26) reduces to

$$G(x) = F(x)[1 - \log F(x)]$$

and the density function of G is

$$g(x) = f(x) \int_x^\infty \frac{f(t)}{F(t)} dt = -f(x) \log F(x).$$

Thus

$$\lambda_G(x) = \lambda_F(x) \left(-\frac{\log F(x)}{1 - \log F(x)} \right)$$

and $\bar{\theta}(x) = -\frac{\log F(x)}{1 - \log F(x)}$ which is decreasing and less than unity. Hence by Theorem $\bar{D}(t)$ is decreasing.

EXAMPLE 5.9. *Let X and Y have power distributions $F_X(x) = \left(\frac{x}{\alpha}\right)^{\beta_1}$ and $F_Y(x) = \left(\frac{x}{\alpha}\right)^{\beta_2}$. Then $\lambda_F(x) = \frac{\beta_1}{x}$ and $\lambda_G(x) = \frac{\beta_2}{x}$ giving $\bar{\theta}(x) = \frac{\beta_2}{\beta_1}$. The divergence function is*

$$\bar{D}(t) = \frac{\beta_2}{\beta_1} - 1 - \log \frac{\beta_2}{\beta_1},$$

a constant, Y being the proportional reversed hazards model of X.

5.3. Renyi's divergence

The divergence measure of order α proposed by Renyi (1961) is defined by

$$
\begin{aligned}
D_\alpha(F, G) &= \frac{1}{\alpha - 1} \log \int \left(\frac{f(x)}{g(x)} \right)^{\alpha - 1} f(x) dx \\
&= \frac{1}{\alpha - 1} \log E_f \left(\frac{f(x)}{g(x)} \right)^{\alpha - 1}, \quad \alpha \neq 1
\end{aligned}
\tag{5.27}
$$

where $\alpha \to 1$, $D_\alpha(F, G)$ reduces to the Kullback-Leibler divergence. Between the residual life distributions of X and Y, (5.27) takes the form

$$
\begin{aligned}
D_\alpha(t) &= \frac{1}{\alpha - 1} \log \int_t^\infty \left[\frac{f(x)}{\bar{F}(t)} \frac{\bar{G}(t)}{g(x)} \right]^{\alpha - 1} \frac{f(x)}{\bar{F}(t)} dx \\
&= \frac{1}{\alpha - 1} \log \frac{1}{\bar{F}(t)} \left[\int_t^\infty \left(\frac{f(x)}{g(x)} \right)^{\alpha - 1} f(x) dx \right] + \log \frac{\bar{G}(t)}{\bar{F}(t)}.
\end{aligned}
\tag{5.28}
$$

The relationship between Renyi entropy and Renyi divergence is very similar to that between Shannon entropy and the Kullback-Leibler divergence. Apart from being a generalization of Kullback-Leibler divergence (5.27) satisfies almost the same set of axioms as (5.1) and enjoys more flexibility with the addition of the parameter α. Particular values of α like zero, $1/2$, 1 and ∞ have special implications. A detailed description of the properties of (5.27) is available in van Erven and Harremoes (2014).

Asadi et al. (2005a) have shown that $D_\alpha(t)$ is independent of t if and only F and G have proportional hazard rates and gave an example in the case of series systems. A more detailed analysis of (5.28) yields some interesting properties of $D_\alpha(t)$. Rewriting (5.27) as

$$
\exp[(\alpha - 1) D_\alpha(t)] \frac{\bar{F}^\alpha(t)}{\bar{G}^{\alpha-1}(t)} = \int_t^\infty \frac{f^\alpha(x)}{g^{\alpha-1}(x)} dx
$$

and differentiating with respect to t, leaves after some algebra

$$
\begin{aligned}
&\bar{F}^\alpha(t) \bar{G}(t) e^{(\alpha-1) D_\alpha(t)} (\alpha - 1) D_\alpha'(t) \\
&\quad + e^{(\alpha-1) D_\alpha(t)} \bar{F}^{\alpha-1}(t) [(\alpha - 1) g(t) \bar{F}(t) - \bar{G}(t) \alpha f(t)] \\
&= -\frac{f^\alpha(t)}{g^{\alpha-1}(t)} \bar{G}^\alpha(t)
\end{aligned}
$$

or,

$$
\exp[(\alpha - 1) D_\alpha(t)][(\alpha - 1) D_\alpha'(t) + (\alpha - 1) h_G(t) - \alpha h_F(t)] = -\frac{h_F^\alpha(t)}{h_G^{\alpha-1}(1)}.
\tag{5.29}
$$

We can express (5.28) in the alternative form

$$(\alpha - 1)D'_\alpha(t)\bar{F}(t)\exp[(\alpha - 1)D_\alpha(t)]$$
$$= [\{\alpha - (\alpha - 1)\theta(t)\}\exp((\alpha - 1)D_\alpha(t)) - \theta^{\alpha-1}(t)]f(t). \tag{5.30}$$

THEOREM 5.8. *The Renyi divergence measure $D_\alpha(t)$ is increasing (decreasing) in t if $\theta(t) \leq (\geq)1$ and $\frac{f(t)}{g(t)}$ is increasing (decreasing).*

PROOF. Equation (5.29) can be restated as

$$\bar{F}(t)e^{(\alpha-1)D_\alpha(t)}D'_\alpha(t)$$
$$= h_F(t)[\{\alpha - (\alpha - 1)\theta(t)\}\exp((\alpha - 1)D_\alpha(t)) - \theta^{\alpha-1}(t)]$$
$$= h_F(t)\left[\{\alpha - (\alpha - 1)\theta(t)\}\frac{\bar{G}^{\alpha-1}(t)}{\bar{F}^\alpha(t)}\int_t^\infty \frac{f^\alpha(x)}{g^{\alpha-1}(x)}dx - -\frac{f^{\alpha-1}(t)\bar{G}^{\alpha-1}(t)}{g^{\alpha-1}(t)\bar{F}^{\alpha-1}(t)}\right]$$
$$= \frac{h_F(t)}{(\alpha - 1)}\frac{\bar{G}^{\alpha-1}(t)}{\bar{F}^{\alpha-1}(t)}\left[\frac{1}{\bar{F}(t)}\left(\int_t^\infty \{\alpha - (\alpha - 1)\theta(t)\}\frac{f^{\alpha-1}}{g^{\alpha-1}(t)}\right)f(x)dx\right]. \tag{5.31}$$

Assume that $\frac{f(x)}{g(x)}$ is increasing in t. Then for $x > t$,

$$\frac{f^{\alpha-1}(x)}{g^{\alpha-1}(x)} \geq \frac{f^{\alpha-1}(t)}{g^{\alpha-1}(t)}.$$

Then (5.31) reveals that $D'_\alpha \geq 0$ if $\frac{\alpha - (\alpha-1)\theta(t)}{\alpha-1} - 1 \geq 0$ or when $\theta(t) \leq \frac{1}{\alpha-1}$, for $\alpha > 1$, since $\theta(t)$ is always positive. □

REMARK 5.7. *Note that $\frac{f(t)}{g(t)}$ is increasing and is equivalent to saying that $X \geq_{lr} Y$.*

Some more interesting characteristics of the Renyi measure are revealed if we consider the distribution function $B(u) = G(Q_X(u))$ defined on $(0,1)$ and its density function $b(u) = \frac{g(Q_X(u))}{f(Q_X(u))}$. Under the additional assumption that $f(x) > 0$ implies $g(x) > 0$, we further have $B(0) = 0$ and $B(1) = 1$ so that the range of B is extended to $[0,1]$. The transformation $x = Q(u)$ in (5.28) now leads to

$$D_{\alpha,Q}(u) = \frac{1}{\alpha - 1}\log\int_u^1\left[\frac{f(Q_X(p))}{1-u}\frac{\bar{G}(Q_X(u))}{g(Q_X(p))}\right]\frac{dp}{(1-u)}$$
$$= \frac{1}{\alpha - 1}\log\frac{1}{(1-u)^\alpha}(1 - B(u))^{\alpha-1}\int_u^1 b^{1-\alpha}(p)dp \tag{5.32}$$

as the quantile-equivalent of (5.28). We denote by $h_B(u) = \frac{b(u)}{1-B(u)}$ and $H_B(u) = \int_0^u h_B(p)dp$ the hazard rate and the cumulative hazard rate of the distribution $B(u)$. We further have

$$(1-u)^\alpha(1-B(u))^{1-\alpha}\exp[(\alpha-1)D_\alpha(u)] = \int_u^1 b^{1-\alpha}(p)dp. \quad (5.33)$$

Writing $P(u) = (1-u)^\alpha \exp[(\alpha-1)D_{\alpha,Q}(u)]$ and differentiating (5.33),

$$(1-B(u))^{1-\alpha}\frac{dP(u)}{du} - (1-\alpha)(1-B(u))^{-\alpha}b(u)P(u) = -b^{1-\alpha}(u)$$

or

$$\frac{dP(u)}{du} - (1-\alpha)h_B(u)P(u) = -h_B^{1-\alpha}(u). \quad (5.34)$$

Multiplying by $e^{-(1-\alpha)H_B(u)}$, the differential equation (5.34) becomes

$$\frac{d}{du}e^{-(1-\alpha)H_B(u)}P(u) = -e^{-(1-\alpha)H_B(u)}h_B^{1-\alpha}(u)$$

and

$$e^{-(1-\alpha)H_B(u)}P(u) = \int_u^1 e^{-(1-\alpha)H_B(p)}h_B^{1-\alpha}(p)dp.$$

Finally

$$\exp[(\alpha-1)D_{\alpha,Q}(u)] = \frac{1}{(1-u)^\alpha}e^{(1-\alpha)H_B(u)}\int_u^1 e^{-(1-\alpha)H_B}h^{1-\alpha}(p)dp. \quad (5.35)$$

Thus we have proved the following theorem.

THEOREM 5.9. *The divergence measure $D_{\alpha,Q}(u)$ is uniquely determined by the hazard rate $h_B(u)$ or equivalently by the distribution function $B(u)$.*

There are some important ramifications for the above result. Firstly it gives a formula to calculate $D_{\alpha,Q}(u)$. Secondly the form of the single distribution function $B(u)$ is sufficient to determine $D_{\alpha,Q}(u)$ which is an advantage over (5.27) that needs the forms of both F and G to calculate the divergence. Thirdly, (5.34) directly relates the hazard rate $h_B(u)$ to the function $D_{\alpha,Q}(u)$. This means that the reliability aspects of $B(u)$ can be employed to distinguish the behaviour of $D_{\alpha,Q}(u)$. We will now explore in detail the above aspects by means of examples and theorems.

EXAMPLE 5.10. *Let G be the proportional hazard rate model of F so that $\bar{G}(x) = \bar{F}(x)$. In terms of the quantile function $Q_X(u)$ of F this relation turns out to be*

$$B(u) = 1 - (1-u)^\theta$$

with

$$h_B(u) = \frac{\theta(1-u)^{\theta-1}}{(1-u)^\theta} = \frac{\theta}{(1-u)} \text{ and } H_B(u) = -\theta\log(1-u).$$

Substituting in (4.33),

$$\exp[(\alpha - 1)D_{\alpha,Q}(u)] = \frac{\theta^{1-\alpha}}{\alpha + \theta - \alpha\theta}$$

giving

$$D_{\alpha,Q}(u) = -\log\theta - \frac{1}{\alpha - 1}\log(\theta + \alpha - \alpha\theta),$$

which is a constant as has been found by Asadi et al. (2005a). It is interesting to note that in the PHM, $B(u)$ is a beta distribution. The hazard rate is reciprocal linear and IFR and further the mean residual life function is linear.

This example raises the interesting question whether the form of $D(u)$ determines the corresponding $B(u)$. Some partial answers seem to be in order.

THEOREM 5.10. *The function $D_{\alpha,Q}(u)$ is a constant if and only if $B(u) = 1 - (1 - u)^\theta$, $\theta > 0$.*

PROOF. Assume that $D_{\alpha,Q}(u)$ is a constant so that $\exp[(\alpha - 1)D_{\alpha,Q}(u)] = K$. Substituting in (4.33) and differentiating we get after some simplifications,

$$K(1 - u)^\alpha(\alpha - 1)h_B^\alpha(u) - K\alpha(1 - u)^{\alpha-1} = -h_B^{1-\alpha}(u)$$

or

$$K(1 - u)^\alpha(\alpha - 1)h_B - K\alpha(1 - u)^{\alpha-1}h_B^{\alpha-1}(u) = -1.$$

Differentiating again

$$(1 - u)^2 h_B(u)h_B'(u) - (1 - u)h_B^2(u) - (1 - u)h_B'(u) + h = 0.$$

This factors into

$$[(1 - u)h_B(u) - 1][(1 - u)h_B'(u) - h_B(u)] = 0$$

leaving two solutions

$$h_B(u) = (1 - u)^{-1} \text{ and } \frac{h'(B)}{h(B)} = \frac{1}{1 - u}.$$

The second one gives $h_B(u) = \frac{\theta}{1-u}$ which subsumes the first solution also. Thus $B(u) = 1 - (1 - u)^\theta$, $\theta > 0$. The second part of the theorem is established in Example 5.10.   □

REMARK 5.8. *When the distribution functions F has a known form it is easy to write down G as F^θ. However when the distribution of X is specified in quantile form with no closed form distribution function the standard form is difficult to find. The above theorem specifies the distribution of Y as*

$$Q_Y(u) = Q_X(1 - (1 - u)^\theta).$$

Table 5.2. Quantile functions of proportional hazards models.

Distribution of X	$Q_X(u)$	Proportional hazards model $Q_Y(u)$
generalized Tukey	$\lambda_1 + \frac{u^{\lambda_3} - (1-u)^{\lambda_4}}{\lambda_2}$	$\lambda_1 + \frac{[1-(1-u)]\lambda_3 - (1-u)^{\theta\lambda_4}}{\lambda_2}$
generalized Tukey lambda family	$\lambda_1 + \frac{1}{\lambda_L}[\frac{u^{\lambda_3}-1}{\lambda_3} - \frac{(1-u)^{\lambda_4}-1}{\lambda_4}]$	$\lambda_1 + \frac{1}{\lambda_L}[\frac{(1-(1-u)^{\theta})-1}{\lambda_3} - \frac{(1-u)^{\theta\lambda_4}-1}{\lambda_4}]$
van Staden-Loots model	$\lambda_1 + \lambda_2[\frac{(1-\lambda_3)u^{\lambda_4}-1}{\lambda_4} - \lambda_3\frac{(1-u)^{\lambda_4}-1}{\lambda_4}]$	$\lambda_1 + \lambda_2[\frac{(1-\lambda_3)}{\lambda_4}(\frac{(1-(1-u)^{\theta})^{\lambda_3}-1}{\lambda_4}) - \frac{\lambda_3(1-u)}{\lambda_4} - 1]$
power-Pareto	$\frac{cu^{\lambda_1}}{(1-u)^{\lambda_2}}$	$\frac{c(1-(1-u)^{\theta})^{\lambda_1}}{(1-u)^{\theta\lambda_2}}$
Govindarajulu	$\sigma[(\beta+1)u^{\beta} - \beta u^{\beta+1}]$	$\sigma[(\beta+1)(1-(1-u)^{\theta})^{\beta} - \beta(1-(1-u)^{\theta})^{\beta+1}]$

Several examples of proportional hazards models when X has a quantile function of the above mentioned nature are given in Table 2.

How do the ageing properties reflect on the monotonicity of $D_\alpha(u)$ is explained in the following theorem.

THEOREM 5.11. *The distribution $B(u)$ is DFR if $P(u)$ is increasing, convex and $\alpha > 1$ or if $P(u)$ is increasing, concave and $\alpha < 1$.*

PROOF. Rewriting (4.32) as

$$h_B^{\alpha-1}(u)P'(u) - (1-\alpha)P(u)h_B^{\alpha}(u) = -1$$

and differentiating

$$h_B(u)P''(u) + (\alpha-1)h_B'(u)P'(u) + (\alpha-1)h_B^2(u)P'(u)$$
$$+ \alpha(\alpha-1)h_B(a)h_B'(u)P(u) = 0$$

or

$$h_B'(u) = -\frac{[(\alpha-1)h_B(u)P'(u) + P''(u)]}{(\alpha-1)(h_B(u)P(u) + P'(u))}.$$

The desired conclusions now follow.   □

REMARK 5.9. *Recalling $P(u) = (1-u)^\alpha e^{(\alpha-1)D_\alpha(u)}$ and noting $(1-u)^\alpha$ is a decreasing function $P(u)$ increases only if $D_\alpha(u)$ increases. Likewise from*

$$P'(u) = (1-u)^{\alpha-1}e^{(\alpha-1)D_\alpha(u)}[(1-u)(\alpha-1)D_\alpha'(u) - \alpha]. \qquad (5.36)$$

$P(u)$ is convex only if (5.36) is increasing which happens when $\alpha > 1$ and $D_\alpha(u)$ is increasing. Thus the conclusions of the theorem boil out to the monotonicity of $D_\alpha(u)$.

Sunoj et al. (2017) considered a quantile version of Renyi divergence same as (5.32), by treating $G(Q_X(u))$ as a quantile function. They reasserted that $D_\alpha(u))$ is a constant if X and Y satisfy the Cox proportional hazard model and also that if Y is exponential, then X is exponential if and only if $D_\alpha(u)$ is constant. The latter result is a restatement of the former. Their other contributions are

(a) If $Q_3(u) = G(Q_X(u))$ is convex

$$D_\alpha(u) \geq (\leq) \log(H_3(u)(1 - Q_3(u))); \ 0 < \alpha < 1, \ (\alpha > 1) \qquad (5.37)$$

where the $H_3(\cdot)$ is the hazard quantile function of $Q_3(u)$.

(b) Let X, Y_1, Y_2 be non-negative random variables with distribution functions F, G_1, G_2 and $G_i F^{-1}$ is concave, $i = 1, 2$

$$D_{\alpha,X,Y_1}(u) \geq D_{\alpha,X,Y_2}(u) + \log \frac{H_{Y_1}(u)}{H_{Y_2}(u)}$$

(c) If $T_1(\cdot)$ and $T_2(\cdot)$ are continuous non-decreasing transformations

$$D_{\alpha,T_1,T_2} = \frac{1}{\alpha - 1} \log \int_u^1 \frac{d}{dp}[T_2(Q_2(p))^{-1} T_1(Q_1(p))]^{1-d} dp$$
$$+ \log(1 - Q_3(u)) - \frac{\alpha}{\alpha - 1} \log(1 - u).$$

Maya and Sunoj (2008) have considered the weighted model

$$F_{X(u)}(x) = \frac{w(x)f(x)}{\mu_w}, \ \mu_w = E(w(X)) < \infty$$

and have concluded that in the equilibrium case the Renyi divergence between X and X_E satisfies $D_{X,X_E,\alpha}(t) = k$, a constant if and only if X is distributed as Pareto I. The divergence function in the general case for residual life is

$$D_{w,\alpha}(t) = \frac{1}{\alpha - 1} \log \int_t^\infty \frac{f^\alpha(x) \ f_w^{1-\alpha}(x)}{\bar{F}^\alpha(t) \ \bar{F}^{1-\alpha}(t)} dx.$$

They have shown that if $X \leq_{\text{LR}} Y$ then

$$D_\alpha(t) \leq (\geq) \frac{\alpha}{\alpha - 1} \log \frac{h_X(t)}{h_Y(t)} \ \text{if} \ \alpha > 1 \ (0 < \alpha < 1) \qquad (5.38)$$

and as a Corollary if $X \leq_{\text{LR}} X_w$,

$$D_\alpha(t) \leq (\geq) \frac{\alpha}{\alpha - 1} \log \frac{E(w(X)|X > t)}{w(t)} \ \text{if} \ \alpha > 1 \ (0 < \alpha < 1) \qquad (5.39)$$

and if $X \leq_{\text{LR}} X_{\bar{F}}$, $D_\alpha(t) \leq (\geq) \frac{\alpha}{\alpha-1} \log(1 + r'(t))$ if α, $(0 < \alpha < 1)$. There is an interesting bound for the divergence in terms of the Renyi entropy

$$D_\alpha(t) \geq - \log h_Y(t) - H_\alpha(t),$$

$\alpha \neq 1$ where $H_\alpha(t) = \frac{1}{1-\alpha} \log \int_t^\infty \frac{f^\alpha(x)}{F^\alpha(t)}$, provided that the density $g(x)$ is decreasing. Similarly if $f_w(x)$ is decreasing,

$$D_w(t) \geq -\log \frac{w(t)h_X(t)}{E(w(X)|X>t)} - H_\alpha(t).$$

If X_1, X_2 and Y be non-negative random variables

$$X_1 \leq_{\mathrm{LR}} X_2 \Rightarrow D_{X_1,Y}(t) \leq (\geq) \frac{\alpha}{\alpha-1} \log \frac{h_{X_1}(t)}{h_{X_2}(t)} + D_{\alpha,X_2,Y}(t), \quad (5.40)$$

$\alpha > 1 \ (0 < \alpha < 1)$ and similarly X, Y_1 and Y_2

$$Y_1 \leq_{\mathrm{LR}} Y_2 \Rightarrow D_{\alpha,X,Y_1}(t) \geq \log \frac{h_{Y_1}(t)}{h_{Y_2}(t)} + D_{\alpha,X,Y_2}(t), \quad (5.41)$$

Some interesting contributions are available in Renyi divergence between past lives defined by (Asadi et al. 2005a)

$$\bar{D}_\alpha(t) = \frac{1}{\alpha-1} \log \int_0^t \frac{f^\alpha(x)g^{1-\alpha}(x)}{F^\alpha(t)G^{1-\alpha}(t)} dx \quad (5.42)$$

and its weighted version (Maya and Sunoj (2008))

$$\bar{D}_{\alpha,X,X_w}(t) = \frac{1}{\alpha-1} \log \int_0^t \frac{f^\alpha(x)f_w^{1-\alpha}(x)}{F^\alpha(t)F_w^{1-\alpha}(t)} dx. \quad (5.43)$$

If the weight function $w(t)$ is increasing in $t > 0$,

$$\bar{D}_{\alpha,X,X_w}(t) \geq (\leq) \frac{\alpha}{\alpha-1} \log \frac{\lambda(t)}{\lambda_w(t)}.$$

On the other hand when $w(t)$ is increasing (decreasing) with $E(w(X)|X \leq t)/w(t)$ is increasing (decreasing) then D_{α,X,X_w} is increasing (decreasing) in the special case $\alpha = 1$. Sunoj and Linu (2012a) further have

$$X \leq_{\mathrm{lr}} Y \Rightarrow \bar{D}_\alpha(t) \geq (\leq) \frac{\alpha}{\alpha-1} \log \frac{\lambda_X(t)}{\lambda_Y(t)}, \quad \alpha > 1 \ (0 < \alpha < 1),$$
$$\quad (5.44)$$

$$X \leq_{\mathrm{lr}} X_w \Rightarrow \bar{D}_\alpha(t) \geq (\leq) \frac{\alpha}{\alpha-1} \log \frac{E(w(X)|X \leq t)}{w(t)}, \quad (5.45)$$

if $g(x)$ is increasing in x,

$$\bar{D}_\alpha(t) \geq -\log \lambda_Y(t) - \bar{D}(t)$$

and

$$\bar{D}_{X,X_w}(t) \geq -\log \frac{w(t)\lambda_X(t)}{E(w(X)|X \leq t)}.$$

The quantile version of (5.42) given in Sunoj et al. (2017) reads

$$\bar{D}_\alpha(u) = \frac{1}{\alpha - 1} \log \left[\frac{Q_3^{\alpha-1}(u)}{u^2} \right] \int_o^u q_3^{1-\alpha}(p) dp$$

with $Q_3(u) = G(Q_X(u))$ and q_3 is the corresponding quantile density function. If $\lambda_{Q_3}(u)$ is the reversed hazard quantile function of Q_3,

$$\bar{D}_\alpha(u) = \frac{1}{\alpha - 1} \log \int_0^u (p\lambda_{Q_3}(p))^{\alpha-1} dp + \log Q_3(u) - \frac{\alpha}{\alpha - 1} \log u.$$

For non-negative random variables X and Y, $\bar{D}_\alpha(u)$ is a constant if Y is the proportional reversed hazard rate model of X. Further if Q_3 is concave

$$\bar{D}_\alpha(u) \geq (\leq) \log(H_3(u)Q_3(u)), \ 0 < \alpha < 1 \ (\alpha > 1)$$

where $H_3(u) = [(1 - u')q_3(u)]^{-1}$, the hazard quantile function of Q_3.

Hooda and Saxena (2011) have some alternative conditions that provide bounds for $\bar{D}_\alpha(t)$. If $\frac{f(t)}{g(t)}$ is increasing in t or $Z \geq_{lr} Y$, then

$$\bar{D}_\alpha(t) \leq \left(\frac{\lambda_X(t)}{\lambda_Y(t)} \right)^{\beta-1} \log \frac{\lambda_X(t)}{\lambda_Y(t)}.$$

This shows by the choice of

$$F(x) = \begin{cases} e^{-\frac{1}{2}-\frac{1}{x}}, & 0 < x \leq 1 \\ e^{-2+\frac{x^2}{2}}, & 1 < x \leq 2 \end{cases} \text{ and } G(x) = \begin{cases} \frac{x^2}{4}, & 0 < x \leq 2 \\ 1, & x \geq 2 \end{cases}$$

that $\bar{D}_\alpha(t)$ need not be monotone.

5.4. Varma divergence

A simple extension of the Renyi divergence by the introduction of an additional parameter leads to the Varma's measure (Varma (1966)) given by

$$D_{\alpha,\beta} = \frac{1}{\alpha - \beta} \log \int_0^\infty \left(\frac{f(x)}{g(x)} \right)^{\alpha+\beta-2} f(x) dx, \qquad (5.46)$$
$$\alpha \neq \beta, \ \beta \geq 1, \ \beta - 1 < \alpha < \beta.$$

It gives the Renyi divergence as $\beta \to 1$ and the Kullback-Leibler divergence as $\alpha, \beta \to 1$. A reparametrization $\delta = \alpha + \beta - 1$, reduces (5.46) to

$$D_{\alpha,\beta} = \frac{1}{\alpha - \beta} \log \int_0^\infty f^\delta(x) g^{1-\delta}(x) dx \qquad (5.47)$$

so that the expression under the integral sign is that in (5.28) where α is replaced by δ. For discriminating the residual and past lives we have

$$D_{\alpha,\beta}(t) = \frac{1}{\alpha - \beta} \log \int_t^\infty \left(\frac{f(x)}{\bar{F}(t)} \right)^\delta \left(\frac{g(x)}{\bar{G}(t)} \right)^{1-\delta} dx \qquad (5.48)$$

and

$$\bar{D}_{\alpha,\beta}(t) = \frac{1}{\alpha - \beta} \log \int_0^t \left(\frac{f(x)}{F(t)} \right)^\delta \left(\frac{g(x)}{\bar{G}(t)} \right)^{1-\delta} dx. \qquad (5.49)$$

By means of the similarity in the expressions (5.48) and (5.49) with those in (5.28) and (5.42) all the algebraic manipulations for the Varma divergence remains the same for results analogous to (5.38)-(5.45) if we make the following changes in them. Replace $\alpha - 1$ in the above cases $\alpha - \beta$ and α by δ. For example

$$X \leq_{\mathrm{lr}} Y \Rightarrow D_{\alpha,\beta}(t) \geq (\leq) \frac{\delta - 1}{\alpha - \beta} \log \frac{h_X(t)}{h_Y(t)}, \ \delta > 1, (\delta < 1)$$

and

$$X \leq_{\mathrm{lr}} X_W \Rightarrow D_{\alpha,\beta}(t) \geq (\leq) \frac{\delta - 1}{\alpha - \beta} \log \frac{E(w(X)|X > t)}{w(t)}, \ \delta > (<)1.$$

In view of this, we do not proceed further to elaborate the results for the Varma divergence. Notice that $D_{\alpha,\beta}(t)$ $(\bar{D}_{\alpha,\beta}(t))$ will be independent of t if and only if Y is the proportional hazards (reversed) hazard model of X, provided that, $(\theta - 1)(1 - \delta) + 1 > 0$.

5.5. Csiszar's family

Let X and Y be absolutely continuous random variables with support $(0, \infty)$ for both. The Csiszar's family of divergence measures has the form

$$D_\phi(F, G) = \int_)^\infty g(x)\phi\left(\frac{f(x)}{g(x)} \right) dx. \qquad (5.50)$$

Some members of this family given in Table 5.3.

Table 5.3. Members of the Csiszar's family.

$\phi(x)$	Measure
$(1 - \sqrt{x})^2$	Hellinger
$x \log x$ or $x \log x + 1 - x$	Kullback-Leibler $D(F, G)$
$\frac{1}{2}(1 - x)^2$	Pearson chi-square
$x^{1+a} - (1 + \frac{1}{a})x^a + \frac{1}{a}, a \neq 0$	BHHJ Power divergence
$\frac{x^{a+1} - x - a(x-1)}{a(a+1)}, a \neq 0, -1$	Cressie and Read
$-\log x + x - 1$ or $-\log x$	Kullback-Leibler $D(G, F)$
$(1 - \sqrt{x})^2$	Matusita divergence
$\mathrm{sgn}\,(\alpha - 1)x^\alpha$	Renyi

Vonta and Karagrigoriou (2010) have made a detailed study of (5.49) with reference to the residual and past lives of X and Y represented as

$$D_\phi(t) = \int_t^\infty \frac{g(x)}{\bar{G}(t)} \phi\left(\frac{f(x)/\bar{F}(t)}{g(x)/\bar{G}(t)}\right) dx \qquad (5.51)$$

and

$$\bar{D}_\phi(t) = \int_0^t \frac{g(x)}{G(t)} \phi\left(\frac{f(x)/F(t)}{g(x)/G(t)}\right) dx \qquad (5.52)$$

in which $\phi(x)$ is continuous, differentiable and convex for $x \geq 0$, $\phi(1) = 0$ and $\phi'(1) = 0$. This means that $\phi(x) \geq 0$ for $x > 0$, $\phi'(x) > 0$ for $x > 1$ and $\phi'(x) < 0$ for $x < 1$.

Arising from the property that for every convex functions ϕ, the inequality $\phi(t^*) \leq \phi(0) + t^* \lim_{r\to\infty} \frac{\phi(r)}{r}$, we have the upper bounds

$$D_\phi(t) < \phi(0) + \lim_{r\to\infty} \frac{\phi(r)}{r}$$

and

$$\bar{D}_\phi(t) < \phi(0) + \lim_{r\to\infty} \frac{\phi(r)}{r}.$$

If t_0 be a point such that $\frac{f(t_0)}{\bar{F}(t)} = \frac{g(t_0)}{\bar{G}(t)}$ and $t_0 < t$, there also exist lower bounds

$$D_\phi(t) \geq \phi\left(\frac{h_X(t)}{h_Y(t)}\right) \geq 0$$

and for $t_0 > t$,

$$\bar{D}_\phi(t) \geq \phi\left(\frac{\lambda_X(t)}{\lambda_Y(t)}\right) \geq 0.$$

For a bijective transformation $T(\cdot)$,

$$D_{\phi,T(X),T(X^*)}(t) = D_{\phi,X,X^*}(T^{-1}(t))$$

and

$$\bar{D}_{\phi,T(X),T(X^*)}(t) = \bar{D}_{\phi,X,X^*}(T^{-1}(t)).$$

Among three random variables X_1, X_2 and Y such that $\frac{f_1(x)}{g(x)}\left(\frac{f_2(x)}{g(x)}\right)$ is increasing in x and $X_{2,t} \geq_{st} X_{1,t}$ ($X_{1,t} \geq_{st} X_{2,t}$) the inequality

$$D_{\phi,X_1,Y}(t) \leq (\geq) D_{\phi,X_2,Y}(t)$$

holds where $X_{i,t} = [X_i | X_i > t]$, $i = 1, 2$. Similarly if $\frac{f_1(x)}{g(x)} \left(\frac{f_2(x)}{g(x)} \right)$ is increasing in x and $X_2^{(t)} \geq_{\mathrm{st}} X_1^{(t)}$ $(X_1^{(t)} \geq_{\mathrm{st}} X_2^{(t)})$ then

$$\bar{D}_{\phi,X_1,Y}(t) \leq (\geq) \bar{D}_{\phi,X_2,Y}(t)$$

and $X_i^{(t)} = [X_i | X_i \leq t]$. On the other hand with X, Y_1 and Y_2 with densities f, g_1 and g_2 satisfying $Y_{1,t} \geq_{\mathrm{lr}} Y_{2,t}$ and $\frac{f(x)}{g_1(x)}$ increasing

$$\bar{D}_{\phi,X,Y_1}(t) < \bar{D}_{\phi,X,Y_2}(t) + \lim_{r \to \infty} \frac{\phi(r)}{r} \left(1 - \frac{\lambda_{Y_2}(t)}{\lambda_{Y_1}(t)} \right), \quad t > 0$$

and if $Y_1^{(t)} \geq_{\mathrm{st}} Y_2(t)$

$$D_{\phi,X,Y_1}(t) < D_{\phi,X,Y_2}(t) + \lim_{r \to \infty} \frac{\phi(r)}{r} \left(1 - \frac{h_X(t)}{h_{Y_1}(t)} \right).$$

When $\lim_{r \to \infty} \frac{\phi(r)}{r} \leq 0$ the last two relations become

$$D_{\phi,X,Y_1}(t) < D_{\phi,X,Y_2}(t) \text{ and } \bar{D}_{\phi,X,Y_1}(t) < \bar{D}_{\phi,X,Y_2}(t).$$

THEOREM 5.12. *The proportional hazards model $\bar{G} = \bar{F}^\theta$ (proportional reversed hazard model $G = F^\theta$) is independent of t and is*

$$D_\phi(t) = \int_0^1 \phi \left(\frac{1}{\theta y^{\theta-1}} \right) dy^\theta \quad \left(\bar{D}_\phi(t) = \int_0^1 \phi \left(\frac{1}{\theta y^{\theta-1}} \right) dy^\theta \right).$$

EXAMPLE 5.11. *If we consider the Cressie and Read divergence (Table 5.1) the measure is*

$$D_\phi(t) = \frac{1}{a(1+a)} \left(\frac{1}{\theta^a (a(1-\theta)+1)} - 1 \right)$$

(same as $\bar{D}_Q$) from which Kullback-Leibler measure arises as $- \log \theta + \theta - 1$ as found earlier.

Vonta and Karagrigoriou (2010) give some results relating to the Cox model $\bar{F}(x) = e^{-\theta H(x)}$ and the frailty model $\bar{G}(x) = e^{G(\theta H(x))}$ where $H(x)$ is the cumulative hazard function of X.

THEOREM 5.13 (Kullback-Leibler). *The discrimination measure between X and Y which follow the Cox proportional hazards model and the frailty model is given by*

$$D = \int_0^\infty e^{-y} (G(y) - \log G'(y)) dy - 1,$$

the generalized measure

$$D_\phi = \int_0^\infty e^{-G(y)} G'(y) \phi \left(\frac{e^{-y}}{e^{-G(y)} G'(y)} \right) dy,$$

and for the residual and past life,

$$D_\phi(t) = \int_{\theta H(t)}^{\infty} \frac{e^{-G(y)}G'(y)}{e^{-G(\theta)H(t)}} \phi\left(\frac{e^{-y}/e^{-\theta H(t)}}{\bar{e}^{G(y)}G'(y)/\bar{e}^{G(\theta)H(t)}}\right) dy$$

and

$$\bar{D}_\phi(t) = \int_0^{\theta H(t)} \frac{e^{-G(y)}G'(y)}{1 - e^{-G(\theta)H(t)}} \phi\left(\frac{e^{-y}/e^{-\theta H(t)}}{\bar{e}^{G(y)}/\bar{e}^{G(\theta)H(t)}}\right) dy.$$

As was done for Renyi divergence, writing $B(u) = F(Q_y(u))$ so that $b(u) = fQ_Y(u)/gQ_Y(u)$ we have

$$D_\phi(u) = \int_u^1 \frac{g(Q_y(p))}{f(Q_Y(p))} \phi\left(\frac{f(Q_y(p))}{\bar{F}(Q_Y(u))} \frac{\bar{G}(Q_y(u))}{g(Q_Y(p))}\right) dp$$

$$= \frac{1}{1-u} \int_u^1 \phi\left(\frac{(1-u)b(p)}{1 - B(u)}\right) dp$$

or

$$\frac{d}{du}(1-u)D_\phi(u) = \int_u^1 \frac{\partial}{\partial u}\phi\phi\left(\frac{(1-u)}{1 - B(u)}\phi(p)\right) dp - \phi\left(\frac{(1-u)}{1 - B(u)}b(u)\right) dp. \tag{5.53}$$

Substituting the expression for $\phi(u)$ in various cases, one can derive the formula for divergence. For illustration in the Kullback-Leibler divergence, $\phi(u) = -\log u + u - 1$ with u replaced by $\frac{(1-u)b(p)}{1-B(u)}$, we have the first term in (5.53) vanishing and the expression in (5.16) is restored. Similar expressions can be worked out for other measures also.

5.6. Chernoff distance

Another divergence measure with a simple structure is the Chernoff (1951) distance defined by

$$C(F, G) = -\log \int f^\alpha(x)g^{1-\alpha}(x)dx, \ 0 < \alpha < 1 \tag{5.54}$$

which is always non-negative. We note that $C = 0$ when $\alpha = 0$ or 1 or when $f(x) = g(x)$. Also

$$\lim_{\alpha \to 1} \frac{C}{1-\alpha} = D(F, G) \text{ and } \lim_{\alpha \to 0} \frac{C}{\alpha} = D(G, F),$$

D, denoting the Kullback-Leibler divergence. The measure of affinity

$$A = \int \sqrt{f(x)g(x)}dx,$$

the Hellinger distance

$$H = \frac{1}{2}\int(\sqrt{f} - \sqrt{g})^2 dx$$

are related to C as

$$C = -\log A = B \text{ and } C = -\log(1 - H)$$

when $\alpha = \frac{1}{2}$, where B is the Bhattacharya coefficient. Nair et al. (2011b) have studied various aspects of (5.54) for the residual life

$$C(t) = -\log \int_t^\infty \left(\frac{f(x)}{\bar{F}(t)}\right)^\alpha \left(\frac{g(x)}{\bar{G}(t)}\right)^{1-\alpha} dx \qquad (5.55)$$

and the past life

$$\bar{C}(t) = -\log \int_t^\infty \left(\frac{f(x)}{F(t)}\right)^\alpha \left(\frac{g(x)}{G(t)}\right)^{1-\alpha} dx. \qquad (5.56)$$

In view of the identity

$$C(t) = (1 - \alpha)D_\alpha(t)$$

where $D_\alpha(t)$ is the Renyi measure in equation (5.28) all the results for Chernoff distance follow from those of Renyi divergence. The same approach holds for the quantile version in Kayal (2018b) and the doubly truncated measure in Kundu (2017). As mentioned above the affinities for discriminating the residual lives of F and G are

$$A(t) = \int_t^\infty \left(\frac{f(x)}{\bar{F}(t)}\frac{g(x)}{\bar{G}(t)}\right)^{1/2} dx$$

is a straight forward deduction from (5.54) as

$$C(t) = -\log A(t)$$

with $\alpha = 1/2$. Hence the details of the discussions in all these cases are not pursued further.

5.7. Lin-Wong divergence

The continuous version of the measure introduced by Lin and Wong (1990) was considered by Khalil et al. (2018). It is given by

$$L(F, G) = \int f(x) \log \frac{2f(x)}{f(x) + g(x)} dx. \qquad (5.57)$$

For past lifetimes (5.56) transforms to

$$\bar{L}(t) = \int \frac{f(x)}{F(t)} \log \frac{2\frac{f(x)}{F(t)}}{\frac{f(x)}{F(t)} + \frac{g(x)}{G(t)}} \qquad (5.58)$$

When $X_1, X_2, \ldots, X_n$ are independent lifetimes a relationship exists between (5.57) and the Fisher information for the past lifetime in the form

$$L(f_t(x;\theta), f_t(x;\theta + \Delta\theta)) \equiv \frac{(\Delta\theta)^2}{8} I_t(\theta)$$

where $f_t = \frac{f(x)}{F(t)}$ and $I_t(\theta)$ is taken with respect to $f_t(x;\theta)$. Further if $X \geq_{\text{lr}} Y$,

$$L(t) \leq \frac{2\lambda_X(t)}{\lambda_X(t) + \lambda_Y(t)}$$

and if X_1, X_2 and Y are such that $X_1 \geq_{\text{lr}} Y$ and $X_2 \geq_{\text{lr}} Y$, then (i)

$$L_{g_1,f}(t) - L_{g_2,f}(t) \leq \log \frac{\lambda_{X_1}(t)}{\lambda_{X_1}(t) + \lambda_Y(t)} - \log \frac{\lambda_{X_2}(t)}{\lambda_{X_2}(t) + \lambda_Y(t)}$$

(ii)

$$L_{f,g_1}(t) - L_{f,g_2}(t) \geq \log \frac{\lambda_{X_2}(t) + \lambda_Y(t)}{\lambda_{X_1}(t) + \lambda_Y(t)}$$

(iii)

$$L_{f,g_1}(t) \geq L_{f,g_2}(t)$$

provided $\lambda_{X_2}(t) \geq \lambda_{X_1}(t)$.

With the Kullback-Leibler divergence $D(t)$, if $X \geq_{\text{lr}} Y$

$$D_{f,g}(t) - L_{f,g}(t) \leq \log \frac{\lambda_X(t) + \lambda_Y(t)}{2\lambda_Y(t)}.$$

CHAPTER 6

Inaccuracy

6.1. Introduction

The concept of inaccuracy was introduced by Kerridge (1961). He argued that an experimenter stating probabilities of outcomes of an experiment can commit two types of lack of precision; vagueness due to non-availability of enough information and incorrect information. The first aspect of vagueness as part of the inaccuracy stated could be dealt with a measure of uncertainty proposed by Shannon while the second also needs consideration for which the measure of inaccuracy was proposed. If an experimenter asserts the probability of the ith event as p_i while the true probability is q_i, Kerridge defined the inaccuracy in the statement as $I = -\sum q_i \log p_i$. With the aid of four assumptions, three of which are restatements of Shannon's postulates, the only measure satisfying them was proved to be I. Kerridge (1961) conceived it as the amount of information required to convey which of a number of alternatives is true to someone who believes the probability of the ith event to be p_i when it is actually q_i.

Nath (1968) was instrumental in extending the measure of inaccuracy to the continuous case also and to generalize it to inaccuracy of order α for both continuous and discrete distributions. Let $\mathcal{A}$ be an abstract space, $\mathcal{B}$ the σ-algebra of subsets of $\mathcal{A}$, μ a σ-finite measure defined over $\mathcal{B}$ and $\mathcal{P}$, the set of probability distributions defined over $(\mathcal{A}, \mathcal{B})$, which are absolutely continuous with respect to $(\mathcal{A}, \mathcal{B})$. For every F in $\mathcal{P}$, there exists a density function f such that $0 < f(x) < \infty$, $f(x) = \frac{dF}{d\mu}$. We define the inaccuracy when F is replaced by G as

$$I_{F,G} = c \int f(x) \log g(x) d\mu, \qquad (6.1)$$

c being a constant. A choice of $c = -1$ gives

$$I_{F,G} = - \int f(x) \log g(x) d\mu. \qquad (6.2)$$

When $f(x) = g(x)$, (6.1) reduces to H_F, the Shannon's entropy. Throughout this chapter we choose μ as the Lebesgue measure. Nath (1968) has proved the following properties of $I_{F,G}$.

(1) $I_{F,G}$ attains its maximum value only when $F = G$

(2) It is not invariant under the transformation of coordinates

(3) $I_{F,G} = H_F + D_{F,G}$ where $D_{F,G}$ is the Kullback-Leibler measure in (4.1)

(4) A necessary condition that I is finite is that $F = G$, but this is not sufficient

(5) $I_{F,G} < \infty$ does not imply $I_{G,F} < \infty$ and also $I_{F,G} < \infty$ and $I_{G,H} < \infty$ does not mean $I_{F,H} < \infty$

(6) $I_{F,G}$ always exists though it may be $-\infty$ or $+\infty$

(7) $I_{F,H} - I_{F,G} = D_{F,H} - D_{F,G}$

(8) If $\{f_n\}$, $n = 1, 2, \dots$ is a sequence of densities converging to f, then $\lim_{n\to\infty} I_{fn,f} = H_f$ and if, $\lim_{n\to\infty} g_n = g$, then $\lim_{n\to\infty} I_{fn,gn} = I_{f,g}$.

The inaccuracy of order α is defined as

$$I_{\alpha,F,G} = I_\alpha = \frac{1}{1-\alpha} \log \int f(x) \log g^{\alpha-1}(x) dx. \tag{6.3}$$

Note that I_α is a decreasing function α and $I_\alpha > I_1$ when $0 < \alpha < 1$ and $I_\alpha < I_1$, when $1 < \alpha < \infty$ where I_1 is the Kerridge inaccuracy (which is of order 1 in (6.3)). Also I_α may be positive or negative or even zero depending on α and may not always exist for some α.

6.2. Inaccuracy of residual lives

Let X and Y be two non-negative continuous random variables with distribution functions F and G and densities f and g. When the true survival function (of X) is $\bar{F}(x)$ and the reference survival function is $\bar{G}(x)$ (of Y) the inaccuracy associated with the residual lives X_t and Y_t is

$$I(t) = - \int_t^\infty \frac{f(x)}{\bar{F}(t)} \log \frac{g(x)}{\bar{G}(t)} dx \tag{6.4}$$

as is evident from (6.1) when μ is the Lebesgue measure. Since for every $t > 0$, (6.4) is an inaccuracy function such that the property (3) above holds with

$$I(t) = H(t) + D(t) \tag{6.5}$$

where

$$H(t) = - \int_t^\infty \frac{f(x)}{\bar{F}(t)} \log \frac{f(x)}{\bar{F}(t)} dx$$

and

$$D(t) = \int_t^\infty \frac{f(x)}{\bar{F}(t)} \log \left(\frac{f(x)/\bar{F}(t)}{g(x)/\bar{G}(t)} \right) dx$$

are respectively the residual entropy (2.7) and the residual divergence (5.3) respectively. We shall refer to (6.4) as the residual inaccuracy.

Rewriting (6.4) as

$$\bar{F}(t)I(t) = - \int_t^\infty f(x) \log g(x) + \bar{F}(t) \log \bar{G}(t) \qquad (6.6)$$

and differentiating

$$I'(t) = h_F(t) \left[I(t) + \log h_G(t) - \frac{h_G(t)}{h_F(t)} \right]. \qquad (6.7)$$

Equations (6.6) and (6.7) contains three unknowns $I(t)$, $\bar{F}(t)$ $(h_F(t))$ and $\bar{G}(t)$ $(h_G(t))$ and therefore characterizing any one of them is an insurmountable problem. To resolve this, assumptions are made about h_G and h_F or $\bar{F}$ and $\bar{G}$ in order to characterize the inaccuracy function. We shall review some of the major results in this connection.

As seen in an earlier chapter one familiar assumption is that G is the proportional hazards model corresponding to F so that $h_G(t) = \theta h_F(t)$ together with a distributional form of $\bar{F}$. Some examples of $I(t)$ satisfying this assumption is given in Table 6.1.

THEOREM 6.1. *Let $\bar{G}$ be the proportional hazards model of $\bar{F}$. Then $I(t)$ is log linear in t if and only if $\bar{F}$ has generalized Pareto form*

$$\bar{F}(x) = \left(1 + \frac{ax}{b} \right)^{-1-\frac{1}{a}}, \ a > -1, \ b > 0$$

(Nair and Gupta (2007)).

REMARK 6.1. *Since $h_G(t) = \theta h_F(t)$, the form of $I(t)$ is*

$$I_{F,G}(t) = \log \frac{at+b}{\theta(a+1)} - \frac{a+(a+1)\theta}{a+1}.$$

A similar result exists when the roles of F and G are interchanged resulting in

$$I_{G,F}(t) = \log \frac{at+b}{a+1} - \frac{(1+2a)b}{\theta(a+1)}.$$

REMARK 6.2. *As $a \to 0$, $\bar{F}(x) = \exp(-\frac{x}{b})$ and $I(t) = \log(\frac{b}{\theta}) - \theta$ which is independent of t. That is, under proportional hazards model $I(t)$ is a constant if and only if $\bar{F}$ is exponential.*

The monotonicity of $I(t)$ is regulated by the conditions in the next theorem which is obvious from (6.7).

Table 6.1. Residual inaccuracy function for some distributions under proportional hazards assumption.

Distribution	$\bar{F}(x)$	$I(t)$
Power	$1 - x^c, 0 \le x \le 1, c > 0$	$\frac{c-1}{c} + \log \frac{1-t^c}{c}$
(uniform $c = 1$)		$\frac{(c-1)t^c}{1-t^c} + c(\theta)$
Generalized Pareto	$(1 + \frac{ax}{b})^{-(1+\frac{1}{a})}, x > 0$	$\log(b + at) + (a - 1)^{-1}$
	$-1 < a, b > 0$	$-\log(a + 1) + c(\theta)$
Burr XII	$(1 + x^c)^{-k}, x > 0, k, c > 0$	$k^{-1}[\log(kc) - \frac{\log(1+t^c)}{c}$
(Pareto II when $k = 1$)		$+\frac{c-1}{c} \sum_{r=1}^{k-1} \frac{(-1)^{r-1}}{(1+t^c)^k} + c(\theta)]$
Exponential geometric	$(1 - p)e^{-\lambda x}(1 - pe^{-\lambda x})^{-1}$	$2 - \log \lambda$
		$+p^{-1}e^{\lambda t} \log((1 - p)e^{-\lambda t}) + c(\theta),$
		$c(\theta) = \theta - 1 - \log \theta$

THEOREM 6.2. *The inaccuracy function is increasing or decreasing according as*

$$I(t) \ge (\le) \frac{h_G(t)}{h_F(t)} - \log G(t).$$

There is some lower bound to $I(t)$ in terms of the Renyi entropy

$$H_{\alpha, F}(t) = (1 - \alpha)^{-1} \log \int_t^\infty \left(\frac{f(x)}{\bar{F}(t)}\right)^\alpha dx$$

defined in (5.6).

THEOREM 6.3.

$$I_{F,G}(t) \ge [H_{2,F}(t) \cdot H_{2,G}(t)]^{\frac{1}{2}}. \tag{6.8}$$

PROOF. If ϕ is a convex function, then by Jensen's inequality $E\phi(X) \ge \phi E(X)$. Invoking the inequality for the conditional expectations,

$$\int_t^\infty \frac{f(x)}{\bar{F}(t)} \log \frac{g(x)}{\bar{G}(t)} dx \le \log \int_t^\infty \frac{f(x)}{\bar{F}(t)} \frac{g(x)}{\bar{G}(t)} dx \tag{6.9}$$

and by Cauchy-Schwartz inequality

$$\left(\int_t^\infty \frac{f(x)}{\bar{F}(t)} \frac{g(x)}{\bar{G}(t)} dx\right)^2 \le \log \int_t^\infty \left(\frac{f(x)}{\bar{F}(t)}\right)^2 dx \int_t^\infty \left(\frac{g(x)}{\bar{G}(t)}\right)^2 dx. \tag{6.10}$$

Inequalities (6.9) and (6.10) combined together gives (6.8). □

We have considered in the previous chapter, several cases of interest in which there is a meaningful relationship between F and G. They continue to be of relevance in the context of the inaccuracy also. Some attention is paid in literature to the case of weighted distributions. Considering $g(x) =$

$\frac{w(x)f(x)}{\mu(x)}$, $\mu_w = E(w(X))$ which is assumed to be finite and non-zero,

$$\bar{G}(x) = \frac{\int_x^\infty w(t)f(t)dt}{\mu_w} = \frac{E(w(X)|X > x)\bar{F}(x)}{\mu_w}$$

and the measure of inaccuracy becomes

$$I_w(t) = -\int_t^\infty \frac{f(x)}{\bar{F}(t)} \log \frac{w(x)f(x)}{\bar{F}(t)E(w(X)|X > t)} dx.$$

Some simplification leads to

$$I_w(t) = -\frac{1}{\bar{F}(t)} \int [w(x)f(x)]f(x)dx - \log \bar{F}(t)E(w(X)|X > t)$$

$$= E[\log w(X)f(X)|X > t] - \log \bar{F}(t)E(w(X)|X > t).$$

The simplest case of $w(x)$ is when $w(x) = x$, resulting in the length-biased case.

We illustrate the above results with a few examples.

EXAMPLE 6.1. *(i) When X and Y have exponential distributions $\bar{F}(x) = \bar{e}^{\alpha x}$ and $\bar{G}(x) = \bar{e}^{\beta x}$, we have*

$$I_{Ex}(t) = -\log \beta - \frac{\beta}{\alpha}; \quad h_F(x) = \alpha, \; h_G(x) = \beta$$

(ii) If we have two Pareto distributions $\bar{F}(x) = \alpha^c(x + \alpha)^{-c}$ and $\bar{G}(x) = \alpha^{c_1}(x + \alpha)^{-c_1}$,

$$I_P(t) = \log \frac{t + \alpha}{c_1} + \frac{1 + c_1}{c}; \quad h_F(x) = \frac{c}{x + \alpha} \text{ and } h_G(x) = \frac{c_1}{x + \alpha}$$

and
(iii) in the case of beta models $\bar{F}(x) = (1 - \frac{x}{R})^c$ and $\bar{G}(x) = (1 - \frac{x}{R})^{c_1}$, $0 \le x \le R$, $c, R > 0$

$$I_B(t) = \log \frac{R - t}{c_1} + \frac{c_1 - 1}{c}; \quad h_F(x) = \frac{c}{R - x}, h_G(x) = \frac{c_1}{R - x}.$$

All the three choices satisfy the proportional hazards model assumption. Since $\log \frac{t+\alpha}{c_1} + \frac{1+c_1}{c} > \frac{c_1}{c} - \log c_1$ by Theorem 6.2 $I_P(t)$ is increasing, which is also evident from the expression for $I_P(t)$. Similarly $I_B(t)$ is decreasing and $I_{Ex}(t)$ is a constant.

6.2.1. Characterizations. There are several types of characterizations of the residual inaccuracy measure than the one mentioned in Section 6.2 that need attention. Taneja et al. (2009) have claimed that if X and Y are two non-negative random variables satisfying proportional hazards model $h_G(x) = \theta h_F(x)$ and residual inaccuracy $I(t) < \infty$ for every $t \ge 0$ then $I(t)$ uniquely determines the survival function $\bar{F}$ of X. It appears that the proof given by them needs modifications. When the proportional hazards model assumption is satisfied we have from (6.7),

$$I'(t) = h_F(t)[I(t) + \log h_F(t) + \log \theta - \theta]. \tag{6.11}$$

For a fixed $t > 0$, (6.11) with $z = h_F(t)$, we write

$$p(z) = z[I(t) + \log z + \log \theta - \theta] - I'(t) \tag{6.12}$$

so that the hazard rate of X is a solution of the equation $p(z) = 0$. Now consider the power distribution $\bar{F}(x) = 1 - x^c$, $0 \le x \le 1$ with hazard rate $h_F(x) = \frac{cx^{c-1}}{1-x^c}$ and its proportional hazard rate model $\bar{G}(x) = (1 - x^c)^\theta$ having hazard rate $h_G(x) = \frac{\theta c x^{c-1}}{(1-x^c)}$. For this model,

$$I(t) = -\log \theta - \log c + \log(1 - t^c) + \theta - 1 + \frac{c-1}{c} + \frac{(c-1)t^c}{1-t^c} \log t$$

and

$$I'(t) = -\frac{t^{c-1}}{1-t^c} + \frac{(c-1)ct^{c-1} \log t}{(1-t^c)^2}.$$

Looking at the behaviour of the function $p(z)$ we see that

$$p(0) = -I'(t) > 0$$

$p(\infty) = \infty$ and $p(z)$ attains its minimum at z_0, which is the solution of

$$p'(z) = \log \frac{1-t^c}{c} + \frac{c-1}{c} + \frac{(c-1)t^c}{1-t^c} \log t + \log z = 0$$

so that

$$z_0 = \frac{c}{(1-t^c)t^{(c-1)}} \frac{t^c}{1-t^c} e^{-\frac{c-1}{c}}.$$

Thus the minimum value of $p(z)$ is

$$p(z_0) = -z_0 - I'(t) \ne 0.$$

By inspection, $z_1 = h_F(t) = \frac{ct^{c-1}}{1-t^c}$ is a solution of $p(z)$ and the minimum value of $p(z)$,

$$z_0 = \frac{ct^{c-1}}{1-t^c} \times e^{-\frac{c-1}{c}} t^{-\frac{(c-1)}{1-t^c}} > h_F(t).$$

Thus $p(z)$ crosses the z-axis at another point $z_2 = h_2(t) > h_F(t)$. The function $h_2(t)$ satisfies $h_2(t) > h_F(t) > 0$ and

$$\int_0^\infty h_2(t)dt > \int_0^\infty h_F(t)dt = \infty$$

and is therefore a hazard rate. Thus $p(z) = 0$ has two hazard rates as solutions which is equivalent to two distributions for $I(t)$.

Nair et al. (2011a) considered an extension of the Kerridge measure in the form

$$I_r(F, G) = \frac{1}{r} \int_0^\infty f(x)(1 - g^r(x))dx, \ r > -1. \qquad (6.13)$$

Some properties of this measure are (i) As $r \to 0$, $I_r(F, G) = I(F, G)$, the Kerridge measure (ii) $I_r(F, G) = H_r(F) + D_r(F, G)$ (iii) $I_r(F, G)$ is minimum, when $f(x) = g(x)$ and in this case $I_r(F, G) = H_r(F)$, the uncertainty in the observations alone. The truncated version of (6.13) is

$$I_r(t) = \frac{1}{r} \int_t^\infty \frac{f(x)}{\bar{F}(t)} \left[1 - \left(\frac{g(x)}{\bar{G}(t)}\right)^r\right] dx \qquad (6.14)$$

representing the residual life. Differentiating (6.14) with respect to t

$$1 - rI_r(t) = \frac{h_F(t)h_G^r(t) - rI_r'(t)}{rh_G(t) + h_F(t)}. \qquad (6.15)$$

When $\bar{G}(x)$ is the proportional hazards model $\bar{F}(x)$, (6.15) reduces to

$$1 - rI_r(t) = \frac{\theta^r h_F^{r+1}(t) - rI_r'(t)}{(1 + r\theta)h_F(t)}.$$

THEOREM 6.4. *For absolutely continuous distribution functions $F(x)$ and $G(x)$ with G as the proportional hazards model of F, the relationship*

$$1 - rI_r(t) = k(h_F(t))^r$$

holds for $r > -1$ if and only iff F is the generalized Pareto distribution having

$$\bar{F}(t) = \left(1 + \frac{at}{b}\right)^{(1+\frac{1}{a})}, \ t > 0, \ a > -1, \ b > 0$$

and k is some positive constant that depends on r and θ.

THEOREM 6.5. *Under conditions of Theorem 6.4 and X and Y have finite means,*

$$1 - rI_r(t) = c(m(t))^{-r},$$

where c is a constant depending on r and θ if and only if $\bar{F}(t)$ is as in Theorem 6.4.

THEOREM 6.6. *If $I_r(t) = c$, a positive constant with $cr < 1$, then F is exponential with parameter $(1 - rc)^{1/r}(1 - r)^{\frac{r-1}{r}}$ if and only if G is exponential with parameter, $\left(\frac{1-rc}{1-r}\right)^{1/r}$.*

Kumar et al. (2010) discussed the properties of the weighted residual in accuracy.

$$I_L(t) = -\int_t^\infty x\frac{f(x)}{\bar{F}(t)} \log \frac{g(x)}{\bar{G}(t)} dx,$$

also written as the identity

$$I_L(t) = t\, I(t) - \int_t^\infty \beta(y,t)dt,$$

where

$$\beta(y,t) = \int_y^\infty \frac{f(x)}{\bar{F}(t)} \log \frac{g(x)}{\bar{G}(t)} dt.$$

Further differentiation yields

$$\frac{d}{dt} I_L(t) = t\frac{d}{dt} I(t).$$

They have proved the following characterization.

THEOREM 6.7. *If X and Y satisfy the proportional hazard rate model and $I_L(t)$ is increasing for all t and $I(t) \geq 0$, then $I_L(t)$ uniquely determines $\bar{F}$.*

Further there exists a lower bound to $I_L(t)$ viz.

$$I_L(t) \geq -E(X|X > t) \log h_G(t)$$

whenever $h_G(t)$ is decreasing in t and $E(X) < \infty$. Kundu (2014) gave an upper bound

$$I_L(t) \leq -E(X|X < t) \log h_G(t) - \frac{1}{\bar{F}(t)} \int_t^\infty x f(x) \log\left(\frac{\bar{G}(x)}{\bar{G}(t)}\right) dx.$$

He has characterized the uniform distribution over (α, β) by the relationship

$$I_L(t) = (1-\theta)E(X \log(X - \alpha)|X > t) - E(X|X > t)\left[\log\frac{t-\alpha}{h_G(t)}\right]$$

when X and Y satisfy the proportional reversed hazard rate model $\lambda_G(t) = \theta\lambda_H(t)$. Similarly the power distribution $F(t) = (\frac{t}{b})^c$, $0 < t < b$, $c > 0$ is characterized by

$$I_L(t) = (1-c\theta)E(X \log(X - d)|X > t) - E(X|X > t)\left[\log\frac{\log t}{h_G(t)}\right],$$

under the proportional reversed hazard rate model assumption. On the other hand when the assumption is the proportional hazard rate model between X and Y the characteristic property turns out to be

$$I_L(t) = \lambda\theta(E(X^2|X > t) - tE(X|X > t)) - E(X|X > t)\log h_G(t)$$

for the exponential law with parameter λ. Some similar results are also proved for Weibull, Pareto I and II laws. See also the discussion along similar lines in respect of doubly truncated random variables by the same author (Kundu (2017)).

A generalized measure of inaccuracy considered in Kayal et al. (2017) has the expression

$$I_\alpha = \frac{1}{\alpha - 1} \log \left[\frac{\int_0^\infty f^\alpha(x) g^{1-\alpha}(x) dx}{\int_0^\infty f^\alpha(x) dx} \right]$$

for which the residual inaccuracy is

$$I_\alpha(t) = \frac{1}{\alpha - 1} \log \left[\frac{\int_t^\infty (\frac{f(x)}{F(t)})^\alpha (\frac{g(x)}{G(t)})^{1-\alpha} dx}{\int_t^\infty (\frac{f(x)}{F(t)})^\alpha dx} \right].$$

They have the following properties.

THEOREM 6.8. *If $I_\alpha(t)$ is independent of t for all $t > 0$, then G is exponential if F is exponential.*

THEOREM 6.9. *If $I_\alpha(t)$ is independent of t and G is the proportional hazards model of F. Then*

$$I_\alpha(t) = \log(B - Ch_F^{\alpha-1}(t))^{\frac{1}{1-\alpha}}$$

where $B = \frac{\theta a}{A(1-\alpha)}$ and $C = \frac{\theta}{1-\alpha}$.

THEOREM 6.10. *If $\frac{h_F(t)}{h_G(t)}$ is increasing in t and both F and G have increasing (decreasing) hazard rates, then for $\alpha < 1$, $I_\alpha(t) \geq (\leq) I_\alpha$.*

THEOREM 6.11. *If $X \leq_{lr} Y$ then*

$$(a)\ I_\alpha(t) \leq \frac{\alpha}{\alpha - 1} \log \frac{h_F(t)}{h_G(t)} + H_\alpha(t),\ \alpha > 1$$

$$(b)\ I_\alpha(t) \geq \frac{\alpha}{\alpha - 1} \log \frac{h_F(t)}{h_G(t)} + H_\alpha(t),\ \alpha < 1.$$

Further if X_1, X_2 and X_3 have hazard rates h_1, h_2 and h_3 and $X_1 \leq_{lr} X_2$, then

$$I_{\alpha, X_1, X_3}(t) - I_{\alpha, X_1, X_2}(t) \leq \frac{\alpha}{\alpha - 1} \log \frac{h_1(t)}{h_2(t)} + H_{\alpha, X_1}(t) - H_{\alpha, X_2}(t)$$

if $\alpha > 1$ and the inequality is reversed for $\alpha < 1$.

REMARK 6.3. *Infact Theorem 6.8 offers a characterization in the form G is exponential if and only if F is exponential when $I_\alpha(t)$ is independent of t for all $t > 0$. To see this assume that G is exponential,*

$$\bar{G}(x) = xe^{-\lambda x}, x > 0$$

and also that $\frac{(1-\alpha)\lambda_c}{c\alpha - \alpha \lambda^{1-\alpha}}$ is negative, where $I_\alpha(t) = C$, a constant. Then

$$CG^{1-\alpha}(t) \int_t^\infty f^\alpha(x) dx = \int_t^\infty f^\alpha(x) \lambda^{1-\alpha} e^{-\lambda(1-\alpha)x} dx.$$

Differentiating and simplifying

$$Cf^\alpha(t) + (1-\alpha)\lambda C \int_t^\infty f^\alpha(x)dx = f^\alpha(t)\lambda^{1-\alpha}.$$

Differentiating again

$$c\alpha f'(t) - (1-\alpha)\lambda_c f(t) = \alpha f'(t)\lambda^{1-\lambda}$$

gives,

$$\frac{f'(t)}{f(t)} = \frac{(1-\alpha)\lambda_c}{c\alpha - \alpha\lambda^{1-\lambda}} = -\beta, \ say, \ \beta > 0.$$

The last equation gives $f(t) = \beta e^{-\beta t}$, $\beta > 0$ *and thus* F *is exponential.*

In Remark 6.1, it is stated that that for $I(t)$ to be a constant, Y is the proportional hazards model of X and X is exponential. However, it is not essential that these assumptions hold for $I(t)$ to be independent of t. Moreover, there are characterizations based on the form of $I(t)$ along with the distribution of X as discussed in Smitha (2010). The main results are given in the following theorem.

THEOREM 6.12.   *(1) Given* $I(t)$ *is independent of* t, X *is exponential if and only if* Y *is exponential*
(2) Given $I(t)$ *is independent of* t *then* X *has distribution*

$$\bar{F}(x) = \left(\frac{\log \beta - \log x}{\log \beta - \log K}\right)^\alpha, \frac{\beta}{K} \le x \le \beta, \ \beta, k > 1, \ \alpha > 0$$

if and only if Y *is Pareto I with* $\bar{G}(x) = (\frac{k}{x})^\alpha$, $k, \alpha > 0$
(3) If $I(t)$ *is a linear function* t *of the form* $I(t) = a + bt$, $b \ne 0$, *then* X *has generalized Pareto distribution if and only if* Y *follows exponential distribution*

The measure of inaccuracy can be directly connected with reliability functions like hazard rate and mean residual life. Two results in this connection are given below.

THEOREM 6.13. *If* Y *is the proportional hazards model of* X *then*

$$I(t) = A(\theta) + \log m_F(t)$$

if and only if X *follows a generalized Pareto distribution.*

PROOF. When X has a generalized Pareto distribution

$$I(t) = (a+1)^{-1} - \log(a+1) + \theta - 1\log\theta + \log(a+bt).$$

Since $m(t) = a + bt$ for the distribution, $I(t)$ satisfies the stated property. Conversely if $I(t)$ is as above

$$I'(t) = \frac{m'(t)}{m(t)}.$$

Using $I'(t) = h_F(t)[I(t) + \log \theta]$

$$\frac{m'(t)}{m(t)} = h_F(t)[A(\theta) + \log \theta - \theta + \log(h_F(t)m_F(t))]$$

$$= (1 + m'_F(t))[A(\theta) + \log \theta - \theta + \log(1 + m'_F(t))].$$

Differentiating

$$m''_F(t)(A(\theta) + \log \theta - \theta + \log(1 + m'_F(t))) = 0.$$

Thus either $m''_F(t) = 0$ or $m'_F(t) = e^{-A(\theta) - \log \theta + \theta}$. In both cases $m'_F(t)$ is a constant or $m_F(t) = a + bt$ that characterizes the generalized Pareto law. $\qquad\square$

COROLLARY 6.1. *The relation $I(t) = K(\theta) - \log h_F(t)$, holds if and only iff, F is a generalized Pareto distribution.*

6.2.2. Order statistics. Shannon's measure of uncertainty associated with the ith order statistic $X_{i:n}$ is given by

$$H(X_{i:n}) = - \int_0^\infty f_{i:n}(x) \log f_{i:n}(x) dx. \tag{6.16}$$

Under the transformation $U = F(X)$,

$$H(X_i : n) = H_n(w_i) - E_{g_i}(\log f(F^{-1}(w_i))) \tag{6.17}$$

where

$$H_n(w_i) = \log B(i, n - i + 1) - (i - 1)(\psi(i) - \psi(n + 1)) \\ - (n - i)[\psi(n - i + 1) - \psi(n + 1)], \tag{6.18}$$

$$g_i(w) = \frac{w^{i-1}(1 - w)^{n-i}}{B(i, n - i + 1)} \text{ and } \psi(z) = \frac{d \log \Gamma(z)}{dz}.$$

The Kullback-Leibler measures between the distribution of the ith order statistics and the data distribution is

$$D(f_{i:n}, f_X) = \int_0^\infty f_{i:n}(g) \log \left(\frac{f_{i:n}(x)}{f_X(x)} \right) dx \tag{6.19}$$

$$= \int_0^\infty g_i(w) \log g_i(w) dw. \tag{6.20}$$

Adding (6.16) and (6.19)

$$H(X_{i:n}) + D(f_{i:n}, f_X) = - \int_0^\infty f_{i:n}(x) \log f_X(x) dx. \tag{6.21}$$

Using $U = F(X)$, (6.21) becomes $-E_{g_i}(\log f(F^{-1}(w_i)))$. Adding (6.17) and (6.20)

$$H(X_{i:n}) + D(f_{i:n}, f_X) = -E_{g_i}(\log(f F^{-1}(w_i))).$$

Thapliyal and Taneja (2013a) defined the measure

$$I(f_{i:n}, f) = -\int_0^\infty f_{i:n}(x) \log f(x) dx = -E_{g_i}(\log f(F^{-1}(w_i)))$$

as a measure of inaccuracy associated with the ith order statistic and the parent distribution $f(x)$.

THEOREM 6.14. *The distribution functions F and G belong to the same family of distributions but for a change of location if and only if*

$$I(f_{i:n}, f) = I(g_{i:n}, g)$$

for $n = n_j$, $j \geq 1$ such that $\sum_{j=1}^\infty n_j^{-1}$ is infinite.

Thapliyal and Taneja (2013a) have also proposed some properties of order statistics.

(i) If B_i is the ith term of the binomial probability $B(n-1, p_i)$, $p_i = \frac{i-1}{n-1}$,

$$nB_i(H(X) + I(A)) \leq I(f_{i:n}, f) \leq nB_i[H(X) + \bar{I}(A)]$$

where $I_A(A) = \int_A f(x) \log f(x) dx$, $A = \{x : f(x) \leq 1\}$, $\bar{A} = 1 - A$

(ii) If $M = f(m) < \infty$, where m is the mode of the distribution

$$-\log M \leq I(f_{i:n}, f) \leq nB_i[H(X) + \log M] - \log M$$

(iii) $\frac{1}{2}\sum_{i=1}^n I(f_{i:n}, f) = H(X)$.

Thapliyal and Taneja (2015) proposed

$$I(f_{i:n}, f, t) = -\int_t^\infty \frac{f_{i:n}(x)}{\bar{F}_{i:n}(t)} \log\left(\frac{f(x)}{\bar{F}(t)}\right) dx$$

as the dynamic residual measure of inaccuracy associated with two residual lifetime distributions $F_{i:n}$ and F, where the survival function of $X_{i:n}$ is

$$\bar{F}_{i:n} = 1 - F_{i:n} = \frac{B_x(i, n-i+1)}{B(i, n-i+1)}, \quad B_x(p, q) = \int_x^1 u^{p-1}(1-u)^{q-1} du.$$

They have established that

(i) $I(f_{i:n}, f, t) \geq \log \bar{F}(t) - \log M$

(ii) If $X_1, \ldots, X_n$ are independent and identically distributed random variables representing a series system and their common density f is decreasing then the corresponding inaccuracy is a decreasing function of n

(iii) If f satisfies the Lipschitz condition

$$|f(x, y_1) - f(x, y_2)| \leq k(y_1 - y_2), \quad k > 0$$

for every (x, y_1) and (x, y_2) in $D \subset \mathbb{R}^2$, then $y = d(x)$ that satisfies $y' = f(x, y)$ and $\phi(x_0) = y_0$, is unique. Using this $I(f_{i:n}, f, f, t)$ characterizes f.

(iv) The property $I(f_{i:n}, f, t) = c - \log \lambda_F(t)$ is satisfied for a constant C, then X has exponential distribution iff. $c = \frac{1}{n}$, Pareto distribution iff $c > \frac{1}{n}$ and finite range distribution iff $c < \frac{1}{n}$.

6.2.3. Inaccuracy of past life. The inaccuracy measure pertaining to the past life was proposed by Kumar et al. (2011) in the form

$$\bar{I}(F, G; t) = \bar{I}(t) = - \int_0^t \frac{f(x)}{F(t)} \log \frac{g(x)}{G(t)} dx \qquad (6.22)$$

and obtained the relationship between the reversed hazard rates of F and G as

$$\lambda_F(t) = \frac{\lambda_G(t) - \bar{I}'(t)}{\log \lambda_G(t) + \bar{I}(t)}. \qquad (6.23)$$

EXAMPLE 6.2. *Let* $F(x) = x^\alpha$, $0 \le x \le 1$ *and* $G(x) = x^\beta$, $0 \le x \le \beta$. *Then* $\lambda_F = \frac{\alpha}{x}$ *and* $\lambda_G = \frac{\beta}{x}$

$$\bar{I}(t) = - \log \beta + \log t + \frac{\beta - 1}{\alpha}. \qquad (6.24)$$

When the proportional reversed hazard model $\lambda_G = \theta \lambda_F$ *holds between* X *and* Y, (6.23) *reduces to*

$$\lambda_F(t) = \frac{\theta \lambda_F(t) - \bar{I}'(t)}{\log \lambda_F(t) + \bar{I}(t)}. \qquad (6.25)$$

THEOREM 6.15. *Let* X *and* Y *be non-negative random variables such that* $\lambda_Y(t) = \theta \lambda_X(t)$. *The accuracy measure of past entropy is of the form*

$$\bar{I}(t) = A + \log t, \qquad (6.26)$$

$A = (\frac{\beta-1}{\beta})\theta - \log \beta$, *if and only if* Y *has power distribution* $G(x) = x^\beta$, $\beta > 0$, $0 < x < 1$.

PROOF. When $G(x) = x^\beta$, the distribution function of X is $F(x) = x^{\beta/\theta}$ and hence from (6.3)

$$\bar{I}(t) = - \log \beta + \left(\frac{\beta - 1}{\beta} \right) \theta + \log t$$

which is (6.26).

To prove the sufficiency part assume that (6.25) holds. Then (6.25) yields

$$\lambda_F(t) = \frac{\theta \lambda_F(t) - \frac{1}{t}}{\log \theta \lambda_F(t) + \log t - \log \beta + (\frac{\beta-1}{\beta})\theta}$$

or

$$\theta \lambda_F(t) - \frac{1}{t} = \lambda_F(t) \log \theta \lambda_F(t) + \log t - \log \beta + \left(\frac{\beta-1}{\beta}\right)\theta. \quad (6.27)$$

Differentiating

$$\lambda_F'(t)(\log \theta \lambda_F(t)) + \lambda_F'(t)$$

$$+ \lambda_F'(t)\left(\log t - \log \beta + \frac{\beta-1}{\beta}\theta\right) + \frac{\lambda_F(t)}{t} = \frac{1}{t^2} + \theta \lambda_F'(t).$$

Replacing $\log \theta \lambda_F(t)$ by $\theta - \frac{1}{t\lambda_F(t)} - \log t + \log \beta - \frac{(\beta-1)}{\beta}\theta$ obtained from (6.27),

$$\lambda_F'(t) - \frac{\lambda_F'(t)}{t\lambda_F(t)} + \frac{\lambda_F(t)}{t} = 0$$

which factorizes into

$$\left(\lambda_F(t) - \frac{1}{t}\right)\left(\frac{\lambda_F'(t)}{t\lambda_F(t)} + \frac{1}{t}\right) = 0$$

leaving two solutions $\lambda_F(t) = \frac{1}{t}$ and $\lambda_F(t) = \frac{C}{t}$. The first being a particular case of the second, the general solution is $\lambda_F(t) = \frac{C}{t}$ which corresponding to the power distribution and this completes the proof. $\qquad \square$

REMARK 6.4. *When $\beta = 1$, we have the uniform distribution in $(0,1)$ characterized by the property $\bar{I}(t) = \log t$.*

REMARK 6.5. *A somewhat different scenario is presented if we start with the form of the reversed hazard rate $\lambda_F(t)$ in (6.25). For example, if we assume that X has power distribution with $\lambda_F(t) = \frac{\alpha}{t}$, we have*

$$\frac{\alpha}{t}\log\frac{\theta\alpha}{t} + \frac{\alpha}{t}\bar{I}(t) = \frac{\theta\alpha}{t} - \bar{I}'(t)$$

which is a linear differential equation in $\bar{I}(t)$. Solving

$$\bar{I}(t) = \theta - \log\frac{\theta\alpha}{t} - \frac{1}{\alpha} + ct^{-\alpha}.$$

This will give the expression for $\bar{I}(t)$ of the power law only if $C = 0$.

Considering a parallel system of n components having independent and identity distributed lifetimes, the lifetime of the parallel system is $Y = \max(X_1, X_2, \ldots, X_n)$, then it follows that proportional reversed hazards model with $G(x) = F^n(x)$, with $\lambda_G(t) = n\lambda_F(t)$ and all the above results apply. In particular when $F(x)$ is exponential with parameter λ (Kumar et al. (2010))

$$\bar{I}(t) = n - \log n\theta + \log(1 - e^{-\theta t}) - \frac{\theta t e^{-\theta t}}{1 - e^{-\theta t}}.$$

Kundu and Nanda (2014) have obtained relationships that are necessary and sufficient conditions for certain life distributions to satisfy. Among them are

$$\bar{I}(t) + \log \lambda_F(t) + \lambda m(t) = K\theta$$

with $K(\theta) = \theta - 1 - \log \theta$ for the exponential model with mean λ,

$$\bar{I}(t) + \log \lambda_F(t) = \theta - \log \theta - \frac{1}{c}$$

when $F(x) = (\frac{x}{b})^c$, $0 < x < b$,

$$I(t) + \log \lambda_F(t) = K(\theta) + (\alpha - 1)[E(\log X|X > t) - \log t]$$

and

$$I(t) + \log \lambda_F(t) = K(\theta) + (\alpha + 1)[E(\log X|X > t - \log(t - \mu + \beta)],$$

respectively for the Pareto I, $F(x) = 1 - (\frac{\beta}{x})^\alpha$, $x > \beta$ and the Pareto II $F(x) = 1 - [1 + \frac{t-\mu}{\beta}]^{-\alpha}$, $t > \mu$ distributions respectively. They also give several results on interval inaccuracy.

The weighted inaccuracy measure of past life (Kumar and Taneja (2012)) is

$$\bar{I}_L(t) = - \int_0^t x \frac{f(x)}{F(t)} \log \frac{g(x)}{G(t)} dx. \qquad (6.28)$$

They have proved that

$$\frac{d}{dt}\bar{I}_L(t) = t\frac{d}{dt}\bar{I}(t).$$

If X is uniform over (a, b),

$$\bar{I}_L(t) = \left(\frac{t+a}{2}\right) \log \left(\frac{t-a}{\theta}\right) + (\beta - 1) \left(\frac{t+3a}{4}\right)$$

when the reversed proportionality assumption holds with $g(x) = \frac{\theta(x-a)^{\theta-1}}{(b-a)^\theta}$. Differentiating the expression (6.28)

$$\begin{aligned} I(t)F(t) + I(t)f(t) = \ & tf(t)\log g(t) - F(t)\log G(t)r'(t) \\ & -F(t)\frac{g(t)}{G(t)}r(t) - f(t)\log G(t)r(t) \quad (6.29) \end{aligned}$$

where $r(t) = E(X|X \le t) = \frac{1}{F(t)} \int_0^t xf(x)dx$. Also

$$r(t)f(t) + r'(t)F(t) = tf(t)$$

or

$$r'(t) + r(t)\lambda_F(t) = t\lambda_F(t). \qquad (6.30)$$

From (6.29) and (2.20)

$$I(t)\lambda_F(t) + I'(t) = t\lambda_F(t)\log \lambda_G(t) - \lambda_G(t)r(t).$$

Thus

$$I'(t) = \lambda_F(t)[t \log \lambda_G(t) - I(t)] - \lambda_G(t)r(t).$$

In the proportional reversed hazards case, we have

$$I(t)\lambda_F(t) + I'(t) = t\lambda_F(t) \log \theta\lambda_F(t) - \theta\lambda_F(t)r(t)$$

as a relationship involving inaccuracy and reliability functions.

6.3. Quantile-based inaccuracy

In continuation of our discussions in the previous sections we denote by $Q_X(u)$ and $Q_Y(u)$ the quantile functions of X and Y (or equivalently of F and G). Since the quantile versions of Kerridge measure are required in the sequel we first present it and look at some of its properties. We define distribution function $B(u) = G(Q_X(u))$ defined on $(0,1)$ and denote its density function as $b(u) = \frac{g(Q_X(u))}{f(Q_X(u))}$. Assume that the densities $f(x)$ and $g(x)$ of X and Y satisfy $f(x) > 0$ whenever $g(x) > 0$ so that $B(0) = 0$ and $B(1) = 1$, thereby $B(u)$ becomes a distribution function over $[0,1]$. Then the inaccuracy measure (6.1) reduces to

$$I(Q_X, Q_Y) = -\int_0^1 f(Q_X(u)) \log g(Q_X(u)) dQ_X(u).$$

Writing $q_X(u) = \frac{dQ_X}{du}$ and using $q_X(u)f(Q_X(u)) = 1$,

$$I(Q_X, Q_Y) = -\int_0^1 \log g(Q_X(u)) du. \tag{6.31}$$

The quantile version of Shannon entropy

$$H(Q_X(u)) = \int_0^1 \log q_X(p) dp$$

and that of Kullback-Leibler divergence

$$D(Q_X(u), Q_Y(u)) = \int_0^1 \log \frac{f(Q_X(u))}{g(Q_X(u))} du = -\int_0^1 \log \frac{g(Q_X(u))}{q_X(u)} du$$

show that

$$I(Q_X, Q_Y) = H(Q_X(u)) + D(Q_X(u), Q_Y(u)).$$

Note also that

$$I(Q_X, Q_Y) = -\int_0^1 \log \left(\frac{b(u)}{q_X(u)} \right) du. \tag{6.32}$$

EXAMPLE 6.3. *When $Q_X(u) = -\frac{1}{\lambda_1}\log(1-u)$ and $Q_Y(u) = -\frac{1}{\lambda_2}\log(1-u)$, corresponding to $F(x) = 1 - e^{-\lambda_1 x}$ and $G(x) = 1 - e^{-\lambda_2 x}$*

$$B(u) = G(Q_X(u)) = 1 - e^{-\lambda_2(-\frac{\log(1-u)}{\lambda_1})}$$

$$= 1 - (1-u)^{\frac{\lambda_2}{\lambda_1}}$$

and

$$b(u) = \frac{\lambda_2}{\lambda_1}(1-u)^{\frac{\lambda_2}{\lambda_1}-1}$$

Hence

$$I(Q_X, Q_Y) = \frac{\lambda_2}{\lambda_1} - \log \lambda_2.$$

The expression (6.31) is the same as that in Kayal et al. (2020). They have given the following example to illustrate the necessity of (6.31).

EXAMPLE 6.4. *Given that X has power-Pareto distribution $Q_X(u) = C_1 u^{\lambda_1}(1-u)^{-\lambda_2}$, $c, \lambda_1, \lambda_2 > 0$ and $Q_Y = c_2 p^{1/\alpha}$, the power distribution, it is difficult to use the formula (6.4) since the distribution function of X has no tractable form. Using (6.31)*

$$I(Q_X, Q_Y) = -\int_0^1 \log C_2 \left(C_1 u^{\lambda_1}(1-u)^{-\lambda_2}\right)^\alpha du$$

$$= \alpha \log\left(\frac{C_2}{C_1}\right) + \log\left(\frac{C_1}{\alpha}\right) + (1-\alpha)(\lambda_2 - \lambda_1).$$

Thus more models can be developed to analyze uncertainty if the quantile formulation is available.

In terms of the hazard quantile functions $h_{Q_X}(u)$, $h_B(u)$ and reversed hazard quantile functions $h_{Q_X}(u)$ and $h_B(u)$ we can write

$$I(Q_X, Q_Y) = -\int_0^1 \log\left(\frac{h_{Q_X}(p)}{h_B(p)}\right) dp.$$

The inaccuracy function need not always be monotonic. When $\psi_1(\cdot)$ and $\psi_2(\cdot)$ are two non-decreasing continuous transformations

$$I(\psi_1(X), \psi_2(X)) = -\int_0^1 \log \frac{\frac{d}{dp}[(\psi_2 Q_Y(p))^{-1}\psi_1(Q_X(p))]dp}{\frac{d}{dp}[\psi_1(Q_X(p))]} dp.$$

Further, $I(Q_X, Q_Y)$ may be positive or negative. To see this, let $Q_X(\mu) = \frac{cu^\lambda}{1-u}$ and $Q_Y(u) = au^\beta$, the quantile functions of the power-Pareto and power distributions respectively. Then

$$I(Q_X, Q_Y) = (1 - \beta)(\lambda + 1) - \log \beta - \beta \log \frac{c}{a}$$

which is negative when $\beta = \lambda = 1$ and $\alpha < 1$. Kayal et al. (2020) concludes that if $h_{Q_Y}(u) \geq h_{Q_X}(u)$, then I is positive. They also propose an inaccuracy ratio

$$R(X_1, X_2) = \frac{I(Q_X, Q_Y)}{H(Q_X)}$$

as the indicator of the loss of information when the true distribution Q_X is replaced by Q_Y.

For two non-negative absolutely continuous random variables X and Y, the quantile-based residual inaccuracy is defined as

$$\begin{aligned}
I_Q(u) &= -\int_u^1 \frac{f(Q_X(p))}{\bar{F}(Q_X(u))} \log \frac{g(Q_X(p))}{\bar{G}(Q_X(u))} dQ_X(p) dp \\
&= \log(1 - B(u)) - \frac{1}{1-u} \int_u^1 \log \left(\frac{b(p)}{q_X(p)} \right) dp \qquad (6.33) \\
&= -\frac{1}{1-u} \int_u^1 \log \frac{b(p)}{q_X(p)(1 - B(u))} dp.
\end{aligned}$$

The measure (6.33) also satisfies the property that it is sum of the quantile-based residual entropy and the quantile-based divergence measure of residual life discussed earlier. The corresponding inaccuracy ratio is

$$R(u) = \frac{I_Q(u)}{H_{Q_X}(u)}.$$

Monotonicity of $I_Q(u)$ is ascertained from the following theorem.

THEOREM 6.16. *The measure $I_Q(u)$ is increasing (decreasing) in u according as*

$$I_Q(u) \geq (\leq) \frac{(1-u)b(u)}{1 - B(u)} - \log \frac{b(u)}{q_X(u)(1 - B(u))}.$$

Upper and lower bounds for the residual inaccuracy measure are given by

$$I_Q(u) \geq \log \frac{h_B(u)(1 - B(u))}{h_{Q_X}(u)}$$

provided $G(x)$ is concave and

$$I_Q(u) \leq H_{Q_X}(u) + \log \frac{h_{Q_X}(u)}{h_{Q_Y}(u)}$$

whenever $g(Q_X(u))/g(Q_X(u)) = b(u)$ is increasing in u. If X, Y_1 and Y_2 be non-negative absolutely continuous random variables then

$$I_Q(X, Y_1, u) - I_Q(X, Y_2, u) \geq \log \frac{h_{Q_{Y_1}}(u)}{h_{Q_{Y_2}}(u)}.$$

Differentiating (6.32) after writing it as

$$[I_Q(u) - \log(1 - B(u))](1 - u) = - \int_u^1 \log \left(\frac{b(p)}{q_X(p)} \right) dp$$

leads to

$$(1 - u)I_Q'(u) - I_Q(u) = \log h_B(u) + \log(1 - u)h_{Q_X}(u) - (1 - u)h_B(u).$$
$$(6.34)$$

Thus

$$(1 - u)I_Q'(u) = I_Q(u) + \log h_B(u)(1 - u)h_{Q_X}(u) - (1 - u)h_B(u).$$

The last identity shows that $I_Q(u)$ increases (decreases) in u according as

$$I_Q(u) \geq (\leq)(1 - u)h_Q(u) - \log(1 - u)h_B(u)h_Q(u)$$

which is directly in terms of the hazard function rather than the expression in Theorem 6.16.

Specializing to the proportional hazards model

$$G(x) = \bar{F}^\theta(x) \text{ or } G(Q_X(u)) = 1 - (1 - u)^\theta,$$

equivalent to $\bar{B}(u) = (1 - u)^\theta$ and $b(u) = \theta(1 - u)^{\theta-1}$, the identity (6.34) has the simpler form

$$(1 - u)I_Q'(u) - I_Q(u) = \log \theta - \theta + \log h_Q(u) \qquad (6.35)$$

or

$$I_Q(u) = \frac{1}{1 - u} \int_u^1 (\theta - \log \theta + \log h_Q(p))dp. \qquad (6.36)$$

Some important theorems emerge as a result of (6.35) and (6.36).

THEOREM 6.17. *Under the proportional hazards model assumption, the residual inaccuracy measure $I_Q(u)$ uniquely determines the distribution of X (or Y).*

PROOF. From (6.35)

$$\log h_Q(u) = \theta - \log \theta + \frac{d}{du}(1 - u)I_Q(u). \qquad (6.37)$$

Given $I_Q(u)$, the hazard quantile function which determines Q uniquely is found from the above equation.   □

EXAMPLE 6.5. *For the generalized Pareto distribution*

$$Q_X(u) = \frac{b}{a}\left[(1-u)^{-\frac{a}{a+1}} - 1\right], \ b > 0; \ a > -1, \ 0 \le u \le 1$$

$h_Q(u) = \frac{a+1}{b}(1-u)^{\frac{a}{a+1}}$. *We have*

$$I_Q(u) = \theta - \log\theta - \log\frac{a+1}{b} + \frac{a}{a+1}(\log(1-u)-1). \qquad (6.38)$$

The probability constant is θ. *Substituting (6.38) in (6.37) with*

$$(1-u)I_Q(u) = (1-u)\left(\theta - \log\theta - \log\frac{a+1}{b}\right) + \frac{a}{a+1}(1-u)(\log(1-u)-1)$$

so that

$$\log H_Q(u) = \frac{a+1}{b} + \frac{a}{a+1}\log(1-u)$$

and

$$h_Q(u) = \frac{a+1}{b}(1-u)^{\frac{a}{a+1}}$$

giving $Q_X(u)$ *as the quantile function on using*

$$Q_X(u) = \int_0^u \frac{dp}{(1-p)h_Q(p)}.$$

REMARK 6.6. *It was mentioned in Section 6.2.1 that the result in Taneja et al. (2009) was not established correctly. Theorem 6.17 offers a simple proof of their result using quantile functions.*

The residual inaccuracy functions of many standard life distributions are not in compact forms to verify then empirically through real data sets. Example 6.5 is to be seen also as a result in which one can find simple forms of $I_Q(u)$ and the distributions characterized by them. Note that in (6.38), $I_Q(u) = A + B\log(1-u)$ so that I_Q is a linear function of $\log(1-u)$. When $a \to 0$, $Q_X(u)$ is exponential so that a constant $I_Q(u)$ results when X has an exponential distribution.

Formula (6.36) gives a simple method to compute $I_Q(u)$ in two alternative forms

$$I_Q(u) = \theta - \log\theta - \frac{1}{1-u}\int_u^1 \log h_Q(p)dp$$

or using the quantile density instead of $h_Q(u)$,

$$I_Q(u) = \theta - \log\theta - \frac{1}{1-u}\int_u^1 (\log(1-u) + \log q(p)dp)$$

$$= \theta - \log\theta - \log(1-u) - 1 + \int_u^1 \log q(p)dp.$$

Table 6.2. Values of $I_Q(u)$ for some distributions under proportional hazards assumption.

Distribution	$\bar{F}(x)$	$I_Q(u)$
exponential	$e^{-\lambda_1 x}$	$\theta - \log\theta - \lambda_1$
Pareto II	$(1+\frac{x}{\alpha})^{-c}$	$\theta - \log\theta - \log\frac{c}{\alpha}$ $+\frac{1}{c}(\log(1-u)-1)$
rescaled beta	$(1-\frac{x}{R})^{c}$	$\theta - \log\theta - \log cR^{-1}$ $-\frac{1}{c}(\log(1-u)-1)$
half logistic	$2(1+e^{x/\sigma})^{-1}$	$\theta - \log\theta$ $-\frac{1}{1-u}[(1+u)\log(1+u)-1]$
exponential geometric	$\frac{(1-p)e^{-\lambda x}}{1-pe^{-\lambda x}}$	$\theta - \log\theta - \log\frac{\lambda}{1-p}$ $+\frac{1}{(1-u)p}[(1-p)(\log(1-p)-1)$ $-(1-pu)(\log(1-pu)-1)]$
linear hazard quantile	$\log(\frac{a+bu}{a(1-u)})^{\frac{1}{a+b}}$	$\theta - \log\theta$ $+\frac{1}{(1-u)b}[(a+b)(\log(a+b)-1)$ $-\log(a+bu)(\log(a+bu)-1)]$
linear mean residual quantile	$-(c+\mu)\log(1-u)$ $-2cu$	$\theta - \log\theta - \frac{1}{2}(\mu-2c+u)$ $\times[\log(\mu-2c+\mu)-1]$

The expressions of $I_Q(u)$ of some standard distributions are given in Table 6.2.

EXAMPLE 6.6. *The linear hazard quantile distribution*

$$Q_x(u) = \log\left(\frac{a+bu}{1-u}\right)^{\frac{1}{a+b}}$$

has $h_Q(u) = a + bu$. *Hence*

$$I_a(u) = \theta - \log\theta + \frac{1}{(1-u)b}[(a+b)(\log(a+b)-1)-(a+bu)(\log(a+bu)-1)].$$

EXAMPLE 6.7. *Consider the linear mean residual quantile function distribution specified by*

$$Q_X(u) = -(c+\mu)\log(1-u) - 2cu, \quad -\mu < c < \mu, \quad \mu > 0.$$

For this distribution $h_Q(u)$ *is reciprocal linear,* $h_Q(u) = \frac{1}{\mu-c+2cu}$ *so that*

$$I_Q(u) = \theta - \log\theta - \frac{(\mu-c+2u)\log[(\mu-c+2u)-1]}{2}.$$

Note that for this distribution there is no closed form for $F(x)$ so that the definition (6.4) fails to give an analytical form of $I(t)$.

As seen from Tables 6.1 and 6.2 even for simple forms of distributions the expressions of $I(t)$ appears to be difficult for algebraic manipulations to extract their properties. In the next few theorems some simple forms of $I(t)$ are proposed along with the candidate distributions G that enable the identification of the true distribution F. Kayal et al. (2020) in their Theorem 3.6 have stated that if X is exponential with parameter λ_2, and $I_Q(u) = c + du$ then X has quantile density function

$$q_X(u) = \frac{\log \lambda_2 + c - d + du}{\lambda_2(1 - u)}. \tag{6.39}$$

Since $Q_X(u)$ has to be always non-decreasing (6.39) it cannot produce a quantile function without additional conditions on (6.39). Moreover, the converse of this result is also true. Thus we modify the result in the following manner.

THEOREM 6.18. *Let $I_Q(u)$ be a linear function of the form $I_Q(u) = c + du$. Then X has a linear mean residual quantile distribution if and only if Y is exponential with parameter λ_2.*

PROOF. When Y is exponential with parameter λ_2, $G(x) = e^{-\lambda_2 x}$, $\lambda_2 > 0$ and hence $B(u) = e^{-\lambda_2 Q_X(u)}$ with hazard function $h_B(u) = \lambda_2 q_X(u)$.
Substituting in (6.34), we have

$$q_x(u) = \frac{\log \lambda_2 + c + d}{\lambda_2(1 - u)} - \frac{2d}{\lambda_2}$$

and hence

$$Q_X(u) = \int_0^u q_X(p)dp = -\frac{(\log \lambda_2 + c + d)}{\lambda_2} \log(1 - u) - \frac{2d}{\lambda_2}u$$
$$= -(c_1 + \mu) \log(1 - u) - 2c_1 u, \tag{6.40}$$
$$0 \le u \le 1, \ -\mu < c_1 < \mu; \ \mu > 0$$

which is of the form of the linear mean residual quantile distribution in Example 6.7 with $C_1 = \frac{d}{\lambda_2}$ and $\mu = \frac{\log \lambda_2 + c}{\lambda_2}$. The condition to be satisfied by the parameters are $-(\log \lambda_2 + c) \le d \le (\log \lambda_2 + c)$ and $\log \lambda_2 + c > 0$. To prove them we assume Q_X in (6.40). Then the hazard function of $B(u)$ should satisfy

$$\log h_B(u) - \log(1 - u)h_{Q_X}(u) - (1 - u)h_B(u) = (1 - u)I_Q'(u) - I_Q(u).$$

Using $I_a(u) = c + du$ and

$$q_X(u) = \frac{\log \lambda_2 + c - d + 2du}{\lambda_2(1 - u)},$$

$$d - c - 2dx = \log h_B(u) - (1 - u)\hbar_B(u) + \frac{\log \lambda_2(1 - u)}{\log \lambda_2 + c - d + 2du}.$$

This can be written as

$$\log h_B(u) - \left\{ \log \lambda_2 + \log \frac{\lambda_2 + c - d + 2du}{\lambda_2(1-u)} \right\}$$

$$+ \log \lambda_2 + (1-u)\lambda_2 \frac{c - d + 2du + \log \lambda_2}{\lambda_2(1-u)} = 0$$

or

$$\log h_B(u) - \log \lambda_2 q_X(u) + (1-u)[\lambda_2 q_X(x) - h_B(u)] = 0$$

or

$$\log h_B(u) - (1-u)h_B(u) = \log \lambda_2 q_X(u) + (1-u)\lambda_2 q_X(u)$$

for all u in $[0,1]$ implying $h_B(u) = \lambda_2 q_X(u)$. Hence $B(u) = G(Q_X(u)) = e^{\lambda_2 Q_X(u)}$ or $G(x) = e^{\lambda_2 x}$, $x > 0$, as required.   □

6.4. Past inaccuracy measure

The inaccuracy measure of past life given in (6.22) transforms to the quantile form as

$$\bar{I}_Q(u) = -\int_0^u \frac{f Q_X(p)}{\mu} \log \frac{g Q_X(p)}{G Q_X(u)} dQ_X(p)$$

$$= -\frac{1}{u} \int_0^u \log \frac{g Q_X(p)}{G Q_X(u)} \qquad (6.41)$$

$$= [\log B(u) - \frac{1}{u} \int_0^u \log \left(\frac{b(p)}{q_X(p)} \right)] dp.$$

Differentiating with respect to u,

$$\bar{I}_Q(u) - u\bar{I}'_Q(u) = u\lambda_B(u) - \log u\lambda_Q(u)$$

giving

$$\frac{d}{du} u\bar{I}_Q(u) = u\lambda_B(u) - \log \lambda_B(u) - \log u\lambda_Q(u)$$

or

$$\bar{I}_Q(u) = \frac{1}{u} \int_0^u [p\lambda_B(p) - \log \lambda_B(p) - \log p\lambda_Q(p)] dp. \qquad (6.42)$$

Referring to the proportional reversed hazard model $G(x) = F^\theta(x)$, we have $B(u) = G(Q_X(u)) = u^\theta$ and $b(u) = \theta u^{\theta-1}$, the power distribution with reversed hazard rate $\lambda_B(u) = \frac{\theta}{u}$. This provides a modification to (6.42) as

$$\bar{I}_Q(u) = \frac{1}{u} \int_0^u \left[\theta - \log \left(\frac{\theta}{p} \right) - \log p\lambda_Q(p) \right] dp$$

$$= \theta - \log \theta - \frac{1}{u} \int_0^u \lambda_Q(p) dp$$

Table 6.3. Inaccuracy of past life.

$Q_X(u)$	$\lambda_{Q_X}(u)$	$\bar{I}_Q(u)$
$\alpha u^{1/\beta}$	$\beta(\alpha u^{1/\beta})^{-1}$	$\frac{\beta^2}{2}\frac{u^{\frac{\beta-1}{\beta}}}{\beta-1}$
$[R(1-u^{1/c})]^{-1}$	$Rc(1-u^{1/c})^2 u^{-\frac{1}{c}}$	$Rc^2[\frac{u^{\frac{c-1}{c}}}{c-1}-2u+\frac{u^{\frac{c+1}{c}}}{c+1}]$
$[\alpha(u^{1/c}-1)]^{-1}$	$\alpha c(1-u^{1/c})^2 u^{-\frac{1}{c}}$	$\alpha c^2[\frac{u^{\frac{c-3}{c}}}{c-3}-\frac{2u^{\frac{c-2}{2}}}{c-2}+\frac{u^{\frac{c-1}{c}}}{c-1}]$

or as

$$\bar{I}_Q(u) - u\bar{I}'_Q(u) = \theta - \log\theta - \log\lambda_Q(u). \tag{6.43}$$

From (6.43) the following theorem is immediate.

THEOREM 6.19. *Under the proportional reversed hazards model assumption, $\bar{I}_Q(u)$ determines the distribution of X uniquely.*

We give some examples for $\bar{I}_Q(u)$ and $\lambda_{Q_X}(u)$ that determine each other and the corresponding distributions, given in Table 6.3.

If $T(X)$ is an increasing transformation and $Q_X(u)$ is the quantile function of X, then $T(Q_X(u))$ is the quantile function of $T(X)$. Using this basic property of quantile functions one can compute the residual or past quantile inaccuracy of $T_1(X)$ and $T_2(Y)$ that are increasing, by simply replacing $Q_X(u)$ and $Q_Y(u)$ by $T(Q_X(u))$ and $T(Q_Y(u))$.

The quantile versions of the weighted measures of information do not seem to have been initiated in the literature. This applies to all the measures that have been discussed so far. The weighted measure of inaccuracy

$$I_t(t) = -\int_y^\infty x\frac{f(x)}{\bar{F}(t)}\log\frac{g(x)}{\bar{G}(t)}dx$$

in the quantile format is

$$I_Q^L(t) = I_L(Q_X(u)) = -\int_u^1 Q(p)\frac{f(p)}{1-u}\log\frac{g(Q_X(p))}{\bar{G}(Q_X(u))}d(p)$$

$$= -\frac{1}{1-u}\int_u^1 f(p)Q(p)[\log g(Q_X(p)) - \log\bar{G}(Q_X(u))]dQ(p)$$

$$= -\frac{1}{1-u}\int_u^1 Q_X(p)\log\frac{b(p)}{q_X(p)}dp + \log\bar{B}(u)$$

$$\tag{6.44}$$

Differentiating

$$(I_{Q_X}^L(u) - \log \bar{B}(u))(1 - u) = -\int_u^1 Q_X(p) \log \frac{b(p)}{q_x(p)} dp \qquad (6.45)$$

$$(1 - u)\frac{dI_{Q_X}^L}{du} - I_{Q_X}^L(u) = -\log \bar{B} - (1 - u)h_B$$
$$+ Q_X(u) \log b(u) - Q_X(u) \log q_x(u). \qquad (6.46)$$

Integrating (6.44) by parts,

$$Q_X(u)I_Q(u) + \int_u^1 q_X(p)I_Q(p)dp = I_Q^L(u)$$

where $I_Q(u)$ is the residual inaccuracy obtained in (6.32). Thus

$$Q_X(u)\frac{dI_Q}{du}(u) = \frac{d}{du}I_Q^L(u). \qquad (6.47)$$

Since we are considering only non-negative random variables, $Q_X(u) \geq 0$ so that $I_Q(x)$ and $I_Q^L(u)$ have the same monotonicity. Since

$$(1 - u)I_Q' - I_Q(u) = \log h_B(u) - \log q_x(u) - (1 - u)h_B(u)$$

by virtue of (6.47), the relation magnitudes of I_θ^L and I_Q are obtained from

$$I_Q^L(u) = Q(u)I_Q(u) + (1 - Q(u))[\log \bar{B} + (1 - u)h_B(u)].$$

6.5. Estimation

The application of the theoretical properties discussed so far to real data requires the estimate of the measures of inaccuracy. Rajesh and Abdul-Sathar (2017) considered random samples $X_1, X_2, \ldots, X_n$ and $Y_1, Y_2, \ldots, Y_n$ from the populations with distribution functions F and G censored on the right by random variables R_i and S_i which are independent and identically distributed and also are independent of the X_i's and Y_i's. The distribution functions of R_i and S_i respectively are denoted by $P_1(x)$ and $P_2(x)$ and we let $Z_i = \min(X_i, R_i)$ and $Z_i^* = \min(Y_i, S_i)$. The authors proposed the estimate for the censored data as

$$I_n^*(t) = -\frac{1}{n}\sum_{i=1}^n \log \frac{g_i(Z_i^*)}{1 - G_n(t)}D(Z_n > t)$$

where D denotes the indicator function, $g_n(z_i^*) = (n-1)^{-1}\sum_{j \neq 1}^n K\left(\frac{z_i^* - z_j^*}{h}\right)$ and $1 - G_n(t)$ is the Kaplan-Meier estimator or a kernel estimator and

$h = h_n$ a sequence of positive numbers converging to zero. The kernel density estimator of

$$I(t) = -\frac{1}{\bar{F}(t)} \int_t^\infty f(x) \log g(x)dx + \log \bar{G}(t)$$

obtained from (6.5) is

$$I_n(t) = -\frac{1}{\bar{F}_n(t)} \int_t^\infty f_n(x) \log g_n(x)dx + \log \bar{G}_n(t)$$

in which

$$f_n(t) = \frac{1}{nh} \sum_{i=1}^n \frac{K(x - Z_i)}{1 - P_i(Z_i)}$$

and

$$g_n(x) = \frac{1}{nh} \sum_{i=1}^n \frac{K(x - Z_i)}{1 - P_2(Z_i^*)}.$$

Under some general conditions it has been shown that the bias and mean squares are respectively

$$E(f_n(x) - f(x)) = \frac{f''(x)}{2} \int_{R+} [\alpha^2 K(\alpha)d\alpha]h^2 + o(h^4)$$

and

$$E(f_n(x) - f(x)) = \frac{C_k}{nh^2 - 1} \frac{f(x)}{1 - P_1(x)} \left[\frac{f''(x)}{2} \int_{R+} \alpha^2 K(\alpha)d\alpha \right]^2 h^4 + o(h^5).$$

Further the bias and variance of $f_n(x) \log g_n(x)$ are

$$\text{Bias}\,(f_n(x) \log g_n(x)) = \left[\frac{\log g(x) f''(x)}{2} + \frac{f(x)g''(x)}{g(x)} \right]$$
$$\left[\int_{R+} \alpha^2 K(\alpha)d\alpha \right] h^2 + o\left(\frac{1}{nh}\right) + o(h^4)$$

and

$$\text{Var}\,(f_n(x) \log g_n(x)) = \frac{C_K}{nh^2} \left[\frac{f(x) \log^2 g(x)}{1 - P_1(x)} + \frac{f^2(x)}{g(x)(1 - P_2(x))} \right] + o(h^4),$$

where $C_K = \int_{R+} K^2(\alpha)d\alpha.$

Similarly the bias and mean square error of
$h_n(t) = \int_t^\infty f_n(x) \log g_n(x) dx$ are

$$E[h_n(t) - h(t)] = \left(\int^2 K(\alpha) d\alpha\right) h^2 \int_t^\infty \left[\frac{\log g(x) f''(x)}{2} + \frac{f(x) g''(x)}{2g(x)}\right] dx$$

$$+ o\left(\frac{1}{nh}\right) + o(h^4)$$

and

$$\text{MSE}\,(h_n(t)) = \frac{1}{nh^2} C_k \int_t^\infty \left[\frac{f(x) - \log^2 g(x)}{1 - P_1(x)} + \frac{f^2(x)}{g(x)[1 - P_2(x)]} dx\right]$$

$$+ h^4 \left\{\int_t^\infty \left(\frac{\log g(x) f''(x)}{2} + \frac{f(x) g''(x)}{2g(x)}\right) dx\right\}^2 \left(\int_{R^+} \alpha^2 K(\alpha) d\alpha\right)$$

$$+ o\left(\frac{1}{nh}\right) + o(h^4).$$

They have obtained some asymptotic properties of the estimates. It is shown by simulation that the MSE of $I_n(t)$ is smaller when compared to MSE of $I_n^*(t)$. Some similar estimates of the past inaccuracy measure have been obtained by Abdul-Sathar and Nair (2021).

CHAPTER 7

Cumulative Entropy

7.1. Introduction

The concept of cumulative residual entropy was introduced by Wang et al. (2003) as an alternative to Shannon entropy. While proposing this new measure they argued that the entropy H_F has several limitations including (i) it is defined only for distributions possessing density functions by curtailing its applications to distributions not possessing densities and also to cases where it is more meaningful to employ survival or distribution functions (ii) the definitions of the discrete and continuous cases do not have the same properties (iii) entropy need not always be positive with a provision for any value on the extended real line (iv) it assumes inconsistent values as with the uniform distribution over $[0, a]$ in which $H_F = \log a$ can be positive, zero or negative depending on whether a is > 1, equal to unity or < 1. (v) the entropy decreases on conditioning; if entropy of $X|Y$ equals the entropy of X, then X and Y are independent; conditional entropy of Y given X is zero if X is a function of Y but not conversely, and (vi) Shannon's entropy fails to take into consideration the changes in the information content in some cases. To overcome these difficulties Wang et al. (2003) employed the distribution function of a random variable to define the information content and used it to develop a theory that parallels the Shannon entropy in many respects. In the univariate case, if X is a random variable in $\mathbb{R}$ the cumulative residual entropy (CRE) is defined as

$$C_X = C_F = -\int_{\mathbb{R}^+} P(|X| > x) \log P(|X| > x) dx. \qquad (7.1)$$

It can be seen that the above definition is valid in both the continuous and discrete cases and is mathematically more general than H_F. Note that C_F is always non-negative. An interesting property of C_F is that it can be computed for the sample data utilizing the empirical distribution function and such computations converge asymptotically to the true values. Further the conditional CRE of X given Y is zero if and only if X is a function of Y.

Given a pair of images $a(x_1, y_1)$ and $b(x_2, y_2)$ where $(x_2, y_2) = T(x_1, y_1)$, is an unknown parametric transformation to be determined, the problem in the image alignment problem is to determine T with reference to a measure M that maximizes M over all T. Among various methods available for the purpose, one is to maximize the mutual information using the Shannon entropy. Comparison between mutual information and cross cumulative entropy methods showed that the latter has a significantly better performance than the mutual information-based methods. This adds to the significance of CRE in practical problems. For details see Wang et al. (2003).

Rao et al. (2004) have discussed various properties and examples of CRE. Let X and Y be non-negative random variables with distribution functions F and G. Then

(i) $\max(C_X, C_Y) \leq C_{X+Y}$, when X and Y are independent
(ii) $C_X \geq 0$ with equality holding if and only if $P(|X| = x) = 1$ for some x
(iii) If X_i's are independent

$$C_{\underline{X}} = \sum_i \left(\prod_{i \neq j} E(|X_j|) \right) C_{X_i}, \quad \underline{X} = (X_1, \ldots, X_n)$$

(iv) If X_n converge in the distribution to X, $\lim_{n \to \infty} E\phi(X_n) = E\phi(X)$ for bounded continuous functions in $\mathbb{R}^n_+$
(v) $C_{X|Y} = 0$ if and only if $|X|$ is a function of Y
(vi) For any X and σ-field $\mathcal{F}$, $E(C(X, \mathcal{F})) \leq E(X)$, inequality holds iff X is independent of $\mathcal{F}$, that is every random variable measureable with respect to $\mathcal{F}$
(vii) If $X \geq 0$ with density f, $C_X \geq C \exp\{H_X\}$ where

$$C = \exp\{\int_0^1 \log(x|\log x|)dx\}$$

and

$$E(X|\mathcal{F}) \geq C \exp\left[-\int f(x|\mathcal{F}) \log(x|\mathcal{F})dx\right],$$

in which $P(X > t|\mathcal{F}) = \int_t^\infty f(x|\mathcal{F})dx$
(viii) If X is non-negative and continuous, there exists a function $\phi = \phi_X$ such that the Shannon entropy of $Y = \phi(X)$ satisfies

$$H_Y = \frac{C_X}{E(X)} - \frac{1}{E(X)} \log\left(\frac{1}{E(X)}\right), \quad \phi = F_Y^{-1} F_X(x).$$

Let $X_1, \ldots, X_n$ be positive, independent and identically distributed with distribution function F. If F_n is the empirical distribution function corresponding to F then the CRE of F_n is

$$-C(F_n) = C_{F_n} = \int_0^\infty G_n(x) \log G_n(x) dx, \quad G_n = \bar{F}_n(x).$$

Also

$$\sup_x |F_n(x) - F(x)| = \sup_x |G_n(x) - G(x)| \to 0$$

almost surely as $n \to \infty$. Thus $C_{F_n} \to C_F$ almost surely. This property enables the estimation of C_F using C_{F_n}. Further a non-negative random variable X satisfies

$$C_X \leq C_{X_\lambda}$$

where X_λ is an exponential random variable with mean $E(X^2)/2E(X)$. The CRE of an exponential random variable with mean μ is μ. Some additional properties of CRE are discussed in Rao (2005) When the random variable is non-negative, as the case in the present work,

$$C_X = -\int_0^\infty \bar{F}(t) \log \bar{F}(t) dt. \tag{7.2}$$

If X and Y are identically distributed and non-negative

$$E(|X - Y|) \leq 2C_X \tag{7.3}$$

and

$$E(|X - E(X)|) \leq 2C_X. \tag{7.4}$$

Consider the Orlicz space $L \log^+ L = S$, (Measurable functions f such that $\int_{\mathbb{R}^n} |f(x)| \log^+ |f(x)| dx$ is $< \infty$. Here $\log^+$ is the positive part of the logarithm). Then C_X is finite if and only if X is in S or $E(X \log^+ X) < \infty$. Also

$$C_X = -E[X(1 + \log \bar{F}(X))].$$

An upper bound of C_X in terms of $|X - E(X)|$ is

$$C_X \leq 2E[|X - E(X)| \log |X - E(X)|] + \frac{4}{e}.$$

The normalized CRE of X is defined as

$$\text{NCRE}(X) = [E(X)]^{-1} \int_0^\infty |\bar{F}(t) \log \bar{F}(t)| dt$$

$$= \frac{C_X}{E(X)}.$$

For Weibull distribution $\bar{F}(x) = e^{-x^\lambda \sigma^\lambda}$, NCRE $= \sigma^{-1}$. Also among all positive random variables with given $(r+1)^{\text{th}}$ moment the above Weibull distribution has maximum NCRE. This result contrasts with the maximum entropy distribution as uniform on the interval $[a, b]$ when no constraints are imposed and maximum entropy distribution as a truncated exponential ce^{-kx} where c and k satisfy the conditions

$$c \int_a^b e^{-kx} dx = 1 \text{ and } c \int_a^b x e^{-kx} dx = m.$$

Given a distribution $D(.)$ satisfying, decreasing $D(.)$ with $D(\infty) = 0$, $\int_0^\infty D(t)dt < \infty$ and

$$C_D = -\int_0^\infty D(t)(\log D(t))dt,$$

then

$$C_D = -\int_0^\infty F(s)dD(s), \quad F(s) = \int_0^s \log D(t)dt.$$

For all distributions g satisfying

$$\int_0^\infty g(x)dx \geq \int_0^\infty D(x)dx,$$

$$\int_0^\infty F(x)dg(x)dx = \int_0^\infty F(x)dD(x)dx,$$

we have

$$C_g \leq C_D,$$

and

$$\int g(x) \log g(x)dx = -C_D.$$

Some theoretical results and applications of CRE in non rigid multi model image registration is available in Wang and Vemuri (2007). Examples of some distributions and their cumulative entropies are given in Table 7.1.

7.2. Cumulative entropy of residual life

Asadi and Zohrevand (2007) have studied several properties of CRE and residual life CRE. For a non-negative random variable X with $C_X < \infty$,

$$C_X = E(m_F(X)). \tag{7.5}$$

From this it follows that if X is NBUE (NWUE) then $C(X) \leq (\geq)\mu = E(X)$ giving an upperbound (lower bound) for C_X of life distributions that

Table 7.1. CRE's of some distribution.

Distribution	$\bar{F}(x)$	C_F
exponential	$e^{-\lambda x}, x > 0$	λ^{-1}
uniform	$1 - \frac{x}{a}, 0 \le x \le a$	$\frac{a}{4}$
Pareto II	$\alpha^c(x + \alpha)^{-c}$	$\frac{c\alpha}{(c-1)^2}, c > 1$
normal	$f(x) = \frac{1}{\sqrt{2\pi}\sigma}e^{-\frac{(x-m)^2}{2\sigma^2}}$	$-\int_0^\infty \operatorname{erf} c(\frac{x-m}{\sigma}) \log \operatorname{erf} c(\frac{x-m}{\sigma}) dx,$
		$c(x) = \frac{1}{\sqrt{2\pi}} \int_x^\infty e^{-\frac{t^2}{2}} dt$
beta	$(1 - \frac{x}{R})^d$	$\frac{R(d-1)}{d+2}$
generalized Pareto	$(1 + \frac{ax}{b})^{-(1+\frac{1}{a})}$	$\frac{(1+2a)}{(1+a)^2}$

are NBUE (NWUE). The CRE of the residual life distribution $F_t(x)$ is

$$
\begin{aligned}
C_X(t) &= -\int_t^\infty \frac{\bar{F}(x)}{\bar{F}(t)} \log \frac{\bar{F}(x)}{\bar{F}(t)} dx \\
&= -\frac{1}{\bar{F}(t)} \int_t^\infty \bar{F}(x) \log \bar{F}(x) + m_F(t) \log \bar{F}(t).
\end{aligned}
\tag{7.6}
$$

As in (7.5), we have $C(t) = E[m_F(X)|X \ge t]$ and

$$C'(t) = h_F(t)(c(t) - m_F(t))$$

implying $C(t) \ge (\le)m_F(t)$ whenever $c(t)$ is increasing (decreasing). We call the function $c(t)$ as DCRE, being the dynamic cumulative residual entropy.

THEOREM 7.1. *(a) X has a constant DCRE if and only if it is exponential, (b) X is increasing DCRE (decreasing DCRE) if and only if X is IMRL (DMRL).*

It follows that the rescaled beta model has decreasing DCRE and Pareto II distribution has an increasing DCRE (see Table 7.2 for more details).

THEOREM 7.2. *(a) If $h_F(t) \le h_G(t)$ and $m_F(t)$ is increasing then $C_F(t) \ge C_G(t)$. (b) Let the DCRE's of X and Y be increasing in t, and $C_F(t) = C_G(t)$ for all $t \ge 0$. Then $\bar{F}(t) = \bar{G}(t)$.*

Characterization of life distributions by the functional form of $C_F(t)$ and by relationship with reliability functions can be accomplished as seen below.

THEOREM 7.3. *If X is a non-negative absolutely continuous random variable then*

$$C_F(t) = km_F(t), \ k > 0$$

holds for all t if and only if X is exponential for $k = 1$, rescaled beta for $k < 1$ and Pareto II for $k > 1$.

THEOREM 7.4. *If X and Y are non-negative with absolutely continuous survival functions $\bar{F}$ and $\bar{G}$ and $m_G(t) = K m_F(t)$, $K > 1$, then $C_G(t)$ is increasing if $C_F(t)$ is increasing.*

Besides bounds for $C_F(t)$ are prescribed as

$$C_F(t) \leq \frac{E[(X - t)^2 | X > t]}{2 m_F(t)}$$

with equality holding in the exponential case.

Some additional properties of DCRE were investigated by Navarro et al. (2010). If X and Y have finite means such that $X \leq_{\text{st}} Y$,

$$C_F \leq C_G - E(X) \log \frac{E(X)}{E(Y)}$$

and

$$C_F \leq \frac{E(X^{\beta+1}) \Gamma^\beta \left(1 + \frac{1}{\beta}\right)}{(\beta + 1) E^\beta(X)} \text{ for all } \beta \geq 0.$$

The DCRE satisfies

$$C_F(t) = m_X(t) \log \bar{F}_X(t) - \frac{1}{\bar{F}_X(t)} \int_t^\infty \bar{F}_X(x) \log \bar{F}_X(x) dx.$$

We say that X is increasing (decreasing) DCRE, IDCRE (DDCRE) if $C_F(t)$ is increasing (decreasing) in t. When $C_F(t) = c(t) m_F(t)$ for $t \geq 0$ then

$$m_X(t) = \left[K - \int_0^t (1 - c(x)) e^{c(x)} dx \right] e^{-c(t)}$$

with $K = \mu e^{(0)}$ and $\mu = E(X)$. This generalizes the characterization earlier proved in Theorem 7.3. We also have

REMARK 7.1.

$$\textit{IFR (DFR)} \Rightarrow \textit{DMRL (IMRL)} \Rightarrow \textit{DDCRE (IDCRE)}$$

Navarro et al. (2010) show examples to demonstrate that even when X is not DMRL, it is DDCRE, and also to the effect that DCRE need not uniquely determine the distribution when $C_F(t)$ is increasing as claimed by Asadi and Zohrevand (2007). The DCRE for the weighted random variable with

$$\bar{F}_Y(t) = \frac{E[w(X) | X > t]}{E w(X)} \bar{F}_X(t)$$

the expression

$$C_Y(t) = - \int_t^\infty \frac{E[w(X) | X > x] \bar{F}_X(x)}{E(w(X) | X > t) \bar{F}_X(t)} \log \frac{E[w(X) | X > x] \bar{F}_X(x)}{E(w(X) | X > t) \bar{F}_X(t)} dx.$$

In particular for the equilibrium random variable

$$C_Y(t) = \log m_X(t) + \frac{C_F(t)}{m_X(t)}$$

and

X is DDCRE (IDCRE) $\Leftrightarrow$ Y is DURL (IURL)

THEOREM 7.5. *(a) If $E(w(X)|X > t)$ is decreasing (increasing) in t and $X(Y)$ is IMRL then $C_X(t) \geq (\leq)C_Y(t)$. (b) If $w(t)$ is decreasing (increasing) in t and $X(Y)$ is IMRL then $C_X(t) \geq (\leq)C_Y(t)$.*

THEOREM 7.6. *(a) If Y is the proportional odds model of X, then $C_X(t) \geq (\leq)C_Y(t)$ according as $0 < p < 1$ ($p > 1$). (b) If $X(Y)$ is IMRL then $C_X(t) \geq (\leq)C_Y(t)$ provided $0 < p < 1$ ($p > 1$).*

Some other characterizations were presented by Baratpour (2010) based on order statistics. The CRE of $X_{1:n}$ is given by

$$C_{X_{1:n}} = -n \int_0^\infty \bar{F}^n(x) \log \bar{F}^n(x) dx$$

$$= -n \int_0^1 \frac{u^n \log u}{f(\bar{F}^{-1}(1-u))} du, \quad u = \bar{F}(x).$$

When X is Pareto I with $f(x) = \frac{\alpha\beta^\alpha}{x^{\alpha+1}}$, $x \geq \beta$, $C_F = \frac{\alpha\beta}{(\alpha-1)^2}$, $\alpha > 1$ and $C_{X_{1:n}} = \frac{n\alpha\beta}{(n\alpha-1)^2}$, $\alpha > \frac{1}{n}$. Similarly when X is Weibull with $\bar{F}(x) = e^{-(\lambda x)^\alpha}$, $C_F = \frac{1}{\lambda\alpha}\Gamma(1+\frac{1}{\alpha})$ and $C_{X_{1:n}} = \frac{1}{\lambda\alpha n^{1/2}}\Gamma(1+\frac{1}{\alpha})$.

THEOREM 7.7. *When $X_1, \ldots, X_n$ are positive, independent and identically distributed then F is Weibull if and only if $\frac{C_{X_{1:n}}}{E(X_{1:n})} = k$, $k > 0$ for all $n = n_j$, $j \geq 1$ such that $\sum_{j=1}^\infty n_j^{-1} = \infty$.*

THEOREM 7.8. *Two positive random variables X and Y belong to the same family of distributions but for a change of location if and only if*

$$C_{X_{1:n}} = C_{Y_{1:n}}$$

for $n = n_j$, $j \geq 1$ and $\sum n_j^{-1} = \infty$.

The CRE of the residual life distribution is a series system which can be written as

$$C_{X_{1:n,t}} = -\int_0^\infty \bar{F}_{X_{1:n,t}}(x) \log \bar{F}_{X_{1:n,t}}(x) dx$$

$$= -\int_0^\infty \left(\frac{\bar{F}(x)}{\bar{F}(t)}\right)^n \log \left(\frac{\bar{F}(x)}{\bar{F}(t)}\right)^n dx$$

$$= -\frac{1}{\bar{F}^n(t)} \int_t^\infty \bar{F}^n(x) \log \bar{F}^n(x) dx + n \log \bar{F}(t) m_{X_{1:n}}(t).$$

From this Baratpour (2010) concluded that

$$C_{X_{1:n}} \geq \bar{F}^n(t) C_{X_{1:n},t}$$
$$= n|\log \bar{F}(t)|\bar{F}^n(t) m_{X_{1:n}}(t).$$

Further X and Y as defined above, belong to the same family of distributions except for a change of location if and only if

$$C_{X_{1:n},t}(t) = C_{Y_{1:n},t}(t).$$

Kang (2015b) has given several properties of the DDCRE class of life distributions. The random variable X is DDCRE if and only if

$$\int_t^\infty \bar{F}_t(x) \left[\log \frac{\bar{F}(x)}{\bar{F}(t)} + 1\right] dx \geq 0.$$

Under the linear transformation $Y = aX + b$, $a > 0$, $b \geq 0$,

$$C_Y(t) = aC_X\left(\frac{t-b}{a}\right).$$

Using this, it follows that if X is DDCRE so is $aX + b$. Also if X is exponential it is DDCRE.

A new class named new better (worse) than used in DCRE, NBUBR (NWUBR) is defined through the inequality $C_X(t) \geq (\leq)C_X$. Then the exponential distribution is both NBUBR and NWUBR. In the proportional hazard case if $Y(X)$ is DDCRE $X(Y)$ is also DDCRE according as $\theta \geq 1$ $(0 < \theta \leq 1)$. Some closure properties noted below also hold among these classes (a) X is DDCRE for a random number N (d) $X_{1:N}$ is DDCRE $\Rightarrow X$ is DDCRE (e) For every increasing non-negative convex function $\phi(\cdot)$, X is DDCRE $\Rightarrow \phi(X)$ is DDCRE.

Kapdistria and Psarrakos (2012) investigated some extensions of the residual lifetime and its implications to CRE/DCRE. Let X and Y be two non-negative continuous and independent random variables with survival functions $\bar{F}$ and $\bar{G}$. Writing $X_Y = [X - Y|X > Y]$, X_Y extends the concept of mean residual life $m(t)$ to the mean residual life at random time Y. Note that

$$E(X_Y) = E(X - Y|X > Y) = E(m_X(Y)).$$

If we consider total lifetime of X given $X > Y$, viz. $(X|X > Y)$ its distribution is given by the survival function.

$$\bar{T}_{F,G}(x) = \bar{G}\#\bar{F}(x) = \bar{G}(x) + \bar{F}(x) \int_0^x \frac{g(y)}{\bar{F}(y)} dy \qquad (7.7)$$

which was introduced by Krakowski (1973) called the relevation transform of X and Y. For various reliability applications of the relevation transform

Table 7.2. Monotonicity of CRE.

Distribution	$\bar{F}(x)$	$Q(u)$	montonicity
exponential	$e^{-\lambda x}$	$\lambda^{-1}\log(1-u)$	constant DCRE
Weibull	$e^{-(\frac{x}{\sigma})^\lambda}$	$\sigma(-\log(1-u))^{1/\lambda}$	increasing for $\lambda<1$; decreasing for $\lambda>1$
Pareto II	$\alpha^c(x+\alpha)^{-c}$	$\alpha[(1-u)^{-1/c}-1]$	increasing
rescaled beta	$(1-\frac{x}{R})^c$	$R[1-(1-u)^{1/c}]$	decreasing
half-logistic	$2(1+e^{x/\sigma})^{-1}$	$\sigma\log\frac{1+u}{1-u}$	decreasing
power	$1-(\frac{x}{\beta})^\alpha$	$\alpha u^{1/\beta}$	decreasing for $\beta\geq 1$
Pareto I	$(\frac{x}{\sigma})^{-\alpha}$	$\sigma(1-u)^{-\frac{1}{\alpha}}$	increasing
Burr Type XII	$(1+x^c)^{-k}$	$[(1-u)^{1/k}-1]^{1/c}$	increasing for $c\leq 1$
Gompertz	$e^{-\frac{B(c^x-1)}{\log c}}$	$\frac{1}{\log c}\left[1-\frac{\log c\,\log(1-u)}{B}\right]$	decreasing for $C>1$, increasing for $c<1$
log logistic	$(1+(\alpha x)^\beta)^{-1}$	$\alpha^{-1}(\frac{u}{1-u})^{1/\beta}$	increasing for $\beta\geq 1$
exponential geometric	$(1-p)e^{-\lambda x}(1-pe^{-\lambda x})^{-1}$	$\frac{1}{\lambda}\log\frac{1-pu}{1-u}$	increasing
exponential Weibull	$1-(1-e^{-(\frac{x}{\sigma})^\lambda})^\theta$	$\sigma[-\log(1-u^{1/\theta})]^{1/\lambda}$	increasing (decreasing) $\lambda\leq 1$ $\lambda\theta\leq 1 (\lambda\geq 1, \lambda\theta\geq 1)$
generalized exponential	$1-(1-e^{-x/\sigma})$	$\sigma[-\log(1-u^{1/\theta})]$	decreasing (increasing), $\theta\geq 1(\theta\leq 1)$
extended Weibull	$\frac{\theta\exp(-\frac{x}{\sigma})^\lambda}{1-(1-\theta)e^{-(x/\sigma)^\lambda}}$	$\sigma[\log\frac{\theta+(1-\theta)(1-u)}{1-u}]$	decreasing (increasing) $\theta\geq 1,\lambda\geq 1$ $(\theta\leq 1,\lambda\leq 1)$
exponential power	$\exp[e^{-(\lambda t)}-1]$	$\frac{1}{\lambda}[-\log(1+\log(1-u))]^{-1/\alpha}$	decreasing for $\alpha\geq 1$
modified Weibull	$\exp(-\alpha(e^{(x/\sigma)^\lambda}-1))$	$\sigma[\log 1+\frac{\log(1-u)}{\alpha\sigma}]^{1/\lambda}$	decreasing for $\alpha\geq 1$
log Weibull	$\exp[-\log(1+\rho x)^k]$	$\rho^{-1}[\exp(\log(1-u))^{1/k}-1]$	increasing $0\leq k\leq 1$

see Baxter (1982), and Shantikumar and Baxter (1985). In addition if X and Y are identically distributed (7.7) simplifies to

$$\bar{T}_{F,F}(x) = \bar{F}\#\bar{F}(x) = \bar{F}(x)(1 - \log \bar{F}(x)) = \bar{F}(x)(1 + \Lambda(x)) \quad (7.8)$$

where $\Lambda(x) = -\log \bar{F}(x)$ is the cumulative hazard function of X. In this notation

$$C_F = \int_0^\infty \bar{F}(x)\Lambda(x)dx$$

and

$$C_F(t) = \frac{1}{\bar{F}(t)} \int_t^\infty \bar{F}(x)\Lambda(x)dx - m(t)\Lambda(t). \quad (7.9)$$

Kapdistria and Psarrakos (2012) constructed a sequence $\langle X_n \rangle$, $n = 1, 2, \ldots$ of random variables with distribution of X_n as

$$\bar{F}_{n+1}(x) = \bar{F}_n(x)[1 - \log \bar{F}_n(x)] = \bar{F}_n(x)(1 + \Lambda_n(x))$$

and $\bar{F}_1(x) = \bar{F}(x)$, $\Lambda_n(x)$ being the cumulative hazard function of X_n so that the hazard rate is $\lambda_n(x) = \frac{f_n(x)}{\bar{F}_n(x)}$ and $f_{n+1}(x) = \Lambda_n(x)f_n(x)$ with $f_1(x) = f(x)$. The sequence $\langle X_n \rangle$ represents in physical terms a continuous time stochastic process denoting the successive failures of a component, which on failure is replaced by a component of equal age but the distribution function of the nth component is identical to that of the time until the nth failure. In this setting

$$E\left(\frac{\Lambda_n(X_n)}{\lambda(X_n)}|X_n > t\right) = C_{X_n}(t) + m_n(t)\Lambda_n(t)$$

$$\text{Cov}\,(X_n, \Lambda_n(X_n)|X_n > t) = C_{X_n}(t) + (t + m_n(t) - E(X_n))\Lambda_n(t).$$

In particular

$$C_{X_n} = E\left(\frac{\Lambda(X_n)}{h_n(X_n)}\right)$$

and

$$\text{Cov}\,(X_n, \Lambda_n(X_n)) = C_{X_n}.$$

For any non-negative random variable, there holds

$$E\left(\frac{\Lambda(X)}{h(X)}|X > t\right) = C_X(t) + m(t)\Lambda(t)$$

and

$$\text{Cov}\,(X, \Lambda(X)|X > t) = C_X(t) + (t + m(t) - E(X))\Lambda(t).$$

7.3. Cumulative residual entropy of order n

Another extension of the cumulative entropy from a different direction can be thought of by considering a sequence $\langle X_n \rangle$, $n = 1, 2, \ldots$ of non-negative absolutely continuous random variables with $X_0 = X$ and the distribution of X_n is specified by the survival function

$$S_n(x) = \begin{cases} \frac{1}{\mu_{n-1}} \int_x^\infty S_{n-1}(u)du, & x > 0, \ n = 1, 2, \ldots \\ 0, & x \leq 0 \end{cases} \qquad (7.10)$$

where $\mu_n = \int_0^\infty S_n(x)dx < \infty$, is the mean of X_n and $S_0(x) = \bar{F}(x)$. The distribution (7.10) is called the equilibrium distribution of order n of X and $S_1(x)$ is the equilibrium distribution of X considered earlier

$$S_1(x) = \frac{1}{\mu_0} \int_x^\infty \bar{F}(u)du, \qquad \mu_0 = E(X)$$

is the usual equilibrium distribution of X.

Equation (7.10) was first considered by Harkness and Shantaram (1969) and its properties and applications were further studied by several authors including Pakes (1996), Nanda et al. (1996), Pakes and Navarro (2007), Gupta (2007), Nair and Preeth (2009) and Nair et al. (2012a). A multivariate generalization was also given by Nair and Preeth (2008). It can be seen from (7.10) that

$$S_n(x) = \frac{1}{n!\mu_0\mu_1 \ldots \mu_{n-1}} \int_x^\infty (u - x)^n f_X(u)du \qquad (7.11)$$

and also

$$S_n(x) = \frac{E\left((X - x)^n | X > x\right)}{E(X^n)} \bar{F}(x) = \frac{M_n(x)}{M_n(0)} \bar{F}(x), \qquad (7.12)$$

where $M_n(x)$ is the nth moment of the residual life of X. If $h_n(x)$ denotes the hazard rate function of X_n, and $m_n(x) = E(X_n - x | X > x)$, is the mean residual life of X_n,

$$h_n(x) = -\frac{d \log S_n(x)}{dx}$$

and also

$$h_n(x) = \frac{1 + m'_n(x)}{m_n(x)} \qquad (7.13)$$

$$m_{n-1}(x) = \frac{m_n(x)}{1 + m'_n(x)}, \qquad n = 1, 2, \ldots \qquad (7.14)$$

$$h_n(x) = (m_{n-1}(x))^{-1}, \qquad n = 1, 2, \ldots \qquad (7.15)$$

$$h_{n-1}(x) = h_n(x) - \frac{d \log h_n(x)}{dx}. \qquad (7.16)$$

In terms of the weighted distribution one can write the density function $f_n(x)$ of X_n as

$$f_n(x) = \frac{S_{n-1}(x)}{\mu_{n-1}}, \qquad n = 1, 2, \ldots$$

and

$$h_{n-1}(x) = \frac{f_{n-1}(x)}{\mu_1 f_n(x)}, \qquad n = 1, 2, \ldots \qquad (7.17)$$

The nth order equilibrium distribution is a weighted distribution of $f_{n-1}(x)$ with weight function $[h_{n-1}(x)]^{-1}$ or

$$f_n(x) = \frac{w(x) f_{n-1}(x)}{Ew(X)}, \qquad w(x) = \frac{1}{h_{n-1}(x)}$$

In particular, the first order equilibrium has weight function $\frac{1}{h(x)}$ and the second order has weight function $m(x)/h(x)$.

Analogous to the definition of CRE, we define the cumulative residual entropy of order n, denoted by CRE_n as

$$C_n = -\int_0^\infty S_n(x) \log S_n(x) dx, \qquad n = 0, 1, 2, \ldots \qquad (7.18)$$

for a non-negative, absolutely continuous random variable X with $E(X^n) < \infty$. By virtue of above, the usual CRE is of order zero, c_0. The density function of S_n being f_n, we also have the Shannon entropy of order n,

$$H_n = -\int_0^\infty f_n(x) \log f_n(x) dx \tag{7.19}$$

$$= -\int_0^\infty \frac{S_{n-1}(x)}{\mu_{n-1}} \log \frac{S_{n-1}(x)}{\mu_{n-1}} dx$$

$$= \log \mu_{n-1} - \frac{1}{\mu_{n-1}} \int_0^\infty S_{n-1}(x) \log S_{n-1}(x) dx \tag{7.20}$$

$$= \log \mu_{n-1} - \frac{1}{\mu_{n-1}} C_{n-1}, \quad n = 1, 2, \ldots \tag{7.21}$$

generalizing the relationship, between CRE and Shannon's entropy in Rao et al. (2004). Notice that H_1 is the Shannon's entropy of X

EXAMPLE 7.1. *For the generalized Pareto distribution* $\bar{F}(x) = \left(1 + \frac{ax}{b}\right)^{(1+\frac{1}{\alpha})}$, $x > 0$; $a > -1$, $b > 0$ *we have*

$$S_n(x) = \left(1 + \frac{ax}{b}\right)^{-\frac{1}{a} + n - 1}$$

so that

$$C_n = \frac{[1 - a(n-1)]b}{(1 - an)^2}, \quad n = 0, 1, 2, \ldots$$

The expression for CRE is

$$c_0 = (1 + a)b.$$

At the same time the nth order entropy is from (7.19) and $\mu_n = b/(1 - an)$ *is*

$$H_n = \log \frac{b}{(1 - a(n-1))} + \frac{[1 - a(n-2)]}{[1 - a(n-1)]}.$$

Three special cases arising out of this are the exponential with mean b when $a \to 0$

$$C_n = b$$

for all n, the Pareto II distribution when $a = (c-1)^{-1}$ *and* $b = a\alpha$ *so that* $\bar{F}(x) = \alpha^c (x + \alpha)^{-c}$ *with*

$$C_0 = \frac{c\alpha}{c(c-1)^2}, \quad C_n = \frac{(c-n)\alpha}{(c-1-n)^2}$$

and the rescaled beta $\bar{F}(x) = \left(1 - \frac{x}{R}\right)^d$, *when* $(1+d)^{-1}$, $b = R(1+d)^{-1}$ *giving*

$$C_n = \frac{R(d+n-1)}{(d+n+1)} \text{ and } C_0 = \frac{Rd}{(d+1)^2}.$$

EXAMPLE 7.2. *Consider the generalized mixture of exponentials*

$$\bar{F}(x) = \alpha e^{-\lambda_1} + (1-\alpha)e^{-\lambda_2 x}, \; x \geq 0; \; 0 < \lambda_1 < \lambda_2, \; \alpha \geq 0.$$

In this case

$$S_n(x) = \left(\frac{\alpha}{\lambda_1^n} + \frac{1-\alpha}{\lambda_2^n}\right)^{-1} \left[\frac{\alpha}{\lambda_1^n}e^{-\lambda_1 x} + \frac{1-\alpha}{\lambda_2^n}e^{-\lambda_2 x}\right].$$

We then have

$$C_n = -K \int_0^\infty \left[\frac{\alpha}{\lambda_1^n}e^{-\lambda_1 x} + \frac{1-\alpha}{\lambda_2^n}e^{-\lambda_2 x}\right] \log\left(K\frac{\alpha}{\lambda_1^n}e^{-\lambda_1 x} + \frac{1-\alpha}{\lambda_2^n}e^{-\lambda_2 x}\right) dx$$

where $K = \left(\frac{\alpha}{\lambda_1^n} + \frac{1-\alpha}{\lambda_2^n}\right)^{-1}$. *The integral does not have a simple closed form.*

Starting from

$$H_n = -\int_0^\infty f_n(x) \log f_n(x) dx,$$

on integration by parts

$$H_n = -[\log f_n(0)S_n(0) + \int_0^\infty \frac{f_n'(x)}{f_n(x)}S_n(x)dx].$$

We have

$$f_n(0) = \frac{S_{n-1}(0)}{\mu_{n-1}} = \frac{1}{\mu_{n-1}} \text{ and } f_n'(x) = \frac{d}{dx}\frac{S_{n-1}(x)}{\mu_{n-1}} = -\frac{f_{n-1}(x)}{\mu_{n-1}}$$

and so

$$H_n = \log \mu_{n-1} + \int_0^\infty h_{n-1}(x)S_n(x)dx \tag{7.22}$$

$$= \log \mu_{n-1} + \int_0^\infty \frac{h_{n-1}(x)}{h_n(x)}f_n(x)dx. \tag{7.23}$$

Equations (7.22) and (7.23) give two simple formulas for calculating the entropy H_n. Comparing (7.15) and (7.23) we have a simplified expression for C_n as

$$\frac{1}{\mu_n - 1}C_{n-1} = \int_0^\infty h_{n-1}(x)S_n(x)dx$$

or

$$C_n = \mu_n \int_0^\infty h_n(x)S_{n+1}(x)dx. \tag{7.24}$$

EXAMPLE 7.3. *In the generalized Pareto case in Example,* $h_n(x) = (1 - (n-1)a)/(ax+b)$ *and hence*

$$C_n = \mu_n \int_0^\infty \left(\frac{1-(n-1)a}{ax+b}\right)\left(1+\frac{a}{b}x\right)^{-\frac{1}{a}+n} dx$$

where $\mu_n = \frac{b}{1-a_n}$, we have $C_n = \frac{(1-(n-1)a)^b}{(1-an)^2}$ as obtained earlier.

Further from (7.24), we can also write

$$C_n = \mu_n \int_0^\infty \frac{f_n(x)}{S_n(x)} \frac{S_{n+1}(x)}{f_{n+1}(x)} f_{n+1}(x) dx$$

$$= \mu_n \int_0^\infty \frac{1}{h_{n+1}(x)} \frac{f_n(x)}{S_n(x)} \frac{S_n(x)}{\mu_n} dx$$

$$= E\left[\frac{1}{h_{n+1}(X)}\right], \quad n = 0, 1, 2, \ldots$$

Since the cumulative entropy was introduced as an alternative to the Shannon's measure, it is of some interest to compare the behaviour of the two. The normalized CRE of X_n is $C_n^* = \frac{c_n}{\mu_n}$ and hence (7.21) can be written as

$$H_n = \log \mu_{n-1} + C_{n-1}, \quad n = 1, 2, \ldots.$$

This gives

$$H_{n+1} - H_n = \log \frac{\mu_n}{\mu_{n-1}} + C_n^* - C_{n-1}^*.$$

Thus when $\langle \mu_n \rangle$ is an increasing sequence $C_n^* \geq C_{n-1}^* \Rightarrow H_{n+1} \geq H_n$. Similarly for a decreasing $\langle \mu_n \rangle$ we have $C_n^* \leq C_{n-1}^* \Rightarrow H_{n+1} \leq H_n$. These give the conditions for the same monotonicity for C_n^* and H_{n+1}. Also depending on $\mu_{n-1} > (< 1)$, the relative magnitude $H_n \geq (\leq)C_{n-1}^*$ is also obtained.

EXAMPLE 7.4. *From Example 7.3, for the Pareto distribution, $C_n = \frac{(c-n)\alpha}{(c-1-n)^2}$ and $E(X_n) = \mu_n = \frac{\alpha}{c-1-n}$ we conclude that $H_{n+1} \geq H_n$. Similarly in the case of re-scaled beta law $H_{n+1} \leq H_n$.*

7.4. Dynamic version of C_n

As defined in (7.6) the dynamic version of C_n denoted by $C_n(t)$ of X_n is written as

$$C_n(t) = -\int_t^\infty \frac{S_n(x)}{S_n(t)} \log \frac{S_n(x)}{S_n(t)} dt$$

$$= -\frac{1}{S_n(t)} \int_t^\infty S_n(x) \log S_n(x) - m_n(t) \log S_n(t) \tag{7.25}$$

where $m_n(t) = E(X_n - t|X_n > t)$, the mean residual life of X_n. It follows from (7.5) that

$$C_n(t) = E(m_n(X_n)|X_n > t)$$

$$= E_{X_{n,t}}\left[\frac{1}{h_{n+1}(X_{n+1})}|X_{n+1} > t\right] \tag{7.26}$$

$$= \frac{1}{S_n(t)}\int_t^\infty \frac{1}{h_{n+1}(x)}f_n(x)dx.$$

The above result is made use of in establishing some characterizations.

THEOREM 7.9. *Let X be a non negative random variable with absolutely continuous survival function satisfying $E(X^n) < \infty$. Then*

$$C_n(t) = \left(1 + \frac{at}{b}\right)C_n, \tag{7.27}$$

holds for all $t > 0$ and $n = 1, 2, 3, \ldots$ if and only if X has a generalized Pareto distribution.

PROOF. Assume that X has a generalized Pareto distribution for Example. Then

$$S_n(x) = \left(1 + \frac{ax}{b}\right)^{-\frac{1}{a}+n+1}$$

and $m_n(x) = \frac{ax+b}{1-na}$. From (7.26),

$$C_n(t) = \frac{1}{(1 + \frac{ax}{b})^{-\frac{1}{a}+n+1}}\int_t^\infty \left(\frac{at+b}{1-na}\right)\left(\frac{1}{a} - n + 1\right)\frac{a}{b}\left(1 + \frac{ax}{b}\right)^{-\frac{1}{a}+n-2}dx$$

$$= \frac{1 - a(n-1)}{(1-an)^2}(at+b)$$

$$= \left(\frac{at+b}{b}\right)C_n, \text{ on using Example 7.3.}$$

This proves the 'if' part. To establish the 'only if' part, assume (7.27) to write

$$\frac{1}{S_n}\int_t^\infty m_n(x)f_n(x)dx = \frac{at+b}{b}C_n.$$

Differentiating

$$h_n(t)\left(m_n(t) - \frac{at+b}{b}\right) = \frac{a}{b}C_n$$

or

$$m_n(t) - \frac{at+b}{b} = \frac{a}{b}C_n m_{n-1}(t) = K_n m_{n-1}(t), \ K_n > 0$$

$$m_n(t) - K_n m_{n-1}(t) = \frac{at+b}{b}.$$

Since the right side is linear in t, the leftside must also be linear in t which cannot be so unless $m_n(t)$ is a linear function of t. Thus X_n is a generalized Pareto and hence X is also a generalized Pareto. $\square$

REMARK 7.2. *Some equivalent forms of Theorem 7.9 are given below*

(i) $C_n(t)$ is a linear function of t
(ii) $C_n = C_n(1 - na)m_n(t)$
(iii) $C_n(t) = C_n(1 - (n-1)a)\frac{1}{h_n(t)}$
(iv) $C_n(t) = \frac{C_n(1-na)}{h_{n+1}(t)}$

THEOREM 7.10. *Let X be a non-negative absolutely continuous random variable with $E(X^n) < \infty$. Then*

$$C_n(t) = K_1 + K_2 - K_1 K_2 E(h_n(X_n)|X_n > t) \qquad (7.28)$$

for all $t > 0$ and $n = 1, 2, \ldots$ if and only if X follows a generalized mixture of exponentials

$$f(x) = \alpha\lambda_1 e^{-\lambda_1 x} + (1-\alpha)\lambda_2 e^{-\lambda_2 x}, \ x > 0, \ \alpha \geq 0. \qquad (7.29)$$

$0 < \lambda_1 < \lambda_2$ and $K_i = \frac{1}{\lambda_i}, \ i = 1, 2.$

PROOF. When X is distributed as in (7.29),

$$S(x) = \alpha e^{-\lambda_1 x} + (1-\alpha)e^{\lambda_2 x}$$

and

$$S_n(x) = \alpha_n e^{-\lambda_1 x} + (1-\alpha_n)e^{-\lambda_2 x}$$

where

$$\alpha_n = \frac{\alpha}{\lambda_1^n}\left(\frac{\alpha}{\lambda_1^n} + \frac{1-\alpha}{\lambda_2^n}\right)^{-1}.$$

Direct calculations show that

$$m_n(x) = \frac{\alpha\lambda_1^{-n-1}e^{-\lambda_1 x} + (1-\alpha)\lambda_2^{-n-1}e^{\lambda_2 x}}{\alpha\lambda_1^{-n}e^{-\lambda_1 x} + (1-\alpha)\lambda_2^{-n}e^{-\lambda_2 x}}$$

and

$$h_n(x) = \frac{\alpha\lambda_1^{-n+1}e^{-\lambda_1 x} + (1-\alpha)\lambda_2^{-n+1}e^{\lambda_2 x}}{\alpha\lambda_1^{-n}e^{-\lambda_1 x} + (1-\alpha)\lambda_2^{-n}e^{-\lambda_2 x}}.$$

We also have

$$\frac{\alpha}{\lambda_1^{n+1}}e^{-\lambda_1 x} + \frac{1-\alpha}{\lambda_2^{n+1}}e^{-\lambda_2 x} = \frac{1}{\lambda_1}\left(\frac{\alpha}{\lambda_1^n}e^{-\lambda_1 x} + \frac{1-\alpha}{\lambda_2^{n+1}}e^{-\lambda_2 x}\right)$$

$$+ \frac{1}{\lambda_2}\left(\frac{\alpha}{\lambda_1^n}e^{-\lambda_1 x} + \frac{1-\alpha}{\lambda_2^n}e^{-\lambda_2 x}\right)$$

$$- \frac{1}{\lambda_1\lambda_2}\left(\frac{\alpha}{\lambda_1^{n-1}}e^{-\lambda_1 x} + \frac{1-\alpha}{\lambda_2^{n-1}}e^{-\lambda_2 x}\right).$$

Dividing by $S_n(x)$,

$$m_n(x) = K_1 + K_2 - K_1 K_2 h_n(x).$$

Thus

$$C_n(t) = E(m_n(X_n|X_n > t)) = K_1 + K_2 - K_1 K_2 E(h_n(X_n)|X_n > t)$$

as required. Conversely, if (7.29) holds

$$-\int_t^\infty \frac{S_n(x)}{S_n(t)}\log\frac{S_n(x)}{S_n(t)}dx = K_1 + K_2 - \frac{K_1 K_2}{S_n(t)}\int_t^\infty h_n(x)f_n(x)dx.$$

Differentiating,

$$-\frac{1}{S_n(t)}\int_t^\infty S_n(x)dx = K_1 K_2 h_n(t) - (K_1 + K_2)$$

or

$$\int_t^\infty S_n(x)dx = K_1 K_2 f_n(t) - (K_1 + K_2)S_n(t).$$

Differentiating again leads to the differential equation

$$K_1 K_2 \frac{d^2 S_n(t)}{dt^2} + (K_1 + K_2)\frac{dS_n}{dt} + S_n(t) = 0. \tag{7.30}$$

To solve this, we see that the auxiliary equation is

$$K_1 K_2 m^2 + (K_1 + K_2)m + 1 = 0$$

with roots $-K_1^{-1}$ and $-K_2^{-1}$. Hence the general solution of (7.30) is of the form

$$S_n(t) = \frac{A_n}{K_1}e^{-t/K_1} + \frac{B_n}{K_2}e^{-t/K_2}.$$

As $x \to 0$

$$\frac{A_n}{K_1} + \frac{B_n}{K_2} = 1.$$

setting $n = 0$ and $\frac{A_0}{K_1} = \alpha$ and $\lambda_i = K_i^{-1}$ as assumed,

$$S(x) = \alpha e^{-\lambda_1 x_1} + (1-\alpha)e^{-\lambda_2 x_2},$$

which completes the proof. $\square$

THEOREM 7.11. *For an absolutely continuous random variable X with $E(X) < \infty$,*

$$C(t) = E\left(\frac{1}{h(X) + \alpha}|X > t|\right) \tag{7.31}$$

if and only if

$$\bar{F}(x) = \exp\left[\frac{e^{-\alpha x + \beta}}{\alpha} + \alpha x + \frac{e^\beta}{\alpha}\right], \qquad \alpha > 0; \ x > 0. \tag{7.32}$$

PROOF. For the given distribution

$$h(x) = e^{\alpha x + \beta} - \alpha \text{ and } m(x) = e^{-(\alpha x + \beta)}$$

giving $m(x) = (h(x) + \alpha)^{-1}$ and hence (7.30) is obtained from $c(t) = E(m(X)|X > t)$. Given $\bar{F}(x)$ as in (7.31), we have

$$C(t) = \frac{1}{\bar{F}(t)}\int_t^\infty \frac{f(x)}{h(x) + \alpha}dx = -\frac{1}{\bar{F}(t)}\int_t^\infty \bar{F}(x) \log \frac{\bar{F}(x)}{\bar{F}(t)}dx.$$

This simplifies on differentiation to

$$\frac{\int_t^\infty \bar{F}(x)dx}{\bar{F}(t)} = \frac{1}{h(t) + \alpha}$$

and to

$$(h(t) + \alpha)f(t) + \bar{F}(t)h'(t) = \bar{F}(t)(h(t) + \alpha)^2$$

or

$$(h(t) + \alpha)h(t) + h'(t) = (h(t) + \alpha)^2.$$

This leads to

$$\frac{h'(t)}{\alpha + h(t)} = \alpha.$$

Solving

$$h(t) = e^{\alpha t + C} - \alpha$$

which is the hazard rate function of (7.31) and this completes the proof. $\square$

REMARK 7.3. *From Theorem 7.10, the exponential mixture (7.29) has CRE*

$$C = K_1 + K_2 - K_1 K_2 E(h(X))$$

and the distribution (7.31) has

$$C = E\left(\frac{1}{h(X) + \alpha}\right).$$

Two other functions that are related to $m_n(x)$ are worth mentioning in this context. One is the Bondesson's functions (Stein and Dattero (1999)),

$$r_n(x) = E((X - x)_+^n)$$

for integers $n \geq 1$ and all x such that $\bar{F}(x) < 1$ and $(X - x)_+ = \max(X - x, 0)$. We have

$$m_n(x) = \frac{r_n(x)}{nr_{n-1}(x)}.$$

One can consider a new sequence $\langle S_{(n)}(x) \rangle$ of survival functions satisfying

$$S_{(n)}(x) = \int_x^\infty S_{(n-1)}(u)du$$

with $S_{(1)}(x) = \bar{F}(x)$ and obtain results similar to $\langle S_n(x) \rangle$ obtained earlier. The equilibrium distribution is the asymptotic distribution of the age U_t and residual life V_t of a component in used time T in a renewal process. The conditional distribution of V_T given $U_T > u$ has survival function

$$P(V_T > u|U_T > u) = \frac{\int_{u+v}^\infty \bar{F}(t)dt}{\int_u^\infty \bar{F}(t)dt}$$

and its nth moment is

$$e_n(t) = \frac{E(X - t)^{n+1}|X > t}{(n + 1)E(X - t|X > t)}.$$

Thus we have the identities

$$m_n(t) = \frac{r_n(t)}{nr_{n-1}(t)} = \frac{e_{n-1}(t)}{(n - 1)e_{n-2}(t)}.$$

The quantities $r_1(x)$, $m_n(x)$ and $e_1(x)$ all represent mean residual functions respectively of the original distribution $\bar{F}(x)$, its nth order equilibrium distribution and the residual life distribution of the equilibrium renewal process. See the alternative definition of mean residual life function and its properties in Nair and Sankaran (2010).

We examine the additional properties of $C_n(t)$ based on the ageing properties.

THEOREM 7.12. *The function $C_n(t)$ is increasing (decreasing) if X_{n+1} is IHR (DHR), $f_n = 1, 2, \ldots$.*

PROOF. Differentiating (7.26)

$$C_n'(t) = h_n(t)\left[C_n(t) - \frac{1}{h_{n+1}(t)}\right]. \qquad (7.33)$$

The proof now follows from Theorem 4.3 in Navarro et al. (2010). □

Some bounds for CRE are easily obtained using the ageing properties. Recall that X is NBUFR (NWUFR) if $h(x) \geq h(0)$ for all $x > 0$. Accordingly if X_{n+1} is NBUFR (NWUFR) from (7.26), we can write

$$C_n(t) \leq (\geq) \frac{1}{S_n(t)} \int_t^\infty \frac{1}{h_{n+1}(x)} f_n(t) dx = \frac{1}{h_n(0)}.$$

For distributions with $0 < m(\infty) < \infty$, X is said to be used better (worse) than aged in expectation UBAE (UWAE) if $m(x) \geq (\leq)m(\infty)$. Thus if X_n is UBAE (UWAE)

$$C_n(t) \geq (\leq)m_n(\infty).$$

The nth order Shannon's entropy of residual life is

$$H_n(t) = -\int_t^\infty \frac{f_n(x)}{S_n(t)} \log \frac{f_n(x)}{s_n(t)} dx$$

$$= -\int_t^\infty \left(\frac{S_{n-1}(x)}{\mu_{n-1}(\mu_{n-1})^{-1} \int S_{n-1}(u)du} \right) \log \left(\frac{S_{n-1}(x)}{\mu_{n-1}(\mu_{n-1})^{-1} \int S'_{n-1}(u)du} \right) dx$$

$$= -\int_t^\infty \left(\frac{S_{n-1}(x)}{m_{n-1}(t)S_{n-1}(t)} \right) \log \left(\frac{S_{n-1}(x)}{m_{n-1}(t)S_{n-1}(t)} \right) dx$$

$$= -\frac{1}{m_{n-1}} \left[\int_t^\infty \frac{S_{n-1}(x)}{S_{n-1}(t)} \log \frac{S_{n-1}(x)}{S_{n-1}(t)} - \log m_{n-1}(t) \int_t^\infty \frac{S_{n-1}(x)}{S_{n-1}(t)} dx \right]$$

$$= \frac{C_{n-1}(t)}{m_{n-1}(t)} + \log m_{n-1}(t), \ n = 1, 2, 3, \dots$$

$$= C_{n-1}(t)h_n(t) - \log h_n(t).$$

Differentiating,

$$H'_n(t) = C_{n-1}(t)h'_n(t) + C'_{n-1}(t)h_n(t) - \frac{h'_n(t)}{h_n(t)}$$

$$= C_{n-1}(t)h'_n(t) + h_n(t)h_{n-1}(t) \left(C_{n-1}(t) - \frac{1}{h_n(t)} \right) - \frac{h'_n(t)}{h_n(t)}.$$

From (7.33)

$$H'_n(t) = C_{n-1}(t)[h_n(t)h_{n-1}(t) + h'_n(t)] - \frac{h'_n(t)h_{n-1}(t) + h'_n(t)}{h_n(t)}$$

$$= (h_n(t)h_{n-1}(t) + h'_n(t)) \left[C_{n-1}(t) - \frac{1}{h_n(t)} \right].$$

Thus from Theorem 7.12 we have an implication between monotone Shannon entropy and CRE.

THEOREM 7.13. *The random variable X_n has decreasing (increasing) Shannon's entropy if and only if X_{n-1} has decreasing (increasing) CRE for $n = 1, 2, 3, \dots$*

PROOF.

$$h_n(t)h_{n-1}(t) + h'_n(t) = h_n^2(t)$$

(from (7.16)) and hence using (7.33)

$$H'_n(t) = \frac{h_n^2(t)}{h_{n-1}(t)} C'_{n-1}(t)$$

from which the result follows.   □

REMARK 7.4. *It is not always necessary that $H_n(t)$ is monotone. From the above theorem it follows that $H_n(t)$ is BT (UBT) if and only if $C_{n-1}(t)$ is BT (UBT).*

7.5. Weighted cumulative residual entropy

Given a function $\phi(x) \geq 0$ the weighted cumulative entropy (WCRE) with weight function ϕ is (Suhov and Sekeh (2015)),

$$C_\phi = \int \phi(x)\bar{F}(x) \log \bar{F}(x)dx.$$

EXAMPLE 7.5. *For the uniform random variable over (a, b),*

$$C_\phi = (b-a) \int_{\frac{b-2a}{b-a}}^{-\frac{a}{b-a}} \phi(b-a)(1-x) \log x \; dx$$

and when $\phi(x) = x$,

$$C_\phi = \frac{1}{3b(a-b)} \left[24a^3 \log\left(\frac{2a-b}{a-b}\right) - 15a^2b - a^3 - 5b^3 + 6\log\left(\frac{2a-b}{a-b}\right)bs \right.$$
$$\left. -18ab^2 \log\left(\frac{2a-b}{a-b}\right) + 21ab^2 + 6a^2 \log\left(\frac{a}{a-b}\right) - 18a^2b \log\frac{a}{a-b} \right]$$

If we write $m_\phi(t) = \frac{1}{\bar{F}(t)} \int_t^\infty \phi(x)\bar{F}(x)dx$ we have $C_\phi = E(m_\phi(X))$. Also

$$C_\phi \geq \alpha_\phi \exp(H_F)$$

where

$$\alpha_\phi = \exp\left[E \log \phi(X) + \int_0^1 \log(x|\log x|)dx \right].$$

Further if X and Y are independent and identically distributed

$$2C_\phi \geq E(\psi(X) - E\psi(X)), \quad \psi(x) = \int_0^x \phi(t)dt$$

and

$$C_\phi = E[\psi(0) - \psi(X)(1 + \log \bar{F}(X)].$$

Sekeh (2015) provides a method of estimation of C_ϕ. If $\bar{F}_n^*$ denotes the empirical survival function

$$\bar{F}_n^* = \frac{1}{n} \sum_{i=1}^n I(X_i > x),$$

where I is the indicator function the empirical WCRE is given by

$$C(\bar{F}_n^*) = -\int_0^\infty \phi(x)\bar{F}_n^*(x) \log \bar{F}_n^*(x)dx.$$

Given $0 < a < \infty$, $\int_0^a \phi(x)dx < \infty$ and $\int_a^\infty \phi(x)x^{-p}dx < \infty$, then the empirical WCRE converges to the CRE of X.

Mirali et al. (2017) and Kayal and Moharana (2017) considered the weighted cumulative entropy of the form

$$C_w = -\int_0^\infty x\bar{F}(x) \log \bar{F}(x)dx$$

as a special care of the WCRE mentioned above. They have many results in common.

EXAMPLE 7.6.

(i) $\bar{F}_X(x) = e^{-\frac{x}{\lambda}}$, $\quad\quad\quad\quad C_w = 2\lambda^2$

(ii) $\bar{F}(x) = 1 - \frac{x}{b}$, $\quad\quad\quad C_w = \frac{5}{36}b^2$

(iii) $\bar{F}(x) = \frac{\beta^\alpha}{(x+\beta)^\alpha}$, $\quad C_w = \alpha\beta^2 \left[\frac{1}{(\alpha-2)^2} - \frac{1}{(\alpha-1)^2} \right]$

The WCRE, C_w satisfies the following properties

(i) $C_w = E(M_0(X))$, $M_0(t) = \frac{\int x\bar{F}(x)dx}{\bar{F}(t)}$

(ii) $C_w(aX + b) = a^2 C_w(X) + abC_w(X)$ Misagh (2016)

(iii) $C_w = E(T(X))$, $T(X) = \int_0^x y \int_0^y h_F(t)dtdy$

(iv) $X \leq_{st} Y \Rightarrow C_w(X) \leq C_w(Y) + \frac{E(X^2)}{2} \log \frac{E(Y^2)}{E(X^2)}$

(v) $C_w(X) \leq \frac{E(X^3)}{3} - \frac{E(X^2)}{2}(1 - \frac{2}{E(X^2)})$

(vi) $C_w(X) \leq \frac{E(X^3)}{3} - \frac{E(X^2)}{2} \left(\frac{4E(X^3)}{3E^2(X^2)} - 1 \right)$

(vii) $C_w(X) \geq D \exp \left(\frac{H_w(X)}{E(X)} \right)$, where

$$D = E(X) \exp \left[\int_0^1 \bar{F}_X^{-1}(u) \log \left(\bar{F}_X^{-1}(u)u| \log u| \right) du/E(X) \right]$$

(viii) For proportional hazards model $\bar{F}_Y(x) = [\bar{F}_X(x)]^\theta$,

$$\theta \geq 1(0 < \theta \leq 1) \Rightarrow C_w(\theta X) \geq (\leq)\theta C_w(Y).$$

Now, the dynamic WCRE (DWCRE) of X_t is

$$C_w(t) = -\int_t^\infty x \frac{\bar{F}(x)}{\bar{F}(t)} \log \frac{\bar{F}(x)}{\bar{F}(t)} dx$$

$$= -\frac{1}{\bar{F}(t)} \int_t^\infty x \bar{F}(x) \log \bar{F}(x) dx + t M_0(t) \log \bar{F}(t).$$

Also,

$$C_w(t) = E(X M_0(X)|X > t)$$

and if $Y = aX + b$,

$$C_{w,Y}(t) = a^2 C_{w,X}\left(\frac{t-b}{a}\right) + ab C_{w,X}\left(\frac{t-b}{a}\right).$$

The property (d) above of WCRE carries over to DWCRE as

$$X \leq_{\text{hr}} Y \Rightarrow C_{w,X}(t) \leq C_{w,Y}(t).$$

Kayal and Moharana (2017) also defined classes of distributions based on DWCRE. The random variable X has increasing (decreasing) DWCRE according as $C_w(t)$ is increasing or decreasing in t. For example, $\bar{F}(x) = e^{-\lambda x}$ has $C(t) = \frac{t}{\lambda} + \frac{2}{\lambda^2}$ which is increasing while $\bar{F}(x) = 1 - \frac{x}{b}$, has $C_w(t) = \frac{b(b-t)}{4 - (b-t^2)/9}$ which is decreasing. The random variable X has increasing (decreasing) DWCRE according as $C_w(t) \geq (\leq) t M_0(t)$ for all $t > 0$. Also a non-negative random variable has increasing (decreasing) DWCRE if and only if

$$\int_t^\infty x \bar{f}(X) \left[1 + \log \frac{\bar{F}(x)}{\bar{F}(t)}\right] dx \geq (\leq) 0.$$

In the proportional hazards model for $\theta \geq 1$ ($0 < \theta \leq 1$) $X(Y)$ has decreasing DWCRE if $Y(X)$ has decreasing DWCRE.

Denote by $M_2(t) = E((X-t)^2|X > t)$, the second moment of residual life. Then

$$C_w = \frac{1}{2} E(M_2(X)) + E(X m(X))$$

Mirali et al. (2017) proved some inequalities and several other results.

(i) If $\bar{F}$ takes values in $[0, b]$, $b > \infty$, then $C_w \leq b C_X$ and

$$C_w(X) \leq \frac{E(X^2)}{2} \left|\log \frac{E(X^2)}{b^2}\right|$$

(ii) $C_w \geq T(\mu)$ where $T(t) = \int_0^t x \log \log \bar{F}(x) dx$

(iii) If $X_1 \leq_{\text{st}} X_2$, $C_{w,X_1} + \frac{1}{2} E(X_1^2) \leq C_{w,X_2} + \frac{1}{2} E(X_2^2)$.

Like maximum entropy models, one can also have max WCRE models under certain constraints.

THEOREM 7.14. *The distribution*

$$\bar{F}(x) = 1 - \exp\left(\sum_{i=1}^{K} \lambda_i x^i\right), \ \lambda_i \geq 0, \ x > 0$$

has max WCRE within the class of continuous random variable Y in $(0, \infty)$ satisfying

$$\sum_{i=0}^{k} \frac{\lambda_i}{i+2} E(X^{i+2}) = \sum_{i=0}^{k} \frac{\lambda_i}{i+2} E(Y^{i+2}), \ \lambda_0 = -1.$$

In particular the exponential distribution with parameter λ maximizes WCRE among continuous distributions satisfying

$$\frac{\lambda^3}{3} E(Y^3) - \frac{\lambda^2}{2} E(Y^2) = 1$$

and Rayliegh with

$$\bar{F}(x) = \exp\left[-\frac{x^2}{2\sigma^2}\right]$$

maximizes WCRE under constraint $\frac{E(X^4)}{4E(X^2)} = \sigma^2$.

Denoting by $U_i = X_{i+1:n}^2 - X_{i:n}^2, i = 1, 2, \ldots, n-1$ the plug-in estimator

$$C_n(X) = -\int_0^\infty x \bar{F}_n^*(x) \log \bar{F}_n^*(x) dx$$

when $\bar{F}_n^*$ is the empirical survival function, can be written as

$$C_0(X) = -\sum_{i=1}^{n-1} \frac{1}{2} U_i \frac{n-i}{n} \log \frac{n-i}{n}.$$

Mirali et al. (2017) have shown that $C_n(X)$ converges to C_X.

A dynamic version of WCRE (DWCRE) was considered by Misagh (2016)

$$C_w(t) = -\int_t^\infty x \frac{\bar{F}(x)}{\bar{F}(t)} \log \frac{\bar{F}(x)}{\bar{F}(t)} dx$$

$$= -\frac{1}{\bar{F}(t)} \int_t^\infty x \bar{F}(x) \log \bar{F}(x) + \log \bar{F}(t) \left[\frac{1}{2} \psi_2(t) - \frac{t^2}{2}\right]$$

where

$$\psi_2(t) = E(X^2 | X \geq t) = 2 \int_t^\infty \frac{x \bar{F}(x)}{\bar{F}(t)} dt + t^2.$$

Also,

$$C_w(t) = E\left[-\int_t^x z \log \frac{\bar{F}(z)}{\bar{F}(t)} dz | X \geq t\right].$$

The DWCRE is increasing or decreasing if and only if

$$C_w(t) \geq (\leq) \frac{1}{2}\psi_2(t) - \frac{t^2}{2}.$$

Another condition for increasing or decreasing $C_w(t)$ is that $\int_t^\infty \frac{x\bar{F}(x)}{\bar{F}(t)}$ is increasing or decreasing in t. To compare the DWCRE of two random variables X and Y in which $X \leq_{\text{hr}} Y$, the criterion for $C_{w,X}(t) \geq C_{w,Y}(t)$ is $\frac{1}{2}[\psi_2(t) - t^2]$ which is increasing in t. A relationship between the DWCRE and the corresponding weighted Shannon entropy H_W of the equilibrium random variable X_E is

$$H_{W,X_E}(t) = \frac{1}{m_X(t)}\left[C_{w,X}(t) + \frac{1}{2}(\psi_2(t) - t^2)\log m_X(t)\right].$$

Characterization of the lifetime distributions can also be accomplished by the form of DWCRE.

THEOREM 7.15. *The random variable X has*

(i) *constant DWCRE if and only if X has a Rayliegh distribution*
(ii) *exponential distribution with parameter $k^{-1/2}$ if and only if*

$$C_w(t) - \frac{1}{2}(\psi(t) - t^2) = k \text{ and } E(X^2) = 2k.$$

The estimation procedure suggested for C_W can be extended to $C_w(t)$ in the following manner, by

$$\hat{C}_{n,w}(t) = -\int_t^\infty \frac{\bar{F}_n^*(x)}{\bar{F}_n^*(t)}\log\frac{\bar{F}_n^*(x)}{\bar{F}_n^*(t)} dx$$

$$= -\frac{1}{2}\sum_{i=j}^{n-1}\frac{n-i}{n-j+1}\log\frac{n-i}{n-j+1}U_i$$

and $(\hat{C}_{n,w}(t))$ converges to $C_w(t)$ almost surely.

Misagh (2016) has shown that

$$C_w = E\left(\frac{X-1}{h(X)}\right)P(x), \quad P(x) = -\int_0^x \log \bar{F}(t)dt$$

and if X has finite support (a,b), $b > a$, then for $\theta \in (0,1]$,

$$C_w \leq \theta\left(\frac{b^2 - a^2}{2} - \log\theta(x)\right).$$

In estimating $C_{n,w}(t)$, he suggested the empirical estimator

$$\hat{C}_{n,w}(t) = \left[\bar{X}^2 - X_{1:n}^2 + \left(1 - \frac{1}{n}\right)S^2\right]\log\sqrt{n}$$

$$- \sum_{j=1}^{n-1} C_J(n-j)\log(n-j)$$

where $\bar{X}$ is the sample mean and $S^2 = \frac{1}{n-1}\sum(X_i - \bar{X})^2$, the sample variance and $C_j = \frac{1}{2}(X_{j+1:n}^2 - X_{j:n}^2)$. Simulated experiments show that the estimates are almost equal to the real values of WCRE and the mean square error values are almost zero.

7.6. Cumulative entropy

Instead of $\bar{F}(x)$ employed in CRE one may choose the distribution function $F(x)$ in defining a similar measure of uncertainty termed as cumulative entropy (CE),

$$\bar{C} = -\int_0^\infty F(x)\log F(x)dx. \tag{7.34}$$

Like the CRE, $\bar{C}$ is also non-negative. Di Cresenzo and Longobardi (2009) have discussed several aspects of the measure (7.34).

EXAMPLE 7.7. *When X is exponential (λ), $\bar{C} = \left(\frac{\pi^2}{6} - 1\right)\lambda$ and when X is Uniform $[0, b]$, then $\bar{C} = \frac{b}{4}$.*

Some properties of $\bar{c}$ are
  (i) If $Y = aX + b$, $\bar{C}(Y) = a\bar{C}(X)$
 (ii) $\bar{C}_X = E(r(X))$, $r(x) = E(x - X|X \le x)$
(iii) $\bar{C}_X = E(\bar{T}_2(X))$, where

$$\bar{T}_2(x) = -\int_x^\infty \log F(u)du = \int_x^\infty \left(\int_z^\infty \lambda(u)du\right)dz$$

(iv) if X and Y are non-negative random variables with finite unequal means and either $X \ge_{st} Y$ or $Y \ge_{st} X$ then

$$\bar{C}_X = E(r(Y)) + E(r'(Z))[E(X) - E(Y)]$$

where Z is an absolutely continuous random variable with probability density function

$$f_Z(x) = \frac{F_Y(x) - F_X(x)}{E(X) - E(Y)}$$

(v) the normalized cumulative entropy is $N\bar{C}_X = \frac{\bar{C}_X}{E(X)}$, which is related to the $N\bar{C}_X = 1 - E(B(F(X)))$ when $B(\cdot)$ is the Bonferroni curve. Note that $r(t) = E(X)B(F(t))$

(vi) $\bar{C} \leq (b - E(X)) \left| \log \left(1 - \frac{E(X)}{b} \right) \right|$, when X has support $[0, b]$, $b < \infty$

(vii) $\bar{C} \geq C_0 e^{H_F}$, $C_0 = \exp \left[\int_0^1 \log x | \log x dx \right] = 0.2065$.

(viii) $\bar{C} \geq \int_0^\infty F(x)\bar{F}(x)dx$ using $-\log u \geq 1 - u$, $u > 0$, $\bar{C} = \frac{1}{2}E|X - Y|$, when X and Y are independent and identically distributed, and $\bar{C} = E(X)E[\bar{F}_E(X)]$, where F_E is the survival function of the equilibrium distribution of X

(ix) if Y is the proportional reversed hazards model of X,

$$\bar{C}(\theta X) \geq \bar{C}(Y)$$

The dynamic version of cumulative entropy, DCE is defined as (Di Cresenzo and Longobardi (2009)) as

$$\bar{C}(t) = -\int_0^\infty \frac{F(x)}{F(t)} \log \frac{F(x)}{F(t)} dx, \ F(t) > 0$$

and

$$\bar{C}(t) = E\left(\bar{T}_2(X, t)|X \leq t\right) = E(r(X)|X \leq t)$$

where

$$\bar{T}_2(X, t) = -\int_x^t \log \frac{F(z)}{\bar{F}(t)} dz, \ t \geq x \geq 0.$$

THEOREM 7.16. *For a non-negative absolutely continuous random variable X,*

(i) $\bar{C}(t) \leq r(t) \log \left(\frac{r(t)}{t} \right)$

(ii) $\bar{C}(t) \geq C_0 e^{\bar{H}_F(t)}$ *and $\bar{H}_F$ is the Shannon's entropy of past life discussed in Chapter 3.*

(iii) $\bar{C}(t) \geq \int_0^t F(x;t)(1 - F(x;t))$, $F(x;t) = \frac{F(x)}{F(t)}$

(iv) $\bar{C}(t) \geq \bar{T}_2(X, t)$

(v) $\bar{C}(t) = a\bar{C}\left(\frac{t-b}{a}\right)$, $t \geq b$.

THEOREM 7.17. *The DCE $\bar{C}(t)$ is increasing in t if and only if $C(t) \leq r(t)$ and if $r(t)$ is increasing then $C(t)$ is also increasing, but increasing $\bar{C}(t)$ does not imply that $r(t)$ is increasing.*

THEOREM 7.18. *The relationship $\bar{C}(t) = cX(t)$ holds iff $F(t) = (\frac{t}{b})^{\frac{c}{1-c}}$, $0 < c < 1$ and $\bar{C}(t) = cm(t)$ if and only if $F(t) = (\frac{t}{b})^{\frac{1-c}{c}}$, $0 < c < 1$.*

The empirical CE is

$$\bar{C}_n = -\int_0^\infty F_n(x) \log F_n(x) dx$$

when $F_n(x)$ is the empirical distribution function of the sample. Then

$$\bar{C}_n = -\sum_{j=1}^{n-1} U_{j+1} \frac{j}{n} \log \frac{j}{n}, \quad U_j = X_{j:n} - X_{j-1:n}$$

converges to $\bar{C}$ and for a random uniform distribution over $[0,1]$

$$E(\bar{C}_n) = \frac{\zeta(-1) - \zeta^*(-1, u)}{n(n+1)} + \frac{(n-1)\log n}{2(n+1)}$$

with $\zeta(s)$ the Riemann zeta function and $\zeta^*(s, a)$ is the partial derivative of the generalized Riemann zeta function with respect to s. The limiting value of $E(\bar{C}_n)$ is 0.25 and $V(\bar{C}_n) \to 0$ as $n \to \infty$.

In the case of the weighted CE,

$$\bar{C}_\phi = -\int_0^\infty \phi(x) F(x) \log F(x) dx$$

an estimate is suggested by Sekeh (2015) as

$$\hat{\bar{C}}_\phi = \psi(x_{n:n} - \bar{\psi}) \log n - \frac{1}{n} \sum_{i=1}^{n-1} \zeta_i \log i$$

where

$$\sum_{i=1}^n \zeta_i = n(\psi(x_{n:n}) - \bar{\psi}),$$

$$\psi(x) = \int_0^x \phi(t) dt,$$

and

$$\bar{\psi} = \frac{1}{n} \sum_{i=1}^n \psi(x_{i:n}).$$

Di Crescenzo and Toomaj (2015) considered the reversed relevation transform

$$T_{\mathrm{G,F}}(x) = G(x) + F(x) \int_x^\infty \frac{dG(t)}{F(t)},$$

The inactivity time of X at a random time Y defined as $X_{(Y)} = [Y - X | X \leq Y]$ and the total time $X(Y) = (X | X \leq Y)$ show that

$$P(X(Y) \leq x) = T_{G,F}(x).$$

In the special case when X and Y are independent and identically distributed

$$T_{G,F}(x) = F(x)[1 - \log F(x)]$$

$$= F(x) \int_x^\infty \lambda(t)dt.$$

When the assumption of proportional reversed hazard model holds $G = F^\theta$ so that

$$T_{G,F}(x) = \frac{1}{\theta - 1}(\theta e^{-B(x)} - e^{-\theta B(x)})$$

where $B(x) = \int_x^\infty \lambda(t)dt$, is the cumulative reversed hazard rate. If a parallel system of n independent and identically distributed lifetimes satisfy the above model we can write

$$T_{G,F}(x) = \frac{\theta F^n(x) - nF^\theta x}{\theta - n}, \quad \theta \neq n.$$

7.7. Quantile-based cumulative entropy

Like the other entropies discussed so far, the cumulative entropy in its residual and past forms also admits a quantile function-wise version which is advantageous in certain circumstances. Under the transformation $x = Q(u)$ in (7.1), Sankaran and Sunoj (2017) obtained the cumulative residual entropy of X as

$$C_Q = -\int_0^1 (1 - u)(\log(1 - u))q(u)du$$

the corresponding DCRE as

$$C_Q(u) = -\int_u^1 \frac{1-p}{1-u} \log \frac{1-p}{1-u} q(p)dp$$

$$= \frac{\log(1 - u)}{1 - u} \int_u^1 (1 - p)q(p)dp \tag{7.35}$$

$$- \frac{1}{1 - u} \int_u^1 (\log(1 - p))(1 - p)q(p)dp.$$

The relationship

$$m_Q(u) = C_Q(u) - (1 - u)C_Q'(u) \tag{7.36}$$

shows that $C_Q(u)$ determines $m_Q(u)$ and hence the distribution of X. Examples of some quantile functions and their $C_Q(u)$ are given Table 7.3.

THEOREM 7.19. *The random variable X satisfies*

(i) $C_Q(u) = au + b$, $a, b > 0$ iff X *follows a family of distributions with* $Q(u) = -(a + b)\log(1 - u) - 4u$

Table 7.3. Quantile-based dynamic cumulative residual entropy.

Distribution	Quantile function	DCRE
exponential	$-\frac{1}{\lambda}\log(1-u)$	$\frac{1}{\lambda}$
uniform	$a + (b-a)u$	$\frac{(b-a)(1-u)}{4}$
Pareto II	$\alpha((1-u)^{-\frac{1}{c}} - 1)$	$\frac{\alpha c}{(c-1)^2}(1-u)^{-\frac{1}{c}}$
beta	$\alpha(1 - (1-u)^{\frac{1}{c}})$	$\frac{\alpha c}{(c-1)^2}(1-u)^{\frac{1}{c}}$
Pareto I	$\alpha(1-u)^{-\frac{1}{c}}$	$\frac{\alpha c}{(c-1)^2}(1-u)^{-\frac{1}{c}}$
linear mean residual	$-(a+b)\log(1-u) - 4u$	$ax + b$
generalized Pareto	$\frac{b}{a}[(1-u)^{-\frac{a}{a+1}} - 1]$	$b(a+1)(1-u)^{-\frac{a}{a+1}}$

(ii) $C_Q(u) = -K\log(1-u)M(u)$ *if and only if*

$$q(u) = \frac{A(K+1)}{c(1-u)^2}(-\log(1-u))^{\frac{1}{c}}.$$

When several units are tested for studying their life lengths, some of the units may fail wholly and others may survive the test period. The sum of all observed and incomplete life lengths is the total time of the test statistic. During an interval $(0, t)$ when n items are tested,

$$\tau(t) = nX_{1:n} + (n-1)(X_{2:n} - X_{1:n}) + \cdots + (n-r+1)(X_{r:n} - X_{r-1:n})$$
$$+ (n-r)(t - X_{r:n})$$
$$= X_{1:n} + X_{2:n} + \cdots + X_{r:n} + (n-r)t.$$

As the number of units on test tends to infinity the limit of this statistic is called the total time on test transform (TTT)

$$A_F^{-1}(u) = \int_0^{F^{-1}(u)} \bar{F}(t)dt$$

where $\bar{F}(t)$ is an absolutely continuous survival function of a lifetime X. In the quantile formulation we write this as

$$A_Q(u) = \int_0^u (1-p)q(p)dp \qquad (7.37)$$

or $A(u)$ when the meaning is clear.

EXAMPLE 7.8. *For the linear hazard quantile distribution*

$$Q(u) = \log\left(\frac{a+bu}{a(1-u)}\right)^{\frac{1}{a+b}}$$
$$q(u) = [(1-u)(a+bu)]^{-1}$$

and hence

$$A_Q(u) = \frac{1}{b} \log \frac{a + bu}{a}.$$

The distribution of X is uniquely determined by $A(u)$ as

$$Q(u) = \int_0^u \frac{A'(p)}{1 - p} dp. \tag{7.38}$$

The TTT is related to reliability functions. For example,

$$A'(u) = \frac{1}{h_Q(u)}$$

$$m_Q(u) = \frac{\mu - A(u)}{1 - u}.$$

Sometimes it is advantageous to use the scaled TTT, $\phi(u) = \frac{A(u)}{\mu}$.

Equation (7.35) gives,

$$C_Q(u) = \frac{\log(1 - u)}{1 - u} \int_u^1 A'(p) dp - \frac{1}{1 - u} \int_u^1 H'(p) \log(1 - p) dp.$$

Differentiating

$$(1 - u)C_Q'(u) - C_Q(u) = -\int_u^1 A'(p) dp = -\frac{1}{1 - u}(A(1) - A(u))$$

$$= \frac{A(u) - \mu}{1 - u} \tag{7.39}$$

THEOREM 7.20. *The functions $C_Q(u)$ and $A(u)$ determine each other and the corresponding distributions uniquely.*

PROOF.

$$\frac{d}{du}(1 - u)C_Q(u) = \frac{A(u) - \mu}{1 - u} \tag{7.40}$$

$$A(u) = \mu + (1 - u)\frac{d}{du}(1 - u)C_Q(u)$$

and

$$C_Q(u) = \frac{1}{1 - u} \int_u^1 \frac{\mu - A(\phi)}{1 - p} dp. \tag{7.41}$$

The Theorem follows from (7.40) and (7.41).   □

For many of the standard distributions, the form of CRE is complicated. An immediate advantage of (7.41) is that we can determine distributions that correspond to simple functional forms of CRE that are useful in modelling problems. We give some examples.

(1) Take $C_Q(u) = \beta u^\alpha$.

Then

$$A(u) = \beta(\alpha u^{\alpha-1} - (2\alpha + 1)u^\alpha + (\alpha + 1)u^{\alpha+1}) + \mu.$$

The quantile density function of X is

$$q(u) = \frac{A'(u)}{1 - u} = (1 - u)^{-1}[\beta\{(\alpha + 1)^2 u^2 - \alpha(2\alpha + 1)u + \alpha(\alpha - 1)\}u^{\alpha-2}].$$

Since $A(u)$ has to be increasing $\beta \geq 0$ and $\alpha \in (-\frac{1}{2}, 0)$. In the special case when $\beta \geq 0$, X is exponential with mean β.

(2) When $C_Q(u) = \alpha(1 - u)^\beta$,

$$A(u) = \mu - \alpha(\beta + 1)(1 - u)^{\beta+1},$$
$$q(u) = \alpha(\beta + 1)^2(1 - u)^{\beta-1}$$

and

$$Q(u) = \frac{\alpha(\beta + 1)^2}{\beta}(1 - (1 - u)^\beta). \tag{7.42}$$

Reparametrizing with $\beta = -\frac{a}{a+1}$ and $\alpha = (a + 1)b$, we have

$$Q(u) = \frac{b}{a}((1 - u)^{-\frac{a}{a+1}} - 1), b > 0, a > 1$$

the quantile function of the generalized Pareto distribution involving the exponential, rescaled beta, uniform and the Pareto models. Notice also that the model (7.42) is characterized by the relationship $C_Q(u) = \mu - KA(u)$, μ is the mean of X and K is a constant.

(3) When $C_\theta(u) = au + b$,

$$A(u) = \mu + (1 - u)(a - b - 2au)$$

and we obtain the linear mean residual quantile distribution

$$Q(u) = -(a + b)\log(1 - u) - 4u$$

as found in Sankaran and Sunoj (2017).

An important aspect of the relationships mentioned above is that inferring the ageing properties of X can be accomplished through the quantile measure $C_Q(u)$. This is explained in the next theorem.

THEOREM 7.21. *A necessary and sufficient condition for X to be*

(a) *IHR (DHR) is that $(1 - u)C_Q^*(u)$ is concave (convex) where $C_Q^* = \frac{d}{du}(1 - u)C_Q(u)$*

(b) *IHRA (DHRA) is that $\frac{(1-u)}{u}C_Q^*(u)$ is decreasing (increasing)*

(c) *DMRL (IMRL) is that $C_Q^*(u)$ is concave (convex)*

(d) *NBUE (NWUE) is that $\bar{C}_Q^*(u) \geq (\leq) - \mu$*

(e) NBUHR (NWUHR) is that $C_Q(u) \geq (\leq)T_0 + (1-T_0)\log(1-u)$ where $T_0 = \mu + C^*(0)$

(f) DVRL (IVRL) is that

$$\int_0^1 \left(\frac{d}{dp}C^*(p)\right)^2 \leq (\geq)(1-u)(C_Q^*(u))^2$$

PROOF. From Barlow and Campo (1975) X is IHR if and only if $\phi(u)$ is concave and IHRA if and only if $\frac{\phi(u)}{u}$ is decreasing. The results in (a) and (b) follows from these two. Klefsjö (1982) has shown that X is DMRL if and only if $\frac{1-\phi(u)}{u}$ is decreasing in u. This means that

$$\frac{\mu - A(u)}{\mu(1-u)} \text{ is decreasing} \Leftrightarrow -\frac{d}{dx}(1-u)C_Q(u) \text{ is concave}$$

$$\Leftrightarrow C_Q^*(u) \text{ is convex.}$$

Further X is NBUE if and only if $\phi(u) \geq u$. It follows that

$$\phi(u) \geq u \Leftrightarrow A(u) \geq \mu u$$

$$\Leftrightarrow \mu + (1-u)C_Q^*(u) \geq \mu u$$

$$\Leftrightarrow C_Q^*(u) \geq -\mu.$$

Now assume that X is NBUHR. This happens if and only if $A'(u) \leq A(0)$ or $A(u) \leq uA(0)$. Thus

$$C_Q(u) = \frac{1}{1-u}\int_u^1 \frac{\mu - A(p)}{1-p}dp$$

$$\geq \frac{1}{1-u}\int_u^1 \frac{\mu - pA(0)}{1-p}dp$$

$$= \frac{1}{1-u}\int_u^1 \frac{(\mu - A(0)) + A_0(1-p)}{1-p}dp$$

$$= A(0) + (\mu - A(0))\log(1-u)$$

as stated in (e). Lastly X is DURL if and only if

$$\int_0^1 \left(\frac{1-\phi(p)}{1-p}\right)^2 \leq \frac{(1-\phi(u))^2}{1-u}$$

$$\Leftrightarrow \int_0^1 \left(\frac{d}{du}(1-p)C_Q(p)\right)^2 \leq (1-u)\left(-\frac{d}{du}(1-u)C_Q(u)\right)^2$$

$$\Leftrightarrow \int_0^1 \left(\frac{dC_Q^*(p)}{dp}\right)^2 \leq (1-u)(C_Q^*(u))^2.$$

Table 7.4. TTT for some life distributions.

Distribution	$Q(u)$	TTT
exponential	$-\frac{1}{\lambda}\log(1-u)$	$\lambda^{-1}u$
Pareto II	$\alpha((1-u)^{-\frac{1}{c}}-1)$	$\frac{\alpha}{c-1}[1-(1-u)^{\frac{c-1}{c}}]$
rescaled beta	$\alpha(1-(1-u)^{\frac{1}{c}})$	$\frac{\alpha_E}{c+1}[1-(1-u)^{\frac{c+1}{c}}]$
half-logistic	$\sigma\log\frac{1+u}{1-u}$	$2\sigma\log(1+u)$
Power	$\alpha u^{1/\beta}$	$\frac{\alpha u^{1/\beta}}{1+\beta}(1+\beta-u)$
Govindarajulu	$\sigma((\beta+1)u^\beta-\beta u^{\beta+1})$	$\sigma(\beta+1)u^\beta$
		$[(1-u)^2+\frac{2u(1-u)}{\beta+1}+\frac{u^2}{(\beta+1)(\beta+2)}]$
generalized lambda	$\lambda_1+\frac{u^{\lambda_3}-(1-u)^{\lambda_4}}{\lambda_2}$	$\lambda_2^{-1}\{\frac{(1+\lambda_3)(1-u)u^{\lambda_3}}{1+\lambda_3}$
		$+\frac{\lambda^4(1-(1-u))^{1+\lambda_u}}{1+\lambda_4}\}$
power Pareto	$\frac{cu^{\lambda_1}}{(1-u)^{\lambda_2}}$	$c[\lambda_1 B_u(\lambda_1,2-\lambda_2)$
		$+\lambda_2 B_u(\lambda_1+1,1-\lambda_2)]$

The results for NBUHR, NBUE and DVRL in terms of $\phi(u)$ assumed above are proved in Nair and Sankaran (2013a) The proofs of DHR, DHRA, IMRL, NWUE, NWUHR and IVRL are similar. □

The condition for monotonicity of $C_Q(u)$ has an elegant expression. From (7.39) $C_Q(u)$ is increasing (decreasing) according as

$$(1-u)C_Q'(u) = C_Q(u) - \frac{\mu-A(u)}{1-u} \geq (\leq)0$$

giving $C_Q(u) \geq (\leq)m_Q(u)$.

The application of the above results in a practical situation demand the estimates of $A(u)$ and $C_Q(u)$. We note that the estimate

$$\frac{1}{n}A(X_{r:n}) = \frac{1}{n}[nX_{1:n} + (n-1)(X_{2:n}-X_{1:n}) + \ldots \\ + (n-r+1)(X_{r:n}-X_{r-1:n})] \tag{7.43}$$

converges uniformly to $A(u)$ as $n \to \infty$ and $\frac{r}{n}$ tends to u. Also

$$\phi_{r:n} = \frac{A(X_{r:n})}{n\bar{X}_n}$$

where $\bar{X}_n$ is the sample mean of the n order statistics, is called the total time on test statistic and $(\frac{r}{n}, \phi_{r:n})$, $r=1,2,\ldots,n$ when connected by consecutive straight lines is called the TTT-plot. Using the above estimate of $A(u)$ we can prescribe an estimate of $C_Q(u)$ as

$$\hat{C}_Q(u) = \frac{1}{1-u}\int_u^1 \frac{\hat{\mu}-\hat{A}(p)}{1-p}dp \tag{7.44}$$

with $\hat{\mu} = \bar{X}_n$ and $\hat{A}(p) = \frac{1}{n}A(X_{r:n})$.

A closely related function to $A(u)$ is the stop-loss transform defined in the quantile format as (Nair et al. (2013b))

$$P_1(u) = \int_u^1 (1-p)q(p)dp \qquad (7.45)$$

which provides the expected loss in a stop-less treaty with infinite cover as a function of priority in risk analysis. It is easy to see that

$$A(u) = \int_0^u (1-p)q(p)dp = \mu - P_1(u). \qquad (7.46)$$

Note that $P_1(u) = M(Q_X(u))$ where $M(x) = E(X-x)^+$, $(X-x)^+ = \max(X-x, 0)$ and as such $M(x)$ has the interpretation as a mean residual life. By virtue of (7.41) we can write

$$C_Q(u) = \frac{1}{1-u}\int_u^1 \frac{P_1(p)}{1-p}dp. \qquad (7.47)$$

THEOREM 7.22. *The functions $C_Q(u)$, $P_1(u)$ determine each other and the distribution of X uniquely.*

PROOF. Equation (7.47) shows that $P_1(u)$ determines $C_Q(u)$ and form

$$P_1(u) = -(1-u)\frac{d}{du}(1-u)C_Q(u)$$

$P_1(u)$ is determined from $C_Q(u)$. That the two specifies $Q(u)$ follow from Theorem 7.19.   □

REMARK 7.5. *In view of the identity (7.46) all results connecting $C_Q(u)$ and $A(u)$ translate into those between $C_Q(u)$ and $P_1(u)$.*

The quantile based cumulative entropy (CE) is

$$\bar{C}_Q = -\int_0^1 (\log p)pq(p)dp$$

and the corresponding measure for past life is

$$\bar{C}_Q = -\int_0^u \frac{p}{u}\log\left(\frac{p}{u}\right)q(p)dp$$

$$= \frac{\log u}{u}\int_0^u pq(p)dp - \frac{1}{u}\int_0^u (\log p)pq(p)dp \qquad (7.48)$$

For the uniform distribution $Q(u) = a + (b-a)u$ this measure is $\bar{C}_G(u) = \frac{(b-a)u}{4}$. Defining the above measures, Sankaran and Sunoj (2017) proved that for a non-negative random variable X the property $\bar{C}_Q(u) = Kr_Q(u)$ where k is a constant holds for all u if and only if

$$Q(u) = \frac{\alpha}{1-\beta} u^{\frac{1}{\beta}-1}, \quad \alpha > 0; \; 0 < \beta < 1$$

the quantile function of the power distribution.

REMARK 7.6. *In the above result the constant K should be less than unity.*

The application of measures of uncertainty in modelling personal incomes has a long history. For example, Ord et al. (1981) suggested the Renyi entropy $\frac{1}{r} \int_0^\infty f(x)(1 - f^r(x))dx$, $-1 < r < \infty$ as measure of inequality in incomes. We examine some basic concepts in income modelling and their relationship with cumulative entropy. Let X be a non-negative random variable representing the income of a community of individuals with absolutely continuous distribution function $F(x)$ and density function $f(x)$. Assuming a poverty line $X = t$ (those with income below t are considered poor), the proportion of poor people is $F(t)$ and their income distribution becomes that $[X|X \leq t]$. Then the income gap ratio of the poor people is defined as

$$\alpha_F(t) = 1 - E\left(\frac{X}{t} | X \leq t\right) = 1 - \frac{\int_0^t xf(x)dx}{tF(t)}. \tag{7.49}$$

Infact $\alpha(t)$ represents the income short fall of the poor and measures the gap between the poverty line t and the mean income among the poor. It is closely related to the mean inactivity time $r_F(t) = E(t - X|X \leq t)$ by means of

$$r_F(t) = t\alpha_F(t). \tag{7.50}$$

Written in terms of the quantile function $Q(u)$ of F,

$$\alpha_a(u) = \alpha_F(Q(u)) = 1 - \frac{\int_0^u Q(p)dp}{uQ(u)} = \frac{\int_0^u pq(p)dp}{uQ(u)} \tag{7.51}$$

and its relationship with the mean inactivity quantile function becomes

$$r_Q(u) = Q(u)\alpha_Q(u). \tag{7.52}$$

We have on differentiating (7.43)

$$u\bar{C}_Q'(u) + \bar{C}_a(u) = \frac{1}{u} \int_0^u pq(p)dp = r_Q(u),$$

$$\frac{d}{du} u\bar{C}_Q(u) = r_Q(u), \tag{7.53}$$

or

$$\bar{C}_Q(u) = \frac{1}{u} \int_0^u r_Q(p) dp. \tag{7.54}$$

Thus the cumulative past entropy and the income gap ratio are related through

$$\bar{C}_Q(u) = \frac{1}{u} \int_0^u Q(p) \alpha_Q(p) dp.$$

Equation (7.46) provides

$$uQ(u)\alpha_Q(u) = \int_0^u pq(p) dp.$$

Differentiating

$$uq(u)\alpha_Q(u) + Q(u)\alpha_Q(u) + uQ(u)\alpha_Q'(u) = uq(u)$$

giving

$$\frac{q(u)}{Q(u)} = \frac{u\alpha_Q'(u) + \alpha_Q(u)}{u(1 - \alpha_Q(u))}.$$

Integrating from 0 to u,

$$Q(u) = \exp\left[-\int_0^u \frac{u\alpha_Q'(p) + \alpha_Q(p)}{p(1 - \alpha_Q(p))} dp \right]$$

so that $\alpha_Q(u)$ determines $Q(u)$ uniquely. This proves the next theorem.

THEOREM 7.23. *The functions $\alpha_Q(u)$, $\bar{C}_Q(u)$ and $Q(u)$ determine each other uniquely.*

7.8. Other topics

So far we have reviewed some main stream topics in cumulative entropy consistent with our deliberations with regard to other measures of information. In this section some topics that depart from these will be discussed. Renjini et al. (2016a) have developed Bayes estimates of the DCRE of the Pareto I distribution $f(x) = \frac{\alpha\beta^\alpha}{x^{\alpha+1}}$, $\alpha, \beta > 0$; $x \geq \beta$. As noted earlier the DCRE of this model is

$$c(t) = \frac{\alpha t}{(\alpha - 1)^2}, \quad \alpha > 1.$$

If $\underline{x} = (x_{(1)}, \ldots, x_{(5)})$ be the first Type II censored sample, the likelihood function is

$$l(\alpha, \beta | \underline{x}) = \frac{n!}{(n-s)!} \alpha^s \exp(\alpha n\beta) \exp(-z) \exp(-\alpha(n-s) \log x_{(s)} + z)$$

with $z = \sum_{i=1}^{S} x_{(i)}$. Using the Jeffrey's prior

$$p_J(\alpha, \beta) = \frac{K}{\alpha} \beta^{\delta\alpha - 1}$$

uniform prior

$$p_U(\alpha, \beta) = K\beta^{\delta\alpha - 1}$$

and the informative gamma prior

$$p_G(\alpha, \beta) = \alpha^{r-1}\beta^{\delta\alpha-1}e^{-\alpha T}, \ \alpha > 0, \beta, \delta, T > 0$$

they have obtained Bayes estimates with squared error, K-loss, generalized entropy, modified squared error and precautionary loss functions for real and simulated data. Using the expected loss, they have compared the relative accuracies of the different estimates. The same authors (Renjini et al. (2016b)) have also found Bayes estimates of the DCRE for the same distribution based on upper record values. For a complete sample $\underline{x}_n = (x_1, x_2, \ldots, x_n)$ the likelihood becomes

$$l(\alpha, \beta | \underline{x}) = \alpha^n \beta^{\text{nd}} \exp[-(\alpha + 1)v],$$

where $v = \sum_{i=1}^{n} \log x_i$. A similar work for this case using the same set of priors as above and also the prior

$$p_E(\alpha.\beta) = \alpha\beta^{\delta\alpha-1} \exp(-\alpha w)$$

was investigated by Renjini et al. (2016a). On the basis of simulation studies they have found that Jeffrey's and gamma priors have smaller posterior risks as compared to the uniform and exponential prior.

Toomaj et al. (2017) have defined an ordering of cumulative entropies by defining X to be smaller than Y in CRE order, $X \leq_{\text{CRE}} Y$ if $C_X \leq C_Y$. They observed that $X =_{\text{CRE}} Y$ does not imply that X and Y have the same distribution and further if $Y = \phi(X)$ where ϕ is strictly increasing on the support of X,

$$C_Y = C_X - \int_0^1 (\phi'(u) - 1)\bar{F}_X(u) \log \bar{F}_X(u) du$$

and hence if $\phi'(u) \geq 1$, then $X \leq_{\text{CRE}} Y$. Also if $P(x)$ is the cumulative hazard function of X, then $C_X = E(V(X))$, where $V(Z) = \int_0^z P(x)dx$. As $V(Z)$ increases, $X \leq_{\text{icx}} Y \Rightarrow X \leq_{\text{CRE}} Y$. Thus if d is the dispersion order (same as hazard quantile function order) defined by $g(F^{-1}(u)) \leq f(\bar{F}^{-1}(u))$ one should have

$$X \leq_d Y \Rightarrow X \leq_{\text{st}} Y \Rightarrow X \leq_{\text{icx}} Y \Rightarrow X \leq_{\text{CRE}} Y.$$

Toomaj et al. (2017) have further studied the CRE for coherent and mixed systems when the components have identically distributed lifetimes and derived bounds for the CRE. They have also proposed a measure to study the closeness of the system with series and parallel systems.

Hashempour and Doostparast (2018) have considered the role of sequential order statistics in multi-component systems and provide characterizations of the baseline distribution using the CRE of sequential order statistics. It shown that the equality of the CRE in the first sequential order statistic determines the distribution of lifetimes. Also a characterization of the Weibull distribution and some bounds for the CRE, were obtained.

Zaradasht (2020) have obtained some interesting results connecting relative mean residual life (RMRL) and relative cumulative residual entropy (RCRE). If $P(x)$ is the cumulative hazard rate, recall that

$$C(t) = -\frac{1}{\bar{F}(t)} \int_t^\infty \bar{F}(x) \log \bar{F}(x) dx - m(t) P(t).$$

The RMRL of distribution F with respect to distribution G is

$$m_{F,G}(t) = \frac{1}{\bar{G}(t)} \int_t^\infty \bar{G}(x) dP_F(x).$$

Similarly, the RCRE is

$$C_{F,G}(t) = -\int_t^\infty \frac{\bar{G}(x)}{\bar{G}(t)} \log \frac{\bar{G}(x)}{\bar{G}(t)} dP_F(x).$$

This definition is motivated by the fact that if F is exponential then $C_{F,G}(t) = C_G(t)$. Notice that

$$C_{F,G}(t) = \frac{1}{\bar{G}(t)} \int_t^\infty \bar{G}(x) P_F(x) P_G(x) - P_G(t) m_{F,G}(t)$$

and

$$C_{F,G}(t) = E(m_{F,G}(Y)|Y > t)$$

where y is the random variable with distribution function G. Zaradasht (2020) has proved the following theorems.

THEOREM 7.24. *If Y is the proportional hazards model of X, then*

(i) *The ratio of hazard rates $\frac{\lambda_F(t)}{\lambda_G(t)}$ is constant if and only if $C_{F,G}(t)$ is constant for all t in $S_E = \{t | \bar{F}(t) > 0; \bar{G}(t) > 0\}$*

(ii) *$C_{F,G}(t)$ is increasing (decreasing) if and only if $m_{F,G}(t)$ is increasing (decreasing)*

Starting from the relationships $C_X = \text{Cov}(X, P(X))$ and $C(t) = \text{Cov}(X_t, P(X_t))$ a second form of RCRE of F with respect to G

can be defined as

$$C^*(X, Y) = \text{Cov}\,(X, G^{-1}(F(X))) = \int_0^1 F^{-1}(u)G^{-1}(u) - E(X_1)E(X_2)$$

and a third form by

$$C^*_{F,G}(t) = \text{Cov}\,(X_{1t}, G^{-1}(F_t(X_{1t})))$$

$$= \int_0^1 F^{-1}(uF(t) + F(t)G^{-1}(u)du) - (t + m_X(t))E(X_2).$$

When G is exponential with parameter unity $C^*(X, Y)$ and $C^*_{F,G}(t)$ reducing to C_X and $C(t)$ respectively.

THEOREM 7.25. *If $X \leq_{dil} Y$, then $\sigma_1 \leq C_{X_1,X_2} \leq \sigma_2$ where σ_1 and σ_2 are the standard deviations of X and Y respectively.*

CHAPTER 8

Generalized Cumulative Entropy and Divergence

8.1. Introduction

In Chapters 2 and 3 we have discussed the Shannon's entropy and modifications of the residual and past lifetimes along with their applications in reliability modelling. This was followed up in the fourth chapter by various generalizations of the Shannon entropy. Some of those generalizations were supported by justifications through modifications of the axioms that generated the Shannon measure while most of the others were mathematical generalizations of the functional forms involved with some having practical applications. Viewing the cumulative entropy as yet another extension by replacing the density function by the survival or distribution function, it is possible to have similar extensions of cumulative entropies with less emphasis on their individual basis either from the axiomatic or application-wise perspectives. Thus the Renyi, Tsallis, Varma and similar types of the Shannon-induced entropies are generalized to their parallel versions by mere replacement of the densities by their survival and distribution functions to evolve various forms of generalized cumulative entropies. The analysis in the form of properties, bounds, characterizations and others also follows the same patterns in their descriptions. The present chapter aims at a review of these developments. Since the formulations are only in their formative stages it is hoped that adequate justifications to these in the form of applications and theoretical discussions will emerge to find them a deserving place in information theory as well as in reliability modelling.

8.2. Survival and failure entropies

Motivated by the definition of CRE, Zografos and Nadaraja (2005) proposed two broad classes of measures of uncertainty of a random vector $\underline{X} = (X_1, X_2, \ldots, X_n)$ based on its survival function $S(\underline{x})$, where $\underline{x} = (x_1, x_2, \ldots, x_n)$. The measures are survival exponential entropy of order α

$$E_\alpha = \left(\int_{R^+} S^\alpha(\underline{x}) dx \right)^{\frac{1}{1-\alpha}}, \ \alpha \geq 0$$

and the generalized survival exponential entropy of order (α, β)

$$E_{\alpha,\beta} = \left(\frac{\int_{R_n^+} S^\alpha(x)dx}{\int_{R_n^+} S^\beta(x)dx} \right)^{\frac{1}{\beta-\alpha}}, \ \alpha, \beta \geq 0, \ \alpha \neq \beta.$$

Zografos and Nadaraja (2005) discussed several properties of these entropies. However entropies of higher dimensional random variables being beyond the scope of the present work, we refer to Zografos and Nadaraja (2005) for further details and concentrate on their one-dimensional versions for a detailed study.

Abbasnejad et al. (2010) considered a non-negative random variable X with absolutely continuous distribution F and defined the survival entropy (SE) of order α of X as

$$S_\alpha(x) = -\frac{1}{\alpha - 1} \log \int_0^\alpha \bar{F}^\alpha(x)dx, \ \alpha > 0, \ \alpha \neq 1. \tag{8.1}$$

They have shown that

(a) $S_{\alpha,Z} = -\frac{1}{\alpha-1} \log \alpha + S_{\alpha,X}, \ Z = aX + b$

(b) $S_{\alpha,Y} = \frac{\theta\alpha - 1}{\alpha - 1} S_{\theta,\alpha X}$, where Y is the proportional hazards model of X

(c) $S_{\alpha,Y} \geq S_{\alpha,X} \geq S_{\alpha,\theta X}, \ \theta \geq 1, \ \alpha > 1$ and $0 < \theta \leq 1, \ 0 < \alpha < 1$ and the inequalities are reversed for $\theta \geq 1, 0 < \alpha < 1$ and $0 < \theta \leq 1,$ $\alpha > 1$

(d) $S_\alpha \geq -\frac{1}{\alpha-1}\mu, \ \mu = E(X)$

$$S_\alpha \leq (\geq)E(m_X(X)), \ \alpha > 1(0 < \alpha < 1).$$

The dynamic survival entropy (DSE) of order α now takes the form

$$S_\alpha(t) = -\frac{1}{\alpha - 1} \log \int_t^\infty \left(\frac{\bar{F}(x)}{\bar{F}(t)} \right)^\alpha dx, \ \alpha > 0; \ \alpha \neq 1. \tag{8.2}$$

We see that

$$S_\alpha(t) \geq -\frac{1}{\alpha - 1} \log m(t)$$

and

$$S_\alpha(t) \leq (\geq)E(m(X)|X \geq t), \ \alpha > 1 \ (0 < \alpha < 1).$$

EXAMPLE 8.1. *(i) Uniform distribution in $(0, \theta)$,*

$$S_\alpha(t) = -\frac{\log(\theta - t)}{(\alpha - 1)} \frac{(\theta - t)}{(\alpha + 1)}.$$

It is increasing (decreasing) in t for $\alpha > 1$ $(0 < \alpha < 1)$.
(ii) Exponential distribution (λ),

$$C_\alpha(t) = \frac{\log \alpha}{\alpha - 1} + \frac{1}{\alpha - 1} \log \lambda$$

which is independent of t.
(iii) Pareto distribution, $\bar{F}(x) = \beta^c (x + \beta)^{-c}$,

$$C_\alpha(t) = -\frac{1}{\alpha - 1} \log \frac{t + \beta}{\alpha c - 1}, \ \alpha > \frac{1}{c}$$

where it is decreasing (increasing) in t for $\alpha > 1$ ($0 < \alpha < 1$).

A classification of distributions is done by defining increasing (decreasing) DSE abbreviated as, IDSE (DDSE) if $S_\alpha(t)$ is an increasing (decreasing) function of t and it is shown that X is ISDE (DDSE) if and only if

$$S_\alpha(t) \geq (\leq)\frac{\log \alpha}{\alpha - 1} \log h(t).$$

Further the distribution F is characterized by the DSE.

Life distributions can be ordered using $S_\alpha(t)$. Abbasnejad et al. (2010) say that X is less than Y in SE ordering, $X \leq_{S_\alpha} Y$ if $S_{\alpha,X} \leq S_{\alpha,Y}$. Further

$$X \leq_{\text{st}} Y \Rightarrow X \leq_{S_\alpha} Y$$
$$X \leq_{\text{hr}} Y \Rightarrow S_{a,X}(t) \leq (\geq)S_{\alpha,Y}, \ \alpha > 1 \ (0 < \alpha < 1).$$

For the equilibrium random variable X_E of X, then the residual entropy $H_E(t)$ of X_E satisfies

$$H_E(t) = \frac{\alpha}{\alpha - 1} \log m(t) + S_\alpha(t)$$

and

$$X \text{ DDSE and DMRL} \Rightarrow Y, \text{ DDRE}, \alpha < 1$$
$$X \text{ DDSE and IMRL} \Rightarrow Y, \text{ DDRE}, 0 < \alpha < 1$$
$$X \text{ is IMRL} \Rightarrow S_{\alpha,X}(t) \geq (\leq)S_{\alpha,Y}(t)$$
$$\text{and } S_{\alpha,X}(t) \leq (\geq)S_{\alpha,Y}(t) \text{ for } 0 < \alpha < 1.$$

THEOREM 8.1. *The relationship $S_\alpha(t) = c + \frac{1}{\alpha-1} \log h(t)$ where C is a constant is satisfied if and only if X is generalized Pareto.*

Abbasnejad (2011) has proved the following characterization theorems.

THEOREM 8.2. *If X and Y are non-negative random variables with distribution functions F and G belonging to the same family but for a change in location and scale*

$$C_\alpha(X_{1:n}) - C_\alpha(X) = C_\alpha(Y_{1:n}) - C_\alpha(Y)$$

for all $n = n_j$, $j \geq 1$ such that $\sum \frac{1}{n_j}$ diverges to $+\infty$. Further if they belong to the same location family $C_\alpha(X_{1:n}) = C_\alpha(Y_{1:n})$ for all $n_j \geq 1$.

The work also contains the failure entropy of order α

$$\bar{S}_\alpha = -\frac{1}{\alpha - 1} \log \int_0^\infty F^\alpha(x)dx \qquad (8.3)$$

and its past lifetime version

$$\bar{S}_\alpha(t) = -\frac{1}{\alpha - 1} \log \int_0^t \frac{F^\alpha(x)}{F^\alpha(t)}dx. \qquad (8.4)$$

If for X, $\bar{C}_\alpha(t)$ is increasing (decreasing) and ϕ is monotonic and convex (concave) then $\phi(X)$ is also increasing (decreasing) for $\alpha > 1$ and ϕ is monotone and concave (convex) then $\phi(X)$ is also increasing (decreasing) for $0 < \alpha < 1$. Further for $X \leq_{\text{rh}} Y$, $\bar{C}_X(t) \leq (\geq) \bar{C}_Y(t)$ for $\alpha > 1$ $(0 < \alpha < 1)$.

THEOREM 8.3. *The identity $\bar{S}_\alpha(t) = K + \frac{1}{\alpha-1} \log \lambda(t)$ is satisfied, if and only if X has a power distribution $F(x) = \left(\frac{x}{\alpha}\right)^\beta$, $0 < x < \alpha$, $\beta > 0$ and $\bar{S}_\alpha(t) = K - \frac{1}{\alpha-1} \log r(t)$ also holds in this case.*

Calling (8.1) as the cumulative Renyi entropy, Sunoj and Linu (2012a) obtained the relationship

$$(1 - \alpha)S'_\alpha(t) = \alpha h(t) - e^{-(1-\alpha)}S_\alpha(t)$$

from which conditions of monotonicity of $S'_\alpha(t)$ can be derived. They have shown that for the

Pareto I: $\bar{F}(x) = \left(\frac{k}{x}\right)^c$, $(1 - \alpha)S_\alpha(t) = \dfrac{\log t}{c\alpha - 1}$,

Weibull: $\bar{F}(x) = e^{-x^p}$, $(1 - \alpha)S_\alpha(t) = \log\left(\dfrac{\alpha^{-\frac{1}{p}}}{pe^{-\alpha t^p}}\right)\Gamma\left(\dfrac{1}{p}, \alpha t^p\right)$,

and the relationship $(1 - \alpha)S_\alpha(t) = \log K(m(t))$ for some constant $K > 0$ characterizes the generalized Pareto law. The weighted analogue of $S_\alpha(t)$,

$$S_{\alpha,w}(t) = \frac{1}{\alpha} \log \int_t^\infty \frac{\bar{F}_w^\alpha(x)}{\bar{F}_w^\alpha(x)}dx$$

with

$$\bar{F}_w(x) = \frac{E[w(X)|X > t]}{E(w(X))} \bar{F}(x)$$

and $w(x) > 0$ is a weight function was discussed in Sunoj and Linu (2012b). The original and weighted versions were related as

$$S_{\alpha,w}(t) \leq (\geq)S_\alpha(t)$$

for $0 < \alpha < 1$ $(\alpha > 1)$ according as $E(w(X)|X > x) \leq (\geq)E(w(X)|X > t)$ for all $x \geq t$. The equilibrium and length-biased distributions appear as special cases and in the former case $S_{\alpha,E}(t) \geq (\leq)S_\alpha(t)$ for $0 < \alpha < 1$ ($\alpha > 1$) if X is IMRL (DMRL).

The generalized survival entropy proposed by Abbasnejad and Borzadaran (2015) has the form

$$
GS_F = \begin{cases} \frac{1}{\beta-\alpha} \log \frac{\int_0^\infty \bar{F}^\alpha(x)dx}{\int_0^\infty \bar{F}^\alpha(x)dx}, & \alpha \neq \beta \\ -\frac{\int_0^\infty \bar{F}^\alpha(x) \log \bar{F}(x)dx}{\int_0^\infty \bar{F}^\alpha(x)dx}, & \alpha = \beta. \end{cases}
$$

Some examples GS_F given by them are

$$\text{Uniform, } F(x) = \frac{x-a}{b-a}, \quad GS_F = \frac{1}{\beta-\alpha} \log \frac{\beta+1}{\alpha+1}$$

$$\text{exponential, } F(x) = 1 - e^{-\lambda x}, \quad GS_F = \frac{1}{\beta-\alpha} \log \frac{\beta}{\alpha}$$

$$\text{Pareto, } F(x) = \lambda^c(x+\lambda)^{-c}, \quad GS_F = \frac{1}{\beta-\alpha} \log \frac{\beta c - 1}{\alpha c - 1}$$

$$\text{Weibull, } \bar{F}(x) = e^{-(\theta x)^\lambda}, \quad GS_F = \frac{1}{\lambda(\beta-\alpha)} \log \frac{\beta}{\alpha}.$$

They show that

(i) GS_F is invariant under linear transformation

(ii) $GS_{Y,\alpha,\beta} = \theta GS_{\theta\alpha,\theta\beta}$ where Y is the proportional hazards model of X.

The dynamic version is given as

$$
GS(t) = \begin{cases} \frac{1}{\beta-\alpha} \log \frac{\int_t^\infty \bar{F}^\alpha(x)dx}{\int_t^\infty \bar{F}^\beta(x)dx} + \log \bar{F}(t), & \alpha \neq \beta \\ \frac{\int_t^\infty \bar{F}^\alpha(x) \log \bar{F}(x)dx}{\int_t^\infty \bar{F}^\alpha(x)dx} + \log \bar{F}(t), & \alpha = \beta. \end{cases}
$$

A characterization of the generalized Pareto distribution is proved by the property $GS(t) = $ a constant. The corresponding failure entropy is defined by

$$
\overline{GS}(t) = \begin{cases} \frac{1}{\beta-\alpha} \log \frac{\int_0^t F^\alpha(x)dx}{\int_0^t F^\beta(x)dx} + \log F(t), & \alpha \neq \beta \\ -\frac{\int_0^t F^\alpha(x) \log F(x)dx}{\int_0^t F^\alpha(x)dx}, & \alpha = \beta. \end{cases}
$$

Some properties that parallel those of $GS(t)$ are also established in this work.

Kundu and Nanda (2016) introduced an ordering among survival entropies of order α by saying that X is smaller than Y in DSE if for all t $S_{\alpha,X}(t) \leq S_{\alpha,Y}(t)$ and denoted it as $X \leq_{\mathrm{DSE}(\alpha)} Y$. Under increasing linear transformations this ordering is closed. Some conditions for establishing ageing properties in terms of monotone DSE are given as follows:

(i) For $0 < \alpha < 1$, if X is decreasing DSE and $S_\alpha(t)$ is convex, then X is IFR and if X is increasing DSE and $S_\alpha(t)$ is concave then X is DFR.

(ii) For $\alpha > 1$, if X is increasing (decreasing) DSE and $S_\alpha(t)$ is concave (convex) then X is IFR (DFR).

If ϕ is any continuous increasing convex (concave) function with $\phi(x) = \infty$, then X has increasing (decreasing) DSE then $\phi(X)$ also has the same type of monotonicity. The generalized Pareto distribution is characterized by the property $S_\alpha(t) = \frac{1}{\alpha-1} \log m(t)$. Considering the past life version of the cumulative entropy of order α, they show that the reversed hazard rate ordering implies $\bar{S}_{\alpha,X}(t) \geq \bar{S}_{\alpha,Y}(t)$ for $0 < \alpha < 1$ and $\bar{S}_{\alpha,X}(t) \leq \bar{S}_{\alpha,Y}(t)$ for $\alpha > 1$. Further, $\bar{S}_\alpha(t)$ uniquely determine the distribution of X for each $\alpha > 0$ and in particular, for the uniform distribution in (a, b),

$$\bar{S}_\alpha(t) = \frac{1}{1-\alpha} \log \frac{t-\alpha}{\alpha+1}, \quad a < t < b.$$

Rajesh et al. (2017) investigated the weighted version of survival entropy of order α (WSE) defined by

$$WS_\alpha = \frac{1}{1-\alpha} \log \int_0^\infty x \bar{F}^\alpha(x) dx, \quad \alpha > 0, \ \alpha \neq 1 \tag{8.5}$$

or

$$\exp[(1-\alpha)WS_\alpha] = \frac{1}{\alpha} \int_0^\infty \bar{F}^\alpha(x) m(x) dx - \frac{\alpha-1}{\alpha} \int_0^\infty x \bar{F}^\alpha(x) m'(x) dx$$

when $m(x)$ is as usual, the mean residual life of X. The properties of WS_α are

(i) if $Y = aX + b, a > 0, b \geq 0$

$$\exp[(1-\alpha)WS_\alpha(Y)] = a^2 \exp[(1-\alpha)WS_\alpha(X)] + ab \exp[(1-\alpha)WS_\alpha(X)]$$

(ii) if Y is the proportional hazards model of X
   (a) $WS_\alpha(Y) = \frac{1-\theta\alpha}{1-\alpha} WS_\alpha(X)$
   (b) $WS_\alpha(Y) \geq WS_\alpha(X) \geq WS_\alpha(X\theta), 0 < \theta \leq 1, 0 < \alpha < 1$ and $\theta \geq 1, \alpha > 1$, and
   $WS_\alpha(Y) \leq WS_\alpha(X) \leq WS_\alpha(X\theta), 0 < \theta \leq 1, \alpha > 1$ and $\theta \geq 1, 0 < \alpha < 1$.

The dynamic version is straightforward

$$WS_\alpha(t) = \frac{1}{1-\alpha} \log \int_t^\infty x \left(\frac{\bar{F}(x)}{\bar{F}(t)} \right)^\alpha dx.$$

For an increasing (decreasing) $WS_\alpha(t)$ there exist the bounds

$$WS_\alpha(t) \geq (\leq) \frac{\log \alpha}{\alpha-1} + \frac{1}{\alpha-1} \log \frac{h(t)}{t}, \quad \alpha > 0, \alpha \neq 1.$$

Finally, if $X \leq_{st} Y$ then X has larger (smaller) $WS_\alpha(t)$ than Y for $\alpha > 1$ $(0 < \alpha < 1)$. The Rayleigh distribution $\bar{F}(x) = \exp[-\frac{x^2}{2\sigma^2}]$, $x > 0$; $\sigma > 0$ is characterized by $WS_\alpha(t) = K$, a constant and also by

$$(1 - \alpha)WS_\alpha(t) = \log \frac{m^*(t)}{\alpha}$$

where

$$m^*(t) = \frac{1}{\bar{F}(t)} \int_t^\infty x\bar{F}(x)dx.$$

A version similar to (8.5) is which $\bar{F}$ is replaced by F called weighted failure entropy of order α (a generalization of $\bar{S}_\alpha$) in (8.3) as discussed in Nair et al. (2017) *viz.*

$$W\bar{S}_\alpha = -\frac{1}{\alpha - 1} \log \int_0^\infty xF^\alpha(x)dx, \ \alpha > 0, \ \alpha \neq 1$$

and its dynamic version

$$W\bar{S}_\alpha(t) = -\frac{1}{\alpha - 1} \log \int_0^t x \left(\frac{F(x)}{F(\alpha)} \right)^\alpha dx.$$

Apart from proving results similar to WSE, they characterize the power distribution $F(x) = x^\lambda$, by the property

$$W\bar{S}_\alpha(t) = C - \frac{1}{\alpha - 1} \log r^*(t), \quad r^*(t) = \frac{1}{F(t)} \int_0^t xF(x)dx.$$

Nourbhakhsh et al. (2020) also have some new results in this direction. The failure entropy of past life in (8.4) has the additional features

(i)

$$\bar{S}_\alpha(t) \leq \frac{\log \alpha + \log \lambda(t)}{1 - \alpha}$$

if and only if $S_\alpha(t)$ is increasing

(ii)

$$\bar{S}_\alpha(t) \leq (\geq)\frac{1}{1 - \alpha} \int_0^t f(x) \log \frac{f(x)}{F^\beta(t)}dx - 1 \quad 0 < \alpha < 1(\alpha > 1)$$

(iii)

$$\bar{S}_{\alpha,X}(t) \geq (\leq)\bar{S}_{\alpha,Y}(t), \quad 0 < \alpha < 1 \ (\alpha > 1)$$

where Y is the proportional reversed hazard rate model of X.

A more general model is obtained when F is replaced by F_w resulting in

$$\bar{S}_{\alpha,w}(t) = \frac{1}{1 - \alpha} \log \int_0^t \frac{F_w^\alpha(\alpha)}{F_w^\alpha(t)}dx, \ \alpha \neq 1, \alpha > 0.$$

This satisfies if $Ew(X)|X < t) \geq E(w(X)|X < s)$ then $\bar{S}_{\alpha,w}(t) \leq (\geq)\bar{S}_{\alpha,w}(s)$, for $0 < \alpha < 1 \ (\alpha > 1)$. Further $E(w(X)|X < t) \geq (\leq)$

$E[w(X)|X < s]$ for $t < s$ and $\bar{S}_\alpha(t)$ is increasing (decreasing) implies $\bar{S}_{\alpha,w}(t)$ is increasing in t. Some result for special cases when $w(x)$ gives the length-biased and equilibrium models are given in the work. The estimation of $\bar{S}_\alpha(t)$ and $\bar{S}_{\alpha,w}(t)$ can be accomplished through the estimates

$$\hat{\bar{S}}_\alpha(t) = \frac{1}{1-\alpha} \log \sum_{j=1}^{n} \left(\frac{j}{n}\right)^\alpha (X_{j+1:n} - X_{j:n})$$

and

$$\hat{\bar{S}}_{\alpha,w}(t) = \frac{1}{1-\alpha} \log \sum_{j=1}^{n} \left(\left(\frac{j \sum_{i=1}^{j} w(X_{i:n})}{n \sum_{i=1}^{n} w(X_{i:n})} \right)^\alpha (X_{j+1:n} - X_{j:n}) \right),$$

for $x_{k:n} \leq t < X_{k+1:n}$.

8.3. Cumulative divergence

Recalling the definition of Kullback-Leibler divergence in (4.1), cumulative divergence measures are obtained by replacing $f(x)$ by $\bar{F}(x)$ in that definition. Thus for two non-negative random variables X and Y the cumulative divergence becomes

$$C_{X,Y} = \int_0^\infty \bar{F}(x) \log \frac{\bar{F}(x)}{\bar{G}(x)} dx - (E(X) - E(Y)). \tag{8.6}$$

Chamany and Baratpour (2014) proposed the above definition following Baratpour and Rad (2012) and their properties along with those of the dynamic version, given by

$$
\begin{aligned}
C_{X,Y}(t) &= \int_t^\infty \left[\frac{\bar{F}(x)}{\bar{F}(t)} \log \frac{\bar{F}(x)/\bar{F}(t)}{\bar{G}(x)/\bar{G}(t)} + \frac{\bar{G}(x)}{\bar{G}(t)} - \frac{\bar{F}(x)}{\bar{F}(t)} \right] dx \\
&= m_G(t) - m_F(t) - C_X(t) - \int_t^\infty \frac{\bar{F}(x)}{\bar{F}(t)} \log \frac{\bar{G}(x)}{\bar{G}(t)} dx.
\end{aligned}
\tag{8.7}
$$

They show that

(i) $C_{X,Y} \geq 0$ and equality holds if and only if $F = G$.
(ii) $C_{X,Y}(t) \neq C_{Y,X}(t)$. A symmetric version is

$$
\begin{aligned}
\hat{C}_{X,Y}(t) &= \frac{1}{2}(C_{XY}(t) + C_{Y,\bar{X}}(t)) \\
&= \frac{1}{2} \int_t^\infty (\bar{F}_t(x) - \bar{G}_t(x)) \log \frac{\bar{F}_t(x)}{\bar{G}_t(x)}.
\end{aligned}
$$

(iii) $C_{X,Y}(t) \geq 0$ and equality holds if and only if $F = G$.
(iv) $C_{X,Y}(t)$ is convex.

(v) $C_{X,Y}(t)$ is non-decreasing (non-increasing) if and only if

$$C_{X,Y}(t) \geq (\leq) \left(1 - \frac{h_G(t)}{h_F(t)}\right)(m_F(t) - m_G(t)).$$

(vi) For the proportional mean residual life model $m_Y(t) = \theta m_X(t)$,

$$C_{X,Y}(t) \leq (\geq) \frac{(1-\theta)^2}{\theta h_F(t)}$$

is a necessary and sufficient condition for $C_{X,Y}(t)$ to be increasing (decreasing).

(vii) For three lifetimes X, Y and Z, the triangular inequality

$$C_{X,Y}(t) + C_{Y,Z}(t) \geq C_{X,Z}(t)$$

if $Y \geq_{hr} X$ and $Y \geq_{hr} Z$ or $Y \leq_{hr} X$ and $Y \leq_{hr} Z$.

(viii) If $X \leq_{hr} Y$, $C_{X,Y}(t) \leq m_G(t) - m_F(t)$. Further for X, Y, Z satisfying $Z \leq_{hr} Y$, $C_{X,Y}(t) \leq E(Y) - E(X)$ and

$$C_{X,Y} - C_{X,Y}(t) \leq E(Y) - E(X) - (m_Y(t) - m_X(t))$$

(ix) For increasing transformation ϕ,

$$aC_{X,Y}(\phi^{-1}(t)) \leq c_{\phi(X),\phi(Y)}(t) \leq bC_{X,Y}(\phi^{-1}(t))$$

if $a < \phi' < b, a, b > 0$.

Some examples of $C_{X,Y}(t)$ are

(a) exponential, $\bar{F}(x) = e^{-\lambda_1 x}; \bar{G}(x) = e^{-\lambda_2 x}$:

$$C_{X,Y}(t) = \frac{(\lambda_1 - \lambda_2)^2}{\lambda_2}, \text{ a constant}$$

(b) Pareto, $\bar{F}(x) = \left(\frac{\alpha}{x+\alpha}\right)^c; \bar{G}(x) = \left(\frac{\alpha}{x+\alpha}\right)^{c_2}$:

$$C_{X,Y}(t) = (\alpha + t)\frac{(c_1 - c_2)^2}{(c_1 - 1)^2(c_2 - 1)}$$

(c) beta, $\bar{F}(x) = (1-x)^{c_1}, \bar{G}(x) = (1-x)^{c_2}$:

$$C_{X,Y}(t) = \frac{(1-t)(c_1 - c_2)^2}{(1+c_1)^2(1+c_2)}.$$

Di Crescenzo and Longobardi (2015) have discussed the case when f and g are replaced by F and G in (4.1) in the form

$$\bar{C}_{X,Y} = \int_t^{\max(r_X, r_Y)} F(t) \log\left(\frac{F(t)}{G(t)}\right) dt + E(X) - E(Y) \qquad (8.8)$$

where r_X and r_Y are the right hand end points of the supports of X and Y respectively. Then

$$\bar{C}_{X,Y} = C_Y - C_X + E(-\log G(Z) - 1)(E(Y) - E(X))$$

in which Z is an absolutely continuous random variable with density $f_Z(z) = \frac{G(z)-F(z)}{E(X)-E(Y)}$, $z > 0$. Further if X and Y have the same left hand end point l,

$$\bar{C}_{X,Y} \leq \int_t^{\max(r_X, r_Y)} F(u) \left[\frac{F(u)}{G(u)} - 1 \right] du + E(X) - E(Y)$$

and if $X \geq_{\text{st}} Y$

$$\bar{C}_{X,Y} \leq \frac{1}{2} \int_t^{\max(r_X, r_Y)} F(u) \left[\frac{F(u)}{G(u)} - 1 \right] \left[3 - \frac{F(u)}{G(u)} \right] du + E(X) - E(Y).$$

(8.9)

When past lifetimes $X_{(t)} = X | X \leq t$ and $Y_{(t)} = Y | Y \leq t$ are considered the measure (8.8) modifies to

$$\bar{C}_{X,Y}(t) = \int_0^t \frac{F(x)}{F(t)} \log \left(\frac{F(x)G(t)}{F(t)G(x)} \right) dx + \mu_x(t) - \mu_Y(t)$$

(8.10)

where $\mu_X(t) = E(X(t))$ and $\mu_Y(t) = E(Y(t))$. The lower bounds analogous to (8.10) and (8.10) are

$$\bar{C}_{X,Y}(t) \geq (t - \mu_X(t)) \log \frac{t - \mu_X(t)}{t - \mu_Y(t)} + \mu_X(t) - \mu_Y(t)$$

and

$$\bar{C}_{X,Y}(t) \geq \frac{1}{2} \int_0^t \left\{ \left(\frac{F(x)}{F(t)} - \frac{G(x)}{G(t)} \right)^2 / \frac{1}{3} \frac{F(x)}{F(t)} + \frac{2}{3} \frac{G(x)}{G(t)} \right\} dx$$

and upper bounds

$$\bar{C}_{X,Y}(t) \leq \int_0^t \left(\frac{F^2(x)G(t)}{G(x)/F^2(t)} - \frac{F(x)}{F(t)} \right) dx + \mu_X(t) - \mu_Y(t)$$

and

$$\bar{C}_{X,Y}(t) \leq \frac{1}{2} \int_0^t \frac{F(x)}{F(t)} \left(\frac{F(x)}{F(t)} \frac{G(t)}{G(x)} - 1 \right) \left(3 - \frac{F(x)}{F(t)} \frac{G(t)}{F(x)} \right) dx$$
$$\mu_X(t) - \mu_Y(t)$$

when $X \geq_{\text{rh}} Y$. Some applications to failure of nano components and image analysis are also discussed in Di Crescenzo and Longobardi (2015).

Park et al. (2012) pointed out that the direct extension $\int_0^\infty \bar{F}(x) \log(\frac{\bar{F}(x)}{\bar{G}(x)})$ of (4.1) does not keep the non-negativity and characterization properties. To avoid this difficulty Baratpour and Rad (2012) have suggested an extension of Kullback-Leibler information as (8.6). In this case $C_{X,Y} \geq 0$ and the equality sign hold if and only if $F = G$.

Based on a random sample $X_1, X_2, \ldots, X_n$ and $\frac{E(X_1^2)}{2E(X_1)} < \infty$, they have suggested a test of the hypothesis $H_0 : F(x) = F_0(x, \lambda) = 1 - e^{-\lambda x}$ against $H_1 : F(x) \neq F_0(x, t)$. They develop a discrimination information statistic

$$
C_{F,F_0} = -C_F - \int_0^\infty \bar{F}(x) \log \bar{F}_0(x, \lambda) dx - E(X) + \lambda \tag{8.11}
$$
$$
= -C_F + \lambda.
$$

C_F is estimated by the CRE of $F_n(x)$, the empirical distribution function. Thus

$$
\hat{C}_F = -\int_)^\infty \bar{F}_n(x) \log \bar{F}_n(x) = -\sum_{i=1}^{n-1} \frac{n-i}{n} \left(\log \frac{n-i}{n} \right) (X_{i+1:n} - X_{i:n}),
$$

and the estimator of (8.10) becomes

$$
\hat{C}_{F,F_0} = \sum_{i=1}^{n-1} \frac{n-i}{n} \left(\log \frac{n-i}{n} \right) (X_{i+1:n} - X_{i:n}) + \frac{\sum_{i=1}^n X_i^2}{2 \sum_{i=1}^n X_i},
$$

since $\lambda = \frac{E(X_1^2)}{2E(X_1)}$. The test statistic used is

$$
T_n = \frac{\hat{C}_{F,F_0}}{\left(\sum_{i=1}^n X_i^2 / 2 \sum_{i=1}^n X_i \right)}
$$

which converges to zero under H_0. Thus H_0 is rejected at a significance level α if $T_n > T_{n,1-\alpha}$ where $T_{n,1-\alpha}$ is the $100(1-\alpha)$ percentile of T_n under H_0, determined by a Monte Carlo simulation. Park et al. (2012) modified the estimator T_n by

$$
T_n^* = 1 - [(\bar{x} + C(F_n))^2 / 4 \frac{\sum_{i=1}^n x_i^2}{2n}].
$$

Sunoj et al. (2018b) have discussed the quantile form of $C_{X,Y}$ through the expression

$$
C_{X,Y} = \int_0^1 \left(\log \frac{1-p}{1 - Q_3(p)} \right) (1-p) q_X(p) + E(Y) - E(X) \tag{8.12}
$$

where $Q_3(u) = Q_Y^{-1} Q_X(u)$. The effect of non-decreasing transformations T_1 and T_2 on $C_{X,Y}$ is

$$
C_{T_1(X),T_2(Y)} = \int_0^1 \left(\log \frac{1-p}{1 - Q_Y^{-1}(T_2^{-1}(T_1(Q_X(p))))} \right) (1-p) dT_1(Q_X(p))
$$
$$
+ \int_0^1 T_2(Q_Y(p)) dp - \int_0^1 T_1(Q_X(p)) dp.
$$

From (8.12) one can get the cumulative divergence measure of X_t and Y_t as

$$C_{Q_X,Q_Y}(u) = \left(\log \frac{1 - Q_3(u)}{1 - u} - 1 \right) m_{Q_X}(u) + m_{Q_Y}(u)$$

$$+ \frac{1}{1 - u} \int_u^1 \left(\log \frac{1 - p}{1 - Q_3(p)} \right) (1 - p) q_X(p) dp.$$

As an example the generalized lambda distribution is the distribution of X,

$$Q_X(u) = \frac{1}{\lambda_1} + \frac{1}{\lambda_2}(u^{\lambda_3} - (1 - u)^{\lambda_4})$$

and exponential law with parameter θ is the distribution of Y provided

$$C_{Q_X,Q_Y} = \frac{\theta}{\lambda_2^2} \left[\frac{\lambda_3 \lambda_4}{1 + \lambda_4} B_u(\lambda_3, 1 + \lambda_4) + \frac{\lambda_3}{1 + \lambda_3} \sum_{t=\lambda_3}^{2\lambda_3} \frac{1 - u^t}{t} - \frac{1 - u^{2\lambda_3}}{2\lambda_3} \right.$$

$$\left. + \frac{\lambda_4}{1 + 2\lambda_4}(1 - u) \right] \frac{\lambda_4(1 - u)^{\lambda_4 - 1}}{(1 + \lambda_3)(\lambda_4 - 1)}$$

$$- \frac{\lambda_4}{1 + \lambda_3} B_u(\lambda_3 + 2, \lambda_4 - 1) - \lambda_4 B_u(1 + \lambda_3, \lambda_4)$$

$$- \frac{1}{\lambda_2} \left[\frac{\lambda_4}{(1 + \lambda_4)^2}(1 - u)^{1 + \lambda_4} - \frac{(1 - u)^{1 + \lambda_3}}{1 + \lambda_3} + \sum_{t=1}^{\lambda_3} \frac{1 - u^t}{t} \right]$$

More properties of the quantile measure are indicated in the following theorems.

THEOREM 8.4. $C_{Q_X,Q_Y}(u)$ *is non-decreasing (non decreasing) if and only if*

$$C(u) \geq (\leq) H_2(u) + \left(\frac{H_3(u)}{1 - Q_3(u)} - 1 \right) m_{Q_X}(u)$$

where $H_2(u) = [(1 - u)q_X(u)]^{-1}$ *and* $H_3(u) = [(1 - u)q_3(u)]^{-1}$.

THEOREM 8.5.

$$(1 - u)C'_{Q_X,Q_Y}(u) = C_{Q_X,Q_Y}(u) + \left(1 - \frac{h_Y(Q_X(u))}{H_{Q_X}(u)} \right) m_{Q_X}(u)$$

$$+ \frac{1}{h_{Q_X}(u)} - \frac{1}{h_{Q_Y}(u)}.$$

As a consequence of the last theorem, if Y *is exponential with parameter* λ_2, *then* X *is exponential if and only if* $C_{Q_X,Q_Y}(u)$ *is a constant. In a similar manner* $C_{Q_X,Q_Y}(u) = a + bu$, *a linear function if and only if*

$$Q_X(u) = -\frac{1}{\lambda_2} \log(1 - u) - \left(\frac{3a - b}{3\lambda_2} \right)^{1/2} - 2u \left(\frac{\lambda_2}{a + b} \right)^{1/2} \log \frac{(\frac{a+b}{\lambda_2})^{1/2} + t}{(\frac{a+b}{\lambda_2})^{1/2} - t}$$

with $t = \left(\frac{3(a+b) - 4b(1-u)}{3\lambda_2} \right)^{\frac{1}{2}}$.

THEOREM 8.6. *If* $X \leq_{hr} Y$ *then*

$$C_{Q_X, Q_Y}(u) \leq m_{Q_Y}(u) - m_{Q_X}(u).$$

Further if X *and* Y *are NWUE and NBUE respectively, also then* $C_{Q_X, Q_Y}(u) \leq E(Y) - E(X)$.

8.4. Cumulative Tsallis entropy

The Tsallis generalized entropy of order α

$$T_\alpha = \frac{1}{\alpha - 1}(1 - \int_0^\alpha f^\alpha(x)dx) \tag{8.13}$$

was discussed at length in Chapter 4 along with its residual and past life forms. From the above representation, the cumulative Tsallis entropy, CRTE was proposed by Sati and Gupta (2015) as

$$C_T = \frac{1}{\alpha - 1}(1 - \int_t^\infty \bar{F}^\alpha(x)dx), \ \alpha > 0; \alpha \neq 1. \tag{8.14}$$

When $\alpha \to 1$ we have the CRE. Further when the random variable X is replaced by its residual life X_t we have the DCRTE,

$$C_T(t) = \frac{1}{\alpha - 1}\left(1 - \int_t^\infty \frac{\bar{F}^\alpha(x)}{\bar{F}^\alpha(t)}dx\right). \tag{8.15}$$

For each $\alpha, C_T(t)$ determines the distribution of X uniquely. Note that

$$(\alpha - 1)C_T'(t) = 1 + \alpha h(t)[(\alpha - 1)C_T(t) - 1].$$

THEOREM 8.7. *The relationship*

$$(\alpha - 1)C_T'(t) = Ch(t)$$

holds for all t *and real constant* C *if and only if*

$$\bar{F}(x) = \exp[-\int_0^x (K - 2\alpha cx)^{-1/2}dx]$$

where K *is a constant of integration. In particular for* $K > 0$ *and* $C = 0$. X *is exponential with parameter* $K^{-1/2}$ *and for* $K = 0$ *and* $C < 0$, X *is Weibull with* $\bar{F}(x) = e^{-(ax)1/2}$, $a > 0$.

THEOREM 8.8. *The relationship* $(\alpha - 1)C_T(t) = 1 - K \, m(t)$ *holds, then* X *is generalized Pareto with special cases, exponential if* $K = \frac{1}{\alpha}$, *Pareto II if* $K < \frac{1}{\alpha}$ *and rescaled beta if* $K > \frac{1}{\alpha}$.

If X and Y are lifetimes satisfying $X \geq_{\text{hr}} Y$, then $C_{X,T}(t) \leq C_{Y,T}(t)$ whenever $\alpha > 1$ $(0 < \alpha < 1)$. This gives an ordering of DCRTEs. Park et al. (2012) have also proposed the weighted DCRTE,

$$C_{w,T}(t) = \frac{1}{\alpha - 1}\left(1 - \int_t^\infty \left(\frac{\bar{F}_w(x)}{\bar{F}_w(x)}\right)^\alpha dx\right) dx$$

$$= \frac{1}{\alpha - 1}\left(1 - \int_t^\infty \left(\frac{E[w(X)]|X > x}{E[w(X)]|X > t}\right)^\alpha \frac{\bar{F}^\alpha(x)}{\bar{F}^\alpha(t)} dx\right).$$

In the case of the equilibrium random variable X_E,

$$C_{X_E,T}(t) \geq (\leq) C_{X,T}(t)$$

when $h_X(x)$ is decreasing for $0 < \alpha < 1$ $(\alpha > 1)$. They have also given an expression for the empirical DCRTE as

$$C_T(\bar{F}_n) = \frac{1}{\alpha - 1}\left(1 - \sum_{j=1}^{n-1}\left(\frac{j}{n}\right)^\alpha (X_{i+1:n} - X_{i:n})\right).$$

Rajesh and Sunoj (2019) chose the expression for Tsallis entropy as $T_\alpha = \frac{1}{\alpha-1}\int_0^\infty (f(x) - f^\alpha(x))$ and used it to write the corresponding cumulative entropy as

$$C_T^* = \frac{1}{\alpha - 1}[\bar{F}(x) - \bar{F}^\alpha(x)]dx$$

$$= \frac{1}{\alpha - 1}\left[\mu + \int_0^\infty \frac{d}{dx}(h(x)\bar{F}(x))\right]\bar{F}^{\alpha-1}(x)dx. \quad (8.16)$$

For the residual life the above measure extends to

$$C_T^*(t) = \frac{1}{\alpha - 1}\int_0^\infty \left(\frac{\bar{F}(x)}{\bar{F}(t)} - \frac{\bar{F}^\alpha(x)}{\bar{F}^\alpha(t)}\right) dx$$

$$= \frac{1}{\alpha - 1}\left[m(t) - \int_0^\infty \left(\frac{\bar{F}(x)}{\bar{F}(t)}\right)^\alpha dx\right]. \quad (8.17)$$

Two simple bounds are obtained in $C_T^*(t) \geq (\leq)\frac{m(t)}{\alpha}$ when X is DMRL (IMRL) and $C_T^*(t) \geq (\leq)\frac{\mu}{\alpha}$ when X is NBUE (NWUE). If X and Y satisfies $X \leq_{\text{disp}} Y$ then $C_{T,Y}^* \geq C_{T,X}^*$ and if X is DMRL, the equilibrium random variable satisfies $(\alpha - 1)C_{T,X_E}^* \leq C_{T,X}^*(t)$.

THEOREM 8.9. *The relationship* $C_X^*(t) = A(t)m(t)$ *holds for all* $t > 0$ *then*

$$m(t) = (p_\alpha(t))^{\frac{1}{\alpha-1}}\left[\frac{K}{p_\alpha(0)} + \int_0^t (p_\alpha(x))^{-\frac{\alpha}{\alpha-1}}(A(x) - p_\alpha(x))dx\right]$$

where $p_\alpha(x) = 1 - (\alpha - 1)A(x)$. *In particular when* $A(t) = B$, *a constant* X *follows the generalized Pareto distribution.*

The cumulative Tsallis entropy for the past life is discussed in Kumar (2017), through

$$\bar{C}_T(t) = \frac{1}{1-\alpha} \left(\int_0^t \frac{F^\alpha(x)dx}{F^\alpha(t)} - 1 \right).$$

For each α, $\bar{C}_T(t)$ determines the distribution of X. The relationship

$$\bar{C}_T(t) = \frac{1}{1-\alpha}[c\,r(t) - 1]$$

holds if and only if X has power distribution $F(x) = (\frac{x}{a})^\theta$. The first order statistic $X_{1:n}$ and the largest $X_{n:n}$ have respective cumulative Tsallis residual entropies

$$C_{T,X_{1:n}} = \frac{1}{1-\alpha} \left[\int_0^\infty \bar{F}^{n\alpha}(x)dx - 1 \right]$$

and

$$C_{T,X_{n:n}} = \frac{1}{1-\alpha} \left[\int_0^\infty F^{n\alpha}(x)dx - 1 \right].$$

If X and Y have the same support $(0, \infty)$ then for $X_{1:n}$ and $Y_{1:n}$, X and Y belong to the same family of distributions if and only if

$$\frac{(1-\alpha)C_{T,X_{1:n}} + 1}{E(X_{1:n})} = \frac{(1-\alpha)C_{T,Y_{1:n}} + 1}{E(Y_{1:n})}$$

and they belong to the same location family if and only if

$$C_{T,X_{1:n}} = C_{T,Y_{1:n}},$$

provided $\sum_{j=1}^\infty n_j^{-1} = \infty$.

Khammar and Jahanshahi (2018a) point out that in the context of theoretical neurobiology, measures of uncertainty based on weighted entropy have been used to balance the amount of information and degree of homogeneity associated with the partition of data in classes. They defined the weighted cumulative Tsallis entropy as

$$C_{T,w} = \frac{1}{\alpha - 1} \left[1 - \int_0^\infty x \bar{F}^\alpha(x)dx \right], \quad \alpha > 0, \; \alpha \neq 1$$

or alternatively as

$$C_{T,w} = \frac{1}{\alpha - 1} \left[1 - \frac{1}{\alpha} \int_0^\infty \bar{F}^\alpha(x)m(x)dx - \frac{\alpha - 1}{\alpha} \int_0^\infty x \bar{f}^\alpha(x) \right] m'(x)dx$$

It is shown that

(i) If $Y = aX + b$,

$$C_{T,w}(Y) = \frac{1 - b - a}{\alpha - 1} + aC_{T,w}(X) + bC_T(X)$$

(ii) In the proportional hazards model Y,

$$C_{T,w}(Y) \geq C_{T,w}(X) \geq C_{T,w}(\theta X)$$

(iii) $C_{T,w}(X) \geq \frac{1}{\alpha-1}[1 - m_w(0)]$ and

$$C_{T,w}(X) \geq (\leq)wS_\alpha - m_w(t), \ \alpha > 1 \ (0 \leq \alpha < 1)$$

where wS_α is the weighted survival entropy of order α in Rajesh et al. (2017) given in equation (8.5). The dynamic version for residual and past lifetimes are respectively

$$C_{T,w}(t) = \frac{1}{\alpha - 1} \left[1 - \int_t^\infty \frac{x\bar{F}^\alpha(x)}{\bar{F}^\alpha(t)} dx \right]$$

and

$$\bar{C}_{T,w}(t) = \frac{1}{\alpha - 1} \left[1 - \int_0^x \frac{xF^\alpha(x)}{F^\alpha(t)} dx \right].$$

Khammar and Jahanshahi (2018a) have proved the following properties

(a) $C_{T,w}(t) \geq \frac{1}{\alpha-1}[1 - m_w(t)]$, $\alpha > 0$ and $C_{T,w}(t) \leq (\geq)wS_\alpha(t) \leq (\geq)wS(t)$, $\alpha > 1 \ (0 < \alpha < 1)$ where $wS(t)$ is the weighted version of DCRE.

(b) If $X \leq_{st} Y$ then $X \geq_{WCRE} (\leq)Y$, where $\geq_{WCRE}$ represents the weighted cumulative residual entropy order stating that X is smaller than Y in WCRE order if $wS_X(t) \leq wS_Y(t)$.

(c) Similarly defining $X \leq_{DWCRE} Y$ as $C_{T,w}^X(t) \leq C_{T,w}^Y(t)$ where $C_{T,w}^X(t)$ is the $C_{T,w}$ of X we have

$$X \geq_{hr} Y \Rightarrow X \leq (\geq)_{DWCRE}Y.$$

(d) For increasing transformation $\phi(.)$ with $\lim_{t \to \infty} \phi(t) = \infty$,

$$C_{T,w}^Y(t) = C_{T,w}^X(\phi(X); \phi^{-1}(t)).$$

In particular, if $\phi(0) = 0$ and $\phi'(0) > 1$,

$$X \leq (\geq)Y_{DCRE} \Rightarrow X \leq_{DWCRE} Y, \ 0 \leq \alpha < 1 \ (\alpha > 1)$$

(e) If $C_{T,w}^Y(t) = C_{T,w}^X(t)$ then $\bar{F}(t) = \bar{G}(t)$, where $\bar{G}$ is the survival function of Y.

(f) The relationship $C_{T,w}(t) = k$ holds for some constant K if and only if X has Rayleigh distribution $\bar{F}(x) = e^{-\frac{x^2}{2\sigma^2}}$, $\sigma > 0$; $x > 0$. The same distribution is also characterized by the property

$$1 - (\alpha - 1)C_{T,w}(t) = \frac{m_w(t)}{\alpha}$$

An estimator for $C_{T,w}$ is suggested as

$$\hat{C}_{T,w}(t) = \frac{1}{\alpha - 1}\left[1 - \sum_{i=1}^{n-1}\left(1 - \frac{i}{n}\right)^{\alpha}(X_{i:n}^2 - X_{i-1:n}^2)\right].$$

Using the familiar methods Sunoj et al. (2018a) find the quantile-based cumulative Tsallis entropy (QCRTE) as

$$\begin{aligned}
C_{Q,T} &= \frac{1}{\alpha - 1}\left(1 - \int_0^1 (1-p)^{\alpha}q(p)dp\right) \\
&= \frac{1}{\alpha - 1}\left(1 - \int_0^1 \frac{(1-p)^{\alpha-1}}{h_Q(p)}dp\right).
\end{aligned} \tag{8.18}$$

EXAMPLE 8.2. *The Govindarajulu distribution* $Q(u) = \theta + \sigma((\beta - 1)u^{\beta} - \beta u^{\beta+1})$, *has*

$$C_{Q,T} = \frac{1}{\alpha - 1}\left[1 - \frac{\alpha\beta(\beta+1)\Gamma(2+\alpha)\Gamma(\beta)}{\Gamma(2+\alpha+\beta)}\right].$$

When the residual life is considered the form (8.18) *transforms to*

$$\begin{aligned}
C_{Q,T}(u) &= \frac{1}{\alpha - 1}\left(1 - \frac{1}{(1-u)^{\alpha}}\int_u^{\infty}(1-p)^{\alpha}q(p)dp\right) \\
&= \frac{1}{\alpha - 1}\left(1 - m_Q(u) + \frac{\alpha - 1}{(1-u)^{\alpha}}\int_u^1 (1-p)^{\alpha-1}m_Q(p)dp\right).
\end{aligned}$$

Differentiating,

$$q(u) = (\alpha - 1)C'_{Q,T}(u) + \frac{\alpha}{1-u}(1 - (\alpha - 1)C_Q(u)),$$

an explicit formula for the determination of the distribution of X in terms of the function $C_{Q,T}(u)$ is obtained. From the above it is seen that when $C_{Q,T}(u)$ is increasing (decreasing)

$$C_{Q,T}(u) \geq (\leq)\frac{1}{\alpha - 1}\left(1 - \frac{\alpha - 1}{h_Q(u)}\right), \quad \alpha > 1 \ (0 < \alpha < 1).$$

Further $C_{Q,T}(u)$ is constant if and only if X is exponential. A sufficient condition for increasing (decreasing) $C_{Q,T}(u)$ can also be stated as X is IFR (DFR).

THEOREM 8.10. *The relationship $C_{Q,T}(u) = a + bu$, $a, b > 0$ holds if and only if X follows the distribution*

$$Q(u) = A(\alpha^2 - 1)u + \alpha(B(\alpha - 1) - 1)\log(1 - u)$$

for $A, B > 0$.

Other properties of interest are

(i) If $X \leq_{HQ} Y$ then $C_{Q_X,T}(u) \geq (\leq)C_{Q_Y,T}(u)$, $\alpha > 1$ ($0 < \alpha < 1$)
(ii) If X has increasing $C_{Q,T}(u')$ and ϕ is a non-negative increasing function, then $\phi(X)$ also has an increasing $C_{Q,T}(u)$
(iii) An alternative form of the cumulative Tsallis entropy based on (8.17) is

$$C_{Q,T}^*(u) = \frac{1}{\alpha - 1}\left(m_Q(u) - \frac{1}{(1-u)^\alpha}\int_0^1 (1-p)^\alpha q(p)dp\right)$$

$$= \frac{1}{\alpha - 1}(m_Q(u) - 1 + (\alpha - 1)C_Q(u)$$

(iv) $C_{Q,T}^*(u) = \frac{1}{(1-u)^\alpha}\int_0^1 (1-p)^{\alpha-1}m_Q(p)dp$ and hence
(v) $C_{Q,T}^*(u) = K M(u)$, $K > 0$ holds if and only if X is generalized Pareto
(vi) The Tsallis entropy of the ith order static is

$$T_\alpha(X_{i:n}) = \frac{1}{\alpha - 1}\left(1 - \int_0^1 \frac{1}{B(i, n-i+1)}p^{\alpha(i-1)}(1-p)^{\alpha(u-1)}(q(p))^{1-\alpha}dp\right)$$

and hence

$$C_{Q,T}^*X_{1:n} = \frac{1}{\alpha - 1}\left(1 - \int_0^\infty \left(\frac{B_u(i, n-i+1)}{B(i, n-i+1)}\right)^\alpha q(u)du\right).$$

Some results similar to those of C_Q are also derived in Sunoj et al. (2018a).

Cali et al. (2019) found that the cumulative entropy $\bar{C}$ and the cumulative residual entropy are related through

$$C_X + \bar{C}_X = \int_{-\infty}^\infty h(x)dx$$

where $h(x) = -[F(x)\log F(x) + \bar{F}(x)\log \bar{F}(x)]$. They proposed cumulative Tsallis entropy as

$$\bar{C}_{\alpha,T}^X = \frac{1}{\alpha - 1}\left(\int_0^\alpha (F(x) - F^\alpha(x))\right)dx$$

and proved some properties listed below.

(i) $\bar{C}^*_{\alpha,T} = E(r(X)F^{\alpha-1}(X))$ so that $\bar{C}^*_{\alpha,T} \leq (\geq)E(r(X))$, $\alpha \geq 1$ $(0 < \alpha < 1)$. Also $\bar{C}^*_{\alpha,T} = 0$ if and only if X is degenerate and $\bar{C}^*_{\alpha,T} > 0$ for non-negative absolutely continuous random variables

(ii) If $Y = \phi(X)$ is strictly increasing

$$\bar{C}^*_Y = \frac{1}{\alpha - 1} \int_{\max(0,\phi^{-1}(0))}^{\infty} (F(x) - \bar{F}^\alpha(x))\phi'(x)dx.$$

(iii) $\bar{C}^*_{\alpha,T} \leq (\geq)\bar{C}^*_T$ if $\alpha \geq 1$ $(0 \leq \alpha \leq 1)$

(iv) If $X \leq_{\text{st}} Y$,

$$|\bar{C}^*_{\alpha,X} - \bar{C}^*_{\alpha,Y}| \leq E(Y) - E(X)$$

For independent and identically distributed random variables

(a) $\bar{C}^*_\alpha(X_{n:n}) \leq nE(X)$

(b) $\bar{C}^*_{\alpha,T}(X_{1:n}) \leq E(X)$.

(v) When Y is the proportional reversed hazards model of X

$$(\alpha - 1)\bar{C}_{\alpha,T}(Y) = (\alpha\theta - 1)\bar{C}_{\alpha,T}(X) - (\theta - 1)\bar{C}_{\alpha,T}(X), \ \alpha > 1.$$

The corresponding dynamic version is

$$\bar{C}^*_{\alpha,T}(t) = \frac{1}{\alpha - 1} \int_0^t \left(\frac{F(x)}{F(t)} - \frac{F^\alpha(x)}{\bar{F}^\alpha(t)} \right) dx$$

$$= \frac{1}{\alpha - 1} \left(r(t) - \frac{\int_0^t F^\alpha(x)dx}{F^\alpha(t)} \right) \quad (8.19)$$

satisfying $\bar{C}^*_{\alpha,T}(t) = E(r(X)F^{\alpha-1}(x)|X < t)/F^{\alpha-1}(t)$ and hence $\bar{C}^*_{\alpha,T}(t) \leq (\geq)\frac{r(t)}{\alpha}$, $\alpha > 1$ $(0 < \alpha < 1)$. Further

$$Y \geq_{\text{disp}} (\leq_{\text{disp}})X \Rightarrow \bar{C}^*_{\alpha,Y}(t) \geq (\leq)\bar{C}^*_{\alpha,X}(t).$$

Some characterization results are given below

THEOREM 8.11. *Let X have support $[0, b]$, $b < \infty$. Then for $\alpha > 1$*

$$\bar{C}^*_{\alpha,T}(t) = \frac{K}{\alpha}r(t) \Leftrightarrow F(t) = \left(\frac{t}{b}\right)^{\frac{b}{(1-b)}}, \ b = \frac{K}{\alpha - K(\alpha - 1)}$$

$$\bar{C}^*_{\alpha,T}(t) = \frac{K}{\alpha}m(t) \Leftrightarrow F(t) = \left(\frac{t}{b}\right)^{\frac{1-b}{b}}, \ b = \frac{K\alpha}{\alpha + K(\alpha - 1)}.$$

Krishnan et al. (2019) studied the quantile form of $\bar{C}_{\alpha,T}(u)$ as

$$\bar{C}_{Q,T}(u) = \frac{1}{\alpha - 1} \left(1 - (\alpha - 1)\bar{C}_Q(u) - (\alpha - 1)\bar{C}'_Q(u) \right).$$

Some examples are

power-Pareto $\quad Q(u) = \dfrac{cu^{\lambda_1}}{(1-u)^{\lambda_2}}$

$$\bar{C}_{Q,T}(u) = \frac{1}{\alpha - 1}\left[1 - \frac{c\lambda_1 B_u(\alpha + \lambda_1, 1 - \lambda_2)}{u^\alpha} \right.$$
$$\left. - \frac{c\lambda_2 B_u(\alpha + \lambda_1, 1 - \lambda_2)}{u^\alpha} \right]$$

skew-lambda $\quad Q(u) = \delta u^\lambda (1-u)^\lambda$

$$\bar{C}_{Q,T}(u) = \frac{1}{\alpha - 1}\left[1 - \frac{\lambda}{u^\alpha}\left(\frac{\alpha u^{\alpha + \lambda}}{\alpha + \lambda} \right) + B_u(\alpha + 1, \lambda) \right]$$

$\bar{C}_{Q,T}(u)$ decreasing (increasing) if

$$\bar{C}_{Q,T}(u) \geq (\leq) \frac{1}{\alpha - 1}\left(1 - \frac{1}{\alpha\lambda(u)} \right), \quad \alpha > 1, \ (0 < \alpha < 1)$$

where $\lambda(u)$ is the reversed hazard quantile function and if ϕ is non-negative increasing and convex and $\bar{C}_Q(u)$ is increasing then $\phi(x)$ has also an increasing $\bar{C}_Q(u)$. We have also two characterization theorems.

THEOREM 8.12. *(i) The property $\bar{C}_{Q,T}(u) = K\,r_Q(u)$ is satisfied if and only if X has quantile function*

$$Q(u) = \frac{\alpha}{1 + K\alpha(\alpha - 1)} + 2u + cK(\alpha - 1)^{-1 - K\alpha(\alpha - 1)}(1 + \log u), \ K > 0$$

(ii) $\bar{C}_{Q,T}(u) = \frac{1}{\alpha - 1}(1 - K\,r_Q(u))$ if and only if X has power distribution $Q(u) = Au^b$.

Krishnan et al. (2019) also considered the alternative form

$$\bar{C}^*_{Q,T}(u) = \frac{1}{\alpha - 1}\left[\int_0^u \left(\frac{p}{u} - \frac{p^\alpha}{\mu^\alpha} \right) q(p)dp \right]$$
$$= \frac{1}{\alpha - 1}\left[r_Q(u) - \frac{1}{u^\alpha}\int_0^u p^\alpha q(p)dp \right]$$

arising from (8.17). This new version is increasing (decreasing) if and only if

$$\bar{C}^*_{Q,T}(u) \leq (\geq) \frac{r_Q(u)}{\alpha}, \quad \alpha > 1 \ (0 < \alpha < 1).$$

The function $\bar{C}^*_Q(u)$ for the ith order statistic is

$$\bar{C}^*_{Q,X_{i:n}}(u) = \frac{1}{\alpha - 1}\left(1 - \int_0^1 \frac{B_u(i, n - i + 1)}{B(i, n - i + 1)}q(p)dp \right).$$

They have also discussed various other properties of this measure.

Another topic of interest in Tsallis type entropy is the Tsallis divergence of order α discussed in Kayal and Tripathi (2018). Staring from the Tsallis divergence

$$T_\alpha(X,Y) = \frac{1}{\alpha - 1}\left[\int_0^\infty f^\alpha(x)g^{1-\alpha}(x)dx - 1\right]$$

one can define the same in terms of quantiles as

$$T_{\alpha,Q_X,Q_Y} = \frac{1}{\alpha - 1}\left(\int_0^1 q_X(p)f_Y(Q_X(p))dp - 1\right).$$

The case of T_{α,Q_X,Q_Y} for doubly truncated random variables and various properties are also presented in Kayal and Tripathi (2018).

Applications of the cumulative Tsallis entropy to test dilation order is discussed in Zaradasht (2020). A method based on multi-scale permuted distributed version has been proposed by Zang et al. (2019) to measure the complexities of dissimilarities of sequences. See also Zhang et al. (2020) for models to measure dissimilarities between time series.

8.5. Cumulative Varma's entropy

In Chapter 4 we have defined Varma's entropy of order α and type β as

$$H_{\alpha,\beta} = \frac{1}{\beta - \alpha}\log\int_0^\infty f^{\alpha+\beta-1}(x)dx$$

and its generalization as a cumulative measure

$$C_{\alpha,\beta} = \frac{1}{\beta - \alpha}\log\int_0^\infty \bar{F}^{\alpha+\beta-1}dx \tag{8.20}$$

was considered in Kumar and Taneja (2011b).

EXAMPLE 8.3. *When*

$$\bar{F}(x) = \frac{1}{1+\theta x}, \quad C_{\alpha,\beta} = \frac{1}{\beta - \alpha}\log\frac{1}{\theta(\alpha - \beta)}.$$

Its dynamic version now becomes

$$C_{\alpha,\beta}(t) = \frac{1}{\beta - \alpha}\log\left[\int_t^\infty \frac{\bar{F}^{\alpha+\beta-1}(x)}{\bar{F}^{\alpha+\beta-1}(t)}dx\right]. \tag{8.21}$$

When this measure is finite, it determines $\bar{F}(x)$ uniquely. Two characterization theorems by properties of $C_{\alpha,\beta}(t)$ are given in the next theorems.

THEOREM 8.13. *The property $C_{\alpha,\beta}(t) = \log K + \log m(t)$ holds if and only if X has an exponential distribution when $K = \frac{1}{\alpha+\beta-1}$, Pareto II distribution if $K < \frac{1}{\alpha+\beta-1}$ and rescaled beta if $K > \frac{1}{\alpha+\beta-1}$.*

THEOREM 8.14. *If $C_{\alpha,\beta}(t) = K(t) + \frac{1}{\beta-\alpha}\log m(t)$ holds, then*

$$m(t) = \left[d + \left(\int_0^t \frac{\alpha+\beta-1-e^{-(\beta-\alpha)k(x)}}{2-\alpha-\beta} e^{\frac{(\beta-\alpha)k(x)}{2-\alpha-\beta}} dx\right)\right] e^{\frac{\beta-\alpha}{3-\alpha-\beta}c(t)}$$

where $d = \mu e^{\frac{\beta-\alpha}{3-\alpha-\beta}}$.

THEOREM 8.15. *The relationship $(\beta - \alpha)C_{\alpha,\beta}(t) = K\,A_F(t)$ characterizes the generalized Pareto law.*

A quantile version of (8.20) and (8.21) was given in Baratpour and Khammar (2018) as

$$C_{\alpha,\beta,Q} = \frac{1}{\beta-\alpha}\log\int_0^1 (1-p)^{\alpha+\beta-1}q(p)dp \tag{8.22}$$

and

$$C_{\alpha,\beta,Q}(u) = \frac{1}{\beta-\alpha}\left[\log\int_u^1 (1-p)^{\alpha+\beta-1}q(p)dp - (\alpha+\beta-1)\log(1-u)\right] \tag{8.23}$$

EXAMPLE 8.4. *When $q(u) = Cu^\alpha(1-u)^{-(\alpha+\beta-2)}$,*

$$C_{\alpha,\beta,Q} = (\beta-\alpha)^{-1}[\log C - \log(\alpha+1)(\alpha+2) + \log(1-(\alpha+2)u^{\alpha+1}$$
$$+ (\alpha+1)u^{\alpha+2}) - (\alpha+\beta-1)\log(1-u)]$$

and

$$C_{\alpha,\beta,Q}(u) = \frac{1}{\beta-\alpha}[\log\log\Gamma_{-\log(1-y)}(1-M,2) - (\alpha+\beta-1)\log(1-u)],$$
$$M = \alpha+\beta-2.$$

A condition for $C_{\alpha,\beta,Q}(u)$ to increase (decrease) is that

$$C_{\alpha,\beta,Q}(u) \geq (\leq)\frac{-\log h_Q(u) - \log(\alpha+\beta-1)}{\beta-\alpha}.$$

Further

(i) $X \leq_{\mathrm{HQ}} Y \Rightarrow C^X_{\alpha,\beta,Q}(u) \leq C^Y_{\alpha,\beta,Q}(u)$ (or $X \leq_{C^X_{\alpha,\beta,Q}(u)} Y$)

(ii) if X or Y is DFR, then $X \leq_{\mathrm{hr}} Y \Rightarrow X \leq_{C_{\alpha,\beta,Q}} Y$.

(iii) $\frac{m_{Q_X}(u)}{m_{Q_Y}(u)}$ is increasing $X \leq_{C_{\alpha,\beta,Q}} Y$,

are some sufficient conditions for comparing the cumulative Varma entropies of X and Y. There are some properties that characterize specific distributions.

THEOREM 8.16. *The random variable X follows uniform distribution (exponential distribution, Pareto type I distribution) if and only any one of the following properties are satisfied,*

(1) $(\beta - \alpha)C_{\alpha,\beta Q(u)}(u) = \frac{C}{1-u}$, $C = -1$, $(0, \frac{1}{\alpha})$

(2) $(\alpha + \beta - C)e^{\beta - \alpha}C_{\alpha,\beta Q(u)} = \frac{1}{h_Q(u)}$, $C = 0$, $(1, 1 + \frac{1}{\alpha})$

The corresponding measure in past life discussed in Minimol (2017) is

$$\hat{C}_{\alpha,\beta}(t) = \frac{1}{\beta - \alpha} \log \int_0^t \frac{F^{\alpha+\beta-1}(x)}{F^{\alpha+\beta-1}(t)} dx.$$

If $\beta - 1 < \alpha < \beta$ and $\beta \geq 1$, then $\hat{C}_{\alpha,\beta}(t)$ determines F uniquely. Further X follows the uniform distribution if and only if

$$\hat{C}_{\alpha,\beta}(t) = \frac{\log \lambda(t) + \log(\alpha + \beta)}{\alpha - \beta}$$

and more generally, the power distribution $F(t) = (\frac{t}{b})^c$, $0 < t < b$ if and only if

$$\hat{C}_{\alpha,\beta}(t) = \frac{A(\alpha, \beta) - \log \lambda(t)}{\beta - \alpha}, \quad A(\alpha, \beta) = \log \frac{C}{C(\alpha - \beta + 1) + 1}.$$

Bhat and Baig (2020) have considered a slightly modified version

$$C_{\eta,\beta} = \frac{\eta}{\beta(1 - \eta)} \left(\int_0^\infty \bar{F}^{\eta-\beta+1}(x)dx - 1 \right), \quad 0 < \eta < \beta \leq 1$$

its extension

$$C_{\eta,\beta}(t) = \frac{\eta}{\beta(1 - \eta)} \left[\frac{\int_0^\infty \bar{F}^{\eta-\beta+1}(x)}{\bar{F}^{\eta-\beta+1}(t)} dx - 1 \right], \quad 0 < \eta < \beta \leq 1$$

and the weighted entropy

$$W\,C_{\eta,\beta}(t) = \frac{\eta}{\beta(1 - \eta)} \left[\int_0^\infty x \bar{F}^{\eta-\beta+1}(x)dx - 1 \right]$$

and discuss the properties of these models similar to the models presented earlier.

8.6. Cumulative residual inaccuracy

If $F(x)$ and $G(x)$ are survival functions of lifetime random variables X and Y, the cumulative residual inaccuracy is defined as

$$C_I = - \int_0^\infty \bar{F}(x) \log \bar{G}(x) dx \qquad (8.24)$$

Taneja and Kumar (2012) studied several properties of this measure and its dynamic form. When F and G coincide (8.24) reduces to the cumulative residual entropy (7.6). Moreover if Y is the proportional hazards model of X, then (8.24) reduces to the CRE. The dynamic version (8.24) is

$$C_T(t) = - \int_t^\infty \frac{\bar{F}(x)}{\bar{F}(t)} \log \frac{\bar{G}(x)}{\bar{G}(t)} dx \qquad (8.25)$$

THEOREM 8.17. *Under the proportional hazards model assumption and $C_I(t)$ an increasing function (8.25) uniquely determines the survival function of X.*

THEOREM 8.18. *Under the assumption of proportional hazards, the relationship $C_I(t) = Km_F(t)$ is satisfied if and only if X follows the exponential distribution for $K = \theta$, Pareto II distribution if $K > \theta$ and the rescaled beta if $0 < K < \theta$.*

More discussions on cumulative residual and past inaccuracy measures are covered in Kumar and Taneja (2015). For the residual type if

$$C_T(t) = k(t)m_F(t)$$

then

$$m_F(t) = \left[A + \int_0^t \left(\frac{k(x) - \theta}{\theta} \right) e^{\frac{k(x)}{\theta}} dx \right] e^{-\frac{k(t)}{\theta}}. \qquad (8.26)$$

Similarly for the past life

$$\bar{C}_I(t) = -\int_0^t \frac{F(x)}{F(t)} \log \frac{G(x)}{G(t)} dx \qquad (8.27)$$

if $\bar{C}_I(t)$ is a decreasing function of t and the reversed proportional hazards model holds, then $\bar{C}_I(t)$ determines F uniquely. Analogous to Theorem 8.18, the property $\bar{C}_I(t) = Kr_F(t)$ characterizes the power model $F(x) = \left(\frac{x}{b} \right)^{\frac{c}{\theta - c}}, 0 < c < \theta$. Another result parallel to (8.26) is

$$r_F(t) = \left[\int_0^t \frac{\theta - k(x)}{\theta} e^{\frac{k(x)}{\theta}} \right] e^{-\frac{k(t)}{\theta}}$$

where θ is the proportionality constant among the reversed hazard rates and

$$\bar{C}_I(t) = k(t)r_F(t).$$

It X_E and Y_E are equilibrium random variables corresponding to X and Y (Kundu et al. (2016))

$$C_I = E(X)[H_{X_E, Y_E} - \log E(Y)]$$

where

$$H_{X_E, Y_E} = -\int_0^\infty f_{X_E}(x) \log g_{X_E}(x).$$

Denoting

$$A_F^{(2)}(x) = -\int_0^x \log \bar{F}(t) dt,$$

$$A_{G(x)}^{(2)} = -\int_0^x \log \bar{G}(t) dt,$$

$$\Lambda_F^{(2)}(X) = \int_x^\infty \log F(t)dt$$

and

$$\Lambda_G^{(2)}(x) = -\int_x^\infty \log G(t)dt,$$

we have

$$C = E[\Lambda_G^{(2)}(x)] \text{ and } \bar{C} = E[\Lambda_G^{(2)}(x)].$$

Considering the ratios

$$R_1 = \frac{E(\Lambda_G^{(2)}(X))}{E(\Lambda_F^{(2)}(X))} \text{ and } R_2 = \frac{E(\Lambda_G^{(2)}(X))}{E(\Lambda_F^{(2)}(X))}$$

give the proximity between X and Y. If X and Y have finite means

$$C_I \geq C + E(X) \log \frac{E(X)}{E(Y)}$$

and

$$C_I \geq C + E(X) - E(Y)$$

where C stands for CRE. Assuming $X \leq_{\text{st}} (\geq_{\text{st}})Y$, we have (i) $C_I \leq \min(C_X, C_Y)$, and (ii) $C_I \geq \max(C_X, C_Y)$.

Differentiating $\bar{C}_I(t)$ with respect to t in (8.27), gives

$$\bar{C}_I'(t) = A_F(t)\bar{C}_I(t) - A_F(t)m_F(t)$$

and hence $\bar{C}_I(t)$ is increasing (decreasing) if and only if $\bar{C}_I \geq (\leq)\frac{h_G(t)}{h_F(t)}m_F(t)$. The effect of a linear transformation on X and Y can be evaluated as

$$\bar{C}_{I,aX+b,aY+b}(t) = a\bar{C}_I\left(\frac{t-b}{a}\right).$$

If X, Y and Z are non-negative random variables satisfying $X \leq_{\text{hr}} Y, Z \leq_{\text{hr}} Y$ or $Y \leq_{\text{hr}} X$ and $Y \leq_{\text{hr}} Z$, then

$$\bar{C}_{I,X,Y}(t) + \bar{C}_{I,Y,Z}(t) \geq \bar{C}_{I,X,Z}(t).$$

If the concern is about past life, the cumulative past inaccuracy measure and results parallel to the above can be established. See Kundu et al. (2016) for details. The measure C_I for record values and their properties are discussed in Tahmasebi et al. (2017).

Daneshi et al. (2020) have found an expression for the weighted cumulative residual inaccuracy between $X_{1:n}$ and X

$$\tilde{C}_I = \int_0^\infty x\bar{F}_{X_{1:n}}(x)\Lambda(x)dx = \frac{1}{n}C_{w,X_{1:n}}$$

where

$$C_{w,X_{1:n}} = n \int_0^\infty x \bar{F}^n(x) \Lambda(x) dx.$$

They also showed the dynamic version

$$\tilde{C}_I(t) = - \int_t^\infty x \frac{\bar{F}_{X_{1:n}}(x)}{\bar{F}_{X_{1:n}}(t)} \log \left(\frac{\bar{F}(x)}{\bar{F}(t)} \right) dx \qquad (8.28)$$

and proved that it specifies the distribution of X. The empirical measure corresponding to (8.28) is

$$\hat{\tilde{C}}_I = - \int_0^\infty x (\bar{F}_n(x))^n \log \hat{\bar{F}}_n(x) dx$$

and $\hat{\tilde{C}}_I$ converges to $\tilde{C}_I$ almost surely. In a similar manner the authors define the weighted inaccuracy between $F(x)$ and $F_{X_{n:n}}(x)$ and present several properties of this new measure. Daneshi et al. (2019) have also discussed the cumulative past (residual) inaccuracy for record values.

The generalized cumulative Kerridge inaccuracy considered by Cali et al. (2020) is defined by

$$K_n(Y, X) = \frac{1}{n} \int_0^\infty F(x)(-\log G(x))^n dx$$
$$= E[T_{n,Y}^{(2)}(x)]$$

where

$$T_{n,Y}^{(2)}(X) = \frac{1}{n!} \int_x^\infty (-\log G(z))^n dz.$$

If X and Y an non-negative with $X \leq_{\text{st}} Y$, then

$$K_n(Y, X) \leq \bar{C}_n \leq K_n(X, Y).$$

The authors also propose an estimate for $K_n(X, Y)$.

The weighted residual inaccuracy based on double truncation is discussed in Jalayeri and Zadeh (2017). For a measurable function $\phi(x)$, $\phi(0) = 0$, based on $[X|t_1 < X < t_2]$ and $[Y|t_1 < Y < t_2]$, $F(t_1) < F(t_2)$ and $G(t_1) < G(t_2)$, the weighted interval measure is

$$C_I(t_1, t_2) = - \int_{t_1}^{t_2} \phi(x) \frac{\bar{F}(x)}{\bar{F}(t_1) - \bar{F}(t_2)} \log \frac{\bar{G}(x)}{\bar{G}(t_1) - \bar{G}(t_2)} dx.$$

One can define the past measure by replacing $\bar{F}$ and $\bar{G}$ by F and G. Bounds for both measures, monotonicity and results concerning proportional and proportional reversed hazards models are also presented in the work.

The quantile-based cumulative inaccuracy measure (Kayal (2018c)) is

$$C_{QI} = -\int_0^1 (1-p)\log(1 - Q_2^{-1}Q_1(p))q_1(p)dp.$$

For example when $Q_1(p) = 2p - p^2$ and $Q_2(p) = p$,

$$C_{QI} = -4\int_0^1 (1-p)^2 \log(1-p)dp = \frac{4}{9}$$

and the measure for the equilibrium variable X_E and Y_E is

$$C_{QI,E} = -\int_0^1 \frac{(1-p)}{\mu_X} \log\left(\frac{1 - Q_2^{-1}Q_1(p)}{\mu_2}\right) q_1(p)dp.$$

Similarly for the past lifetime

$$\bar{C}_{QI} = -\int_0^1 p\log(Q_2^{-1}Q_1(p))q(p)dp.$$

The corresponding dynamic versions are

$$C_{QI}(u) = -\frac{1}{1-u}\int_u^1 (1-p)\log\left(\frac{1 - Q_2^{-1}Q_1(p)}{1 - Q_2^{-1}Q_1(u)}\right) q_1(p)dp$$

and

$$\bar{C}_{QI}(u) = -\frac{1}{u}\int_0^u p\log\left(\frac{Q_2^{-1}(Q_1(p))}{Q_2^{-1}(Q_1(u))}\right) q_1(u)du.$$

Various properties of the measures are

(1)
$$\bar{C}_{QI}(u) = \frac{1}{1-u}C_{QI}(u) - \frac{\frac{d}{du}(Q_2^{-1}Q_1(u))}{1 - Q_2^{-1}Q_1(u)}m_{Q_1}(u)$$

(2) $C_{QI}(u)$ is increasing (decreasing) in u if and only if

$$C_{QI}(u) \geq (\leq)\frac{m_{Q_1}(u)}{h_{Q_2^{-1}Q_1}(u)(1 - Q_2^{-1}Q_1(u))}$$

(3) If X has quantile function $Q_2(p) = -\frac{1}{\lambda_2}\log(1-p)$, then X is exponential if and only if $C_{QI}(u)$ is a constant

(4)
$$\bar{C}'_{QI}(u) = \frac{\frac{d}{du}Q_2^{-1}Q_1(u)}{Q_2^{-1}Q_1(u)}r_{Q_1}(u) - \frac{1}{u}\bar{C}_{QI}.$$

Note that Q_1 and Q_2 represent the quantile functions of X and Y and m_Q and r_Q the mean residual quantile function of Q and reversed mean residual quantile function of Q and h_Q the hazard quantile of Q.

8.7. Generalized cumulative residual entropy

In our discussions on the relevation transform, we came across the distribution function

$$\bar{F}^*(t) = \bar{F}(t) + \int_0^t \frac{\bar{G}(t)}{\bar{G}(x)} f(x) dx$$

as the reliability of the relevation process in which a failed unit is replaced (or repaired) by another of the same age. As a special case when $G = F$, the reliability function of the time to failure (Krakowski (1973)) after n replacements, becomes

$$\bar{F}_n(t) = \bar{F}(t) \sum_{k=0}^{n-1} \frac{[A(t)]^k}{k!} = q_n(\bar{F}(t)), \; n = 1, 2 \qquad (8.29)$$

and $A(t) = -\log \bar{F}(t)$. Note $q_n(x) = x \sum_{k=0}^{n-1} \frac{(-\log x)^k}{k!}$ is an increasing function satisfying $q_n(0) = 0$ and $q_n(1) = 1$. the density function of (8.29) is

$$f_n(t) = \frac{A(t)^{n-1}}{(n-1)!} f(t)$$

so that the number of failures form a non-homogeneous, Poisson process with intensity function $h(t) = \frac{f(t)}{\bar{F}(t)}$. Citing these factors Psarrakos and Navarro (2013) defined the generalized cumulative residual entropy (GCRE) as

$$C_n(X) = \int_0^\infty \bar{F}(x) \frac{(A(x))^n}{n!} dx, \; n = 1, 2, \ldots \text{ with } C_0(X) = \mu$$

$$= \int_0^\infty \frac{(A(x))^n}{n!} f(x) \frac{\bar{F}(x)}{f(x)} dx = E\left[\frac{1}{h(X_{n+1})}\right]$$

where X_{n+1} corresponds to $\bar{F}_{n+1}$. It is interesting to note that if X is IFR (DFR) $C_n(X) \geq (\leq) C_{n+1}(X)$ and if $X \leq_{\text{hr}} Y$ either X or Y is DFR then $C_n(X) \leq C_n(Y)$. The dynamic GCRE is (DGCRE)

$$C_n(t) = \frac{1}{n!} \int_t^\infty \frac{\bar{F}(x)}{\bar{F}(t)} \left[-\log \frac{\bar{F}(x)}{\bar{F}(t)}\right]^n dx. \qquad (8.30)$$

We have

$$C_n(t) \geq C_{n+1}(t) \text{ when } X \text{ is IFR or DFR}$$

and

$$C_n(t) = E\left(\frac{1}{h(t + X_{t,n+1})}\right)$$

where $X_{t,n}$ is a random variable with reliability function $\bar{F}_n(t)$ and the reliability function of X_t. The expression (8.30) is equivalent to

$$C_n(t) = \frac{1}{\bar{F}(t)} \sum_{k=1}^{n} \frac{(-1)^{n-k}}{k!(n-k)!} (A(t))^{n-k} \int_t^\infty \bar{F}(x) A^k(x) dx$$

and when $n = 1$, we have the DCRE. An extension of the characterization theorem of DCRE is that for $C > 0$, $C_n(t) = cC_{n-1}(t)$ holds for all t and fixed $n = 1, 2, \ldots$ then X is exponential ($c = 1$), Pareto II ($C > 1$) or rescaled beta ($c < 1$).

Navarro and Psarrakos (2017) considered the questions of characterizing F by $C_n(t)$ and proved the following result.

THEOREM 8.19. *Let X be an absolutely continuous random variable with support included in $[0, \infty)$ and $C_n(t)$ be differentiable. Further assume that $C_n(t) \neq C_{n-1}(t)$ for all $t \geq 0$ such that $F(t) < 1$ and $t_0 \geq 0$. Then $C_i(t_0)$ for $i = 0, 1, \ldots, n-1$ and the function $C_n(t)$ uniquely determine $F(t)$. In particular if $C_1(t) \neq C_0(t) = \mu$ for all $t \geq 0$ and $F(t) < 1$ then $C_1(t)$ and μ determine $F(t)$.*

We have seen in the earlier chapters weighted distributions in various forms and studied the properties of such distributions and their entropies. Psarrakos and Economou (2017) considered a sequence of weighted densities

$$f_{Y_n}(x) = \frac{w_n(x)}{E(w_n(X))} f_X(x)$$

where $w_1(x) = m_X(x)$ and $w_n(x) = E(w_n(X)|X > x)$, $n = 2, 3, \ldots$. From equation (8.30) it follows that

$$C_n(x) = E[C_{n-1}(X)|X > x] = \frac{\int_x^\infty C_{n-1}(t) f(t) dt}{\bar{F}_X(x)}$$

and

$$C_n'(x) = h_X(x)[C_n(x) - C_{n-1}(x)].$$

Some interesting distributional properties are possessed by

$$f_{Y_n}(x) = \frac{C_{n-1}(x) f_X(x)}{C_n(x)}, \quad n = 1, 2, \ldots$$

with survival function

$$\tilde{F}_{Y_n}(x) = \frac{C_n(x) \bar{F}_X(x)}{C_n}$$

and hazard rate

$$h_{Y_n}(x) = \frac{C_{n-1}(x)}{C_n(x)} h_X(x).$$

The extension of the result for $C_n(t)$ to the weighted version is given below.

THEOREM 8.20. *For a fixed $n \in (1, 2, \dots)$ and $k > 0$*

$$f_{Y_n}(x) = \frac{1}{k}(\bar{F}_X(x))^{\frac{1}{k}-1} f_X(x)$$

if and only if X follows exponential for $c = 1$, Pareto II for $c > 1$ and rescaled beta for $c < 1$.

Some other results are

(i) $E(Y_n^k | Y_n > x) = EC_{n-1}(x)E(X^k | X > x)/C_n(x)$

(ii) if n_* is the minimum in the set $(1, 2, \dots)$ such that $C_{n-1}(x)$ is increasing (decreasing) with respect to x, then $X \leq_{\mathrm{lr}} (\geq_{\mathrm{lr}})Y_n$ for any $n \geq n_*$ and also if $C_{n_*}(x)$ is increasing (decreasing) in x then $X \leq_{\mathrm{hr}} (\geq_{\mathrm{hr}})Y_n$ for any $n \geq n_*$.

EXAMPLE 8.5. *Let X follow the Weibull (α, β) distribution*

$$f_X(x) = \alpha\beta x^{\beta-1} e^{-\alpha x^\beta}, \ x > 0.$$

Then

$$f_{Y_n}(x) = \left[n\Gamma_{1-\frac{1}{\beta}}(n, \alpha x^\beta)\alpha\beta x^{\beta-1}e^{-\alpha x^\beta} \right] \Big/ \Gamma\left(n + \frac{1}{\beta}\right)$$

with hazard rate

$$h_{Y_n}(x) = \left[n\Gamma_{1-\frac{1}{\beta}}(n, \alpha x^\beta) \right] \Big/ \Gamma_{1-\frac{1}{\beta}}(n+1, \alpha x\beta)^{\beta-1}\alpha\beta x.$$

EXAMPLE 8.6. *For the Pareto II with $f(x) = \frac{c\alpha^a}{(x+\alpha)^{c+1}}$,*

$$C_n(x) = \frac{c^n}{(c-1)^{n+1}}(x+\alpha)$$

and f_{Y_n} is the Pareto distribution $(c-1, \alpha)$.

Some more properties of $C_n(t)$ and its applications to actuarial sciences were presented by Psarrakos and Toomaj (2017).

(i) $C_n(aX+b) = aC_n(X); C_n(X) = \frac{1}{n!}E(h_n(X)), h_n(x) = \int_0^\infty A^n(t)dt.$

(ii) $C_n \max(X_1 \dots X_n) \leq C_n \max(Y_1, \dots, Y_n)$ when $(X_1, \dots, X_n)$ and $(Y_1, \dots, Y_m)$ are independent and the X_i's independent and identically described and $X_i \leq_{\mathrm{icx}} Y_i$.

(iii) $X \leq_{\mathrm{hr}} Y \Rightarrow \frac{C_n(X)}{E(X)} \leq \frac{C_n(Y)}{E(Y)}.$

(iv) $C_n(X) \geq \frac{1}{n}C^*(X)$ and $C_n(X) \geq \frac{1}{n}(\int_0^\infty F(x)\bar{F}(x)dx)^n.$

(v) $C_n(X) \geq \frac{1}{n!}C^n e^{n H(k)}$ where $C = e^{\int_0^1 \log x| \log x|dx} = 0.2065$ and $H(X)$ is the Shannon entropy of X.

(vi) If X is IFRA (DFRA), then $C_n(X) \leq (\geq)\frac{1}{n!}E[X A^{n-1}(X)].$

(vii) The random variables X and Y are identically distributed but for a change of location and scale if and only if n is fixed,

$$\frac{C_n(X_{1:m})}{E(X_{1:m})} = \frac{C_n(Y_{1:m})}{E(Y_{1:m})}$$

for all $m = m_j$, $j \geq 1$ satisfying $\sum_{j=1}^{\infty} m_j^{-1} = \infty$.

The right tailed deviation risk measure

$$D(X) = \int_0^\infty \sqrt{\bar{F}(t)}dt - E(X)$$

and the standard deviations of X are calculated and it is shown that for some well known $C_n(X)$ they are well related to them, to propose $C_n(X)$ as a measure of risk. Tahmasebi and Eskanderzadeh (2017) considered a sequence $\langle X_n \rangle$, $n = 1, 2, \ldots$ of independent and identically distributed variables and the distribution of $X_{n(k)}$, the kth upper record values of $\langle X_n \rangle$ with density

$$f_{n(k)}(x) = \frac{k^n}{(n-1)!} (F(x))^{k-1} (\Lambda(x))^{n-1} f(x)$$

and its distribution function

$$F_{n(k)}(x) = 1 - (\bar{F}(x))^k \sum_{i=0}^{n-1} \frac{(k\Lambda(x))^n}{i!}.$$

They proposed an extension of CRE in the form

$$C_{n,k} = \int_0^\infty \frac{k^{n+1}}{n!} (\bar{F}(x))^k (\Lambda(x))^n \, dx = E_{X_{n+1}(k)} \left(\frac{1}{h(X)} \right). \tag{8.31}$$

This measure reduces to the GCRE of Psarrakos and Navarro (2013) when $k = 1$ and if in addition $n = 1$ also, we have the CRE. The measure satisfies

(i) X is IFR (DFR) $C_{n,k} \geq (\leq)C_{n+1,k}$, $n = 1, 2, \ldots, k \geq 1$ and $C_{n,k} \leq (\geq)C_{n,k+1}$

(ii) $X \leq_{\text{st}} Y$ and X is DFR implies $C_{n,k}(x) \leq C_{n,k}(y)$.

When X_t is considered instead of X, the dynamic version is obtained as

$$C_{n,k}(t) = \int_t^\infty \frac{k^{n+1}}{n!} \left(\frac{\bar{F}(x)}{\bar{F}(t)} \right)^k \left(-\log \frac{\bar{F}(x)}{\bar{F}(t)} \right)^n dx, \ n = 0, 1, 2, \ldots, \ k \geq 1. \tag{8.32}$$

Notice that

(a) $C_{0,1}(t) = m(t)$

(b) If X is exponential with mean λ, $C_{n,k}(t) = C_{n-1,k}(t) = \lambda$

(c) $C_{n,k}(t) \geq (\leq)C_{n+1,k}(t)$ and $C_{n,k}(t) \leq (\geq)C_{n,k+1}(t)$ if X is IFR (DFR)

(d) $C_{n,k}(t) = kh(t)[C_{n,k}(t) - C_{n-1,k}(t)]$

(e) If X is IFR (DF) then $C_{n,k}(t)$ is decreasing (increasing) for all n and $k \geq 1$

(f) If $C_{n,k}(t) = mkC_{n-1,k}(t)$ for a fixed n and $m > 0$, then X is exponential ($m = 1$), Pareto II ($m > 1$) or rescaled beta ($m < 1$)

EXAMPLE 8.7. If $\bar{F}(t) = \left(\frac{\beta - t}{\beta}\right)^{\alpha}$ then

$$C_{n,k}(t) = \frac{\alpha^n k^{n+1}(\beta - t)}{(\alpha + 1)^n(\alpha k + 1)}.$$

If $\bar{F}(t) = \left(\frac{\beta}{t+\beta}\right)^{\alpha}$ then

$$C_{n,k}(t) = \frac{\alpha^n k^{n+1}(t + \beta)}{(\alpha - 1)^n(\alpha k - 1)}, \quad \alpha k > 1$$

and $\bar{F}(t) = \frac{\alpha \beta^{\alpha}}{x^{\alpha+1}}$ gives

$$C_{n,k} = \frac{k\alpha^n b}{(\alpha K - 1)^{n+1}}, \quad \alpha > \frac{1}{k}$$

and

$$C_{n,k} = \frac{\alpha}{(\alpha K - 1)^{n+1}}C_{n-1,K}.$$

The reversed relevation transform discussed in this section has immediate implications to the generalized cumulative entropy of past life. Recall that the transform is

$$F^*(x) = G(x) + F(x)\int_x^{\infty} \frac{dG(t)}{F(t)}, \quad x > 0$$

and when $F = G$,

$$F^X(x) = F(x)[1 + T(x)], \quad T(x) = \int_x^{\infty} \frac{dF(t)}{F(t)} = -\log F(x).$$

The n fold transform thus becomes

$$F_{(n)}(x) = \begin{cases} F(x) & n = 1 \\ F_{n-1}F(x) & n \geq 2 \end{cases}$$

in which

$$F_{(n)}(x) = F(x)\sum_{k=0}^{n-1} \frac{(T(x))^k}{k!} = q_n(F(x)) \qquad (8.33)$$

when $q : [0, 1] \rightarrow [0, 1]$ is a continuous non-decreasing and piece-wise differentiable function satisfying $q(0) = 0$ and $q(1) = 1$ (called a distortion function). Note that (8.33) coincides with the distribution function of the nth lower record value. If $X(n)$ is the random variable corresponding to

$F(n)$, then $X_{(n)}$ converges in probability to zero as $n \to \infty$. The generalized cumulative entropy

$$\bar{C}_n = \int_0^\infty F(x) \frac{T^n(x)}{n!} dx \qquad (8.34)$$

and its dynamic version

$$\bar{C}_n(t) = \frac{1}{n!} \int_0^t \frac{F(x)}{F(t)} \left[-\log \frac{F(x)}{F(t)} \right]^n dx$$

$$= \frac{1}{n!} \int_0^t \frac{F(x)}{F(t)} [T(x) - T(t)]^n dx$$

has the relationship

$$\bar{C}_n = E(X_{(n)}) - E(X_{(n+1)}), \quad n = 1, 2.$$

Also, if $X \leq_{\text{disp}} Y$ then $\bar{C}_n(x) \leq \bar{C}_n(y)$. For independent random variables with log-concave densities

$$\bar{C}_n(X + Y) \geq \max(\bar{C}_n(X) + \bar{C}_n(Y)).$$

The connection with hazard rates is established through the property that if $X \leq_{\text{hr}} Y$ and X or Y is DFR, then $\bar{C}_n(Y) \leq \bar{C}_n(Y)$. If g is measurable and differentiable twice

$$Eg(X_{(n)}) = E(g\, X_{(n+1)}) + E[g'(Z_n)]\bar{C}_n$$

where Z_n is absolutely continuous with density function

$$f_{Z_n}(x) = \frac{F(x)T^n(x)}{\bar{C}_n n!}.$$

Tahmasebi and Eskanderzadeh (2017) besides establishing the above characterises of $\bar{C}_n$ have some results similar to those of C_n reviewed earlier from Psarrakos and Toomaj (2017).

THEOREM 8.21. *Let X have support $[0, b]$, $b < \infty$. Then $\bar{C}_n(t) = k\bar{C}_{n-1}(t)$ holds for all $t >, 0 < k < 1$, and a fixed n, if and only if $F(x) = \left(\frac{x}{b}\right)^{\frac{k}{1-k}}$.*

They have also proved some special properties regarding the monotonicity of $\bar{C}_n$. If one of them, $\bar{C}_n(t)$ is increasing so is $\bar{C}_{n+1}$. Also F and G belong to the same family of distributions save for a change in location and scale if and only if

$$\bar{C}_n(X_{m:m}, t) = \bar{C}_n(Y_{m:m}, t)$$

for all $m \in M$, $M = \{m_j, j \geq 1\}$ is a strictly increasing sequence of positive integers such that $\sum_{j=1}^\infty m_j^{-1} = \infty$. The result holds for $\bar{C}_n$ also.

Using the empirical distribution function and

$$\hat{\bar{C}}_n(\hat{F}_m) = \frac{1}{n!} \sum_{i=1}^{m-1} \frac{i}{m} \left[-\log \frac{i}{m} \right]^n X_{i+1:n} - X_{i:n}$$

the mean and variance of $\hat{\bar{C}}_n$ for a sample from $U[0,1]$ is

$$E(\hat{\bar{C}}_n) = \frac{1}{n!(m+1)} \sum_{i=1}^{m-1} \frac{i}{m} \left(-\log \frac{i}{m} \right)^n$$

and

$$V(\hat{\bar{C}}_n) = \frac{m}{n!(m+1)^2(m+2)} \sum_{i=1}^{m-1} \left(\frac{i}{m} \right)^2 \left(-\log \frac{i}{m} \right)^{2n}.$$

By virtue of this, $\hat{\bar{C}}_n$ is an unbiased and consistent estimator for $\bar{C}_n$ of a population uniformly distributed in $[0,1]$.

The shift dependent weighted cumulative residual entropy

$$C_{n,W} = \frac{1}{n!} \int_0^\infty x \bar{F}(x)(-\log \bar{F}(x))^n dx$$

and

$$C_{n,W}(t) = \frac{1}{n!} \int_0^\infty x \frac{\bar{F}(x)}{\bar{F}(t)} \left[-\log \frac{\bar{F}(x)}{\bar{F}(t)} \right]^n dx, \ n = 1, 2, \ldots$$

are given special attention in Kayal (2018a). A linear transformation $Y = aX + b$ results in

$$C_{n,Y,W}(t) = a^2 C_{n,X,W}(t) + ab C_n(t).$$

For a DFR lifetime X, $C_{n,W} \le C_{n+1,W}$ a similar relationship holds for $C_{n,W}(t)$. If we write $T_n(t) = \frac{1}{n!} \int_t^x z(-\log \frac{\bar{F}_X(z)}{\bar{F}_X(t)})^n dz$ then

$$C_{n,W}(t) = E[T_n(t) | X > t].$$

There exists the relationship

$$C'_{n,W}(t) = \lambda_X(t)[C_{n,W}(t) - C_{n-1,W}(t)], \ n = 1, 2, \ldots$$

An estimator for C_W is presented as

$$\hat{C}_{n,W} = \frac{1}{n} \sum_{j=1}^{m-1} \left(1 - \frac{j}{m} \right) \left(-\log \left(1 - \frac{j}{m} \right) \right)^n (X_{j+1:n}^2 - X_{j:n}^2).$$

The generalized cumulative entropy (GCE), $\bar{C}_n$ introduced in (8.29) was further studied in Kayal (2016). Some important results include

(1) $\bar{C}_n(t) = E(R_n^{(2)}(X)|X \le t)$ where

$$R_n^{(2)}(x) = \frac{1}{n!} \int_x^t \left(-\log \frac{F_X(x)}{F_X(t)} \right)^n dx.$$

(2) $\bar{C}_n(t) \ge R_n^{(n)}(r(t))$ and

$$\bar{C}_n(t) \le \frac{1}{n!} \int_0^t \left(-\log \frac{F_X(x)}{F_X(t)} \right)^n dx.$$

(3) $\bar{C}_n(t) = \lambda_X(t)[\bar{C}_{n-1}(t) - C_n(t)]$
(4) $\bar{C}_n(t)$ is increasing if and only if $\bar{C}_{n-1}(t) \ge C_n(t)$
(5) An estimator of $\bar{C}_n(t)$ based on order statistic and empirical distribution is

$$\hat{\bar{C}} = \sum_{j=1}^{m-1} \frac{1}{n!} \left(\frac{j}{m} \right) \left(-\log \frac{j}{m} \right)^n (X_{j:n} - X_{j-1:n})$$

for a random sample $X_1, X_2, \ldots, X_m$. Cali et al. (2020) show that if $\bar{C}_n(t) \ne \bar{C}_{n-1}(t)$ then $\bar{C}_i(t_0)$, $i = 0, 1, \ldots, n-1$ and $\bar{C}_n(t)$ uniquely determine $F(x)$. However if the $C_n(t)$ does not depend on n, $F(x)$ is uniquely specified. If T represents the lifetime of a coherent system with m identically distributed components with distortion function q, writing $\phi_n(u) = u[-\log u]^n$, it is shown that if $\phi_n(q(u)) \ge (\le)\phi_n(u)$ then $\bar{C}_{n,T}(t) \ge (\le)C_n(X_1)$. This result extends to two coherent systems of lives T_1 and T_2 having two common distributions F_1 and F_2 and a common copula, then $F_1 \le_{\text{disp}} F_2 \Rightarrow \bar{C}_n(T_1) \le \bar{C}_n(T_2)$. The approach allows to state bounds $B_{1,n}\bar{C}_n(X_1) \le \bar{C}_n(T) \le B_{2,n}\bar{C}_n(X_1)$ where $B_{1,n}$ and B_{2n} are the infimum and supremum of $\phi_n(q(u))/\phi_n(u)$ in $u \in (0, 1)$. Also $f(x) \le M$ for all x, then

$$\bar{C}_n(T) \ge \frac{1}{M(n!)} \int_0^1 \phi_n(q(u))du$$

and

$$\bar{C}_n(t) \ge \frac{1}{L(n!)} \int_0^1 \phi_n(q(u))du$$

when $f(x) \ge L$ for all x.

Cali et al. (2019) have more generalized cumulative entropies. In the past cumulative entropy

$$C(t) = -\int_0^t \frac{F(x)}{F(t)} \log \frac{F(x)}{F(t)} dx$$

in Chapter 7 and the dynamic cumulative Tsallis entropy in (8.19), a new measure is defined using $w(x) = \frac{r(x)}{(F(x))^{\alpha-1}}$ and random variables Y_α with

$$f_{Y_\alpha}(x) = \begin{cases} r(x) \left(\frac{F^{\alpha-1}(x)}{C^*_{\alpha,T}} \right) f(x), & \alpha \neq 1 \\ \frac{r(x)}{C_X} f(x), & \alpha = 1 \end{cases}.$$

For $\alpha > 0$ the distribution functions are

$$F_{Y_\alpha}(x) = \begin{cases} \frac{C^*_{\alpha,T}(x)}{C^*_{\alpha,T}} F^\alpha(x), & \alpha \neq 1 \\ \frac{C(x)}{C_{Xt}} F(x), & \alpha = 1 \end{cases}.$$

The distribution of Y_α gives a family of weighted distributions of order α. We refer to Cali et al. (2019) for distributional properties and stochastic comparisons with some known weighted distributions and characterizations. A shift dependent version of the generalized cumulative residual entropy for the kth upper record and its dynamic versions are discussed in Moharana and Kayal (2019).

Xiong et al. (2019) have introduced CRE of fractional order (FCRE)

$$C_q = \int_0^\infty \bar{F}(x)(-\log \bar{F}(x))^q dx \tag{8.35}$$

where $0 \leq q \leq 1$. From the definition (8.35) it is clear that $C_0 = \mu$ and it becomes CRE when $q = 1$. The measure C_q is non-negative, non-additive and is a concave function of F, but convex in q.

EXAMPLE 8.8. *When X is exponential λ, $C_q = \frac{\Gamma(q+1)}{\lambda}$.*

Some properties of FCRE are

(a) $C_q(Y) = aC_Q(X)$, $Y = qX + b$, $a > 0$, $b \geq 0$
(b) When X and Y are independent $\max(C_q(X), C_Q(Y)) \leq C_Q(X + Y)$ and $C_q(X) \leq [C(X)]^q$
(c) $C_q(X) \geq C(q)e^{H(x)}$
(d) The empirical FCRE is

$$\hat{C}_q(\bar{F}_n) = \int_0^\infty \bar{F}_n(x)(-\log \bar{F}_n(x))^q dx$$

where $\bar{F}_n$ is the empirical distribution function arising from a random sample $X_1, \ldots, X_n$ from F. Now

$$\hat{C}_q(\bar{F}_n) = \sum_{j=1}^{n-1} \left(1 - \frac{j}{n}\right) \left(-\log(1 - \frac{j}{n})\right)^q (X_{j+1:n} - X_{j:n}).$$

Note that $\hat{C}_q$ converges atmost surely to C_q. An application of the entropy is given for financial data.

A shift-dependent measure of generalized CRE is discussed in Tahmasebi et al. (2020) in the form

$$\bar{C}_{w,n} = \frac{1}{n!} \int_0^\infty w(x)F(x)(-\log F(x))^n dx. \qquad (8.36)$$

If X and Y are such that $X \leq_{\text{rh}} Y$ then $C_w(X) \leq C_w(Y)$. The dynamic version is

$$\bar{C}_{w,n}(t) = \frac{1}{n!} \int_0^\infty \frac{w(x)F(x)}{F(t)} [-\log \bar{F}(x) + \log \bar{F}(t)]^n dx$$

The measure is such that
(a)
$$\bar{C}'_{w,n}(t) = \lambda(t)[\bar{C}_{w,n-1}(t) - \bar{C}_n(t)].$$

In particular

$$\bar{C}'_{w,1}(t) = \lambda(t) \left[\frac{1}{F(t)} \int_0^t w(x)F(x)dx = C_{w,1}(t) \right]$$

(b)

$$\bar{C}'_{w,1}(t) = \frac{\bar{C}_{w,j}(t) - \bar{C}_{w,j-1}(t)\frac{\partial}{\partial t}\bar{C}_{w,n}(t)}{C_{w,j}(t) - C_{w,n-1}(t)}, \quad j = 1,2,\ldots,n-1$$

(c) $\bar{C}_{w,n}(t)$ is increasing if X is a decreasing reversed hazard rate average and if $\bar{C}_{w,n-1}(t)$ is increasing, so does $\bar{C}_{w,n}(t)$. For more properties see Daneshi et al. (2019).

Moharana and Kayal (2019) describe the features of the shift dependent generalized cumulative entropy

$$C_n^w = \frac{1}{n!} \int_0^\infty x\bar{F}(x)(-\log \bar{F}(x))^n dx.$$

Under the proportional reversed hazard model Y of X

$$C_n^w(Y) = \frac{\theta^n}{n!} \int_0^\infty xF_X^\theta(x)(-\log F_X(x))^n dx.$$

THEOREM 8.22.   (i) $C_n^W(y) \leq \theta^n C_n^w(X)$ if $\theta \geq 1$ and $C_a^w(Y) \geq \theta^n C_n^w(X)$ if $0 < \theta < 1$
(ii) $C_n^w(X) \geq \frac{1}{n!}[\int_0^\infty x^{\frac{1}{n}} F(x)\bar{F}_X(x)]^n$
(iii) $C_n^w(X) = \frac{1}{n!} \int_0^\infty \lambda_x(t)[\int_0^t xF_X(x)(-\log F(x))^{n-1}dx]dt.$

THEOREM 8.23. *The distribution of X belongs to (i) the inverse Weibull family*

$$F_X(x) = e^{-(\frac{\lambda}{x})^{\frac{1}{k}}}, \quad \lambda, k, x > 0$$

if and only if

$$n! C_n^w(X) = k E[X^2(-\log F(X))^{n-1}]$$

for all $n = n_j$, $j \geq 1$ such that $\sum_{j=1}^{\infty} n_j^{-1} = \infty$ and (ii) the power distribution if and only if

$$n! k C_n^W(X) = E[X^2(-\log F(X))^n], \quad k > 0$$

for all $n = n_j$, $j \geq 1$ such that $\sum_{j=1}^{\infty} n_j^{-1} = \infty$. An estimate of $C_n^W(X)$ is proposed as

$$\hat{C}_n^W(X) = \frac{1}{n!} \sum_{j=1}^{m-1} \left(\frac{j}{m}\right) \left(-\log\left(\frac{j}{m}\right)\right)^n (X_{j+1:n}^2 - X_{j:n}^2)$$

based on a random sample $X_1, X_2, \ldots, X_m$ from $F(x)$.

Another version of weighted measures of generalized CRE given in Tahmasebi (2020) is

$$\tilde{C}_n = \frac{1}{n!} \int_0^{\infty} w(x) \bar{F}(x) A^n(x) dx = E\left(\frac{w(X_{n+1})}{h(X_{n+1})}\right)$$

where X_n has density function

$$f_n(x) = \frac{A^{n-1}(x) f(x)}{(n-1)!}.$$

For random variables X and Y satisfying $X \leq_{hr} Y$, and if $\frac{w(x)}{h_X(x)}$ is increasing $\tilde{C}_n(X) \leq \tilde{C}_n(Y)$ and further if $X \leq_{disp} Y$ and $w(x) = F^{-m}(x)$, we have $\tilde{C}_n(X) \leq \tilde{C}_n(Y)$. We can express $\tilde{C}_n$ as an expected value

$$\tilde{C}_n(X) = \frac{1}{n!} E\left[\int_0^X w(t) \Gamma^n(t) dt\right].$$

A more general form for the residual life is

$$\tilde{C}_n(t) = \frac{1}{n!} \int \frac{w(x) \bar{F}(x)}{\bar{F}(t)} [\Gamma(x) - \Gamma(t)]^n dx$$

with derivative

$$\tilde{C}_n'(t) = h(t)[\tilde{C}_n(t) - \tilde{C}_{n-1}(t)].$$

Some general properties of $\tilde{C}_n(t)$ similar to other weighted functions are also discussed in Tahmasebi (2020). Most of the generalizations discussed here are based on cumulative residual entropy. There is scope for similar more general extensions of the other versions involving various generalizations of cumulative residual entropy.

For a detailed survey of literature on various topics related to reliability modelling and information measures, one can refer to Box and Cox (1964); Daroczy (1967); Smith and Bain (1975); Vasicek (1976); Cox and Oakes

(1984); Salicru et al. (1993); Jiang et al. (2003); Belzunce et al. (2005b); Jain and Srivastava (2007); Kundu et al. (2010); Misagh and Yari (2011, 2012); Nair et al. (2012b, 2013a); Nair and Sankaran (2013b); Midhu et al. (2014); Baratpour and Khammar (2015); Di Crescenzo and Longobardi (2015); Rezaei et al. (2015); Xu and Moura (2017); Kayal (2018a); Nanjundan and Pasha (2018), Nair et al. (2019); Noughabi and Jarrahiferi (2019); Zang et al. (2019); Krishnan et al. (2020); Nair et al. (2021a,b).

References

Aarset, M. V. (1987). How to identify a bathtub hazard rate. *IEEE Transactions on Reliability*, 36: 106–108.

Abbasnejad, M. (2011). Some characterizations based on dynamic survival and failure entropies. *Communications of the Korean Statistical Society*, 18: 787–798.

Abbasnejad, M., Arghami, N. R., Morgenthaler, S., Borzadaran, G. R. M. et al. (2010). On the dynamic survival entropy. *Statistics and Probability Letters*, 80: 1962–1971.

Abbasnejad, M. and Borzadaran, G. R. M. (2015). Some results on dynamic generalized survival entropy. *Communications in Statistics-Theory and Methods*, 44: 1653–1668.

Abdul-Sathar, E. A. and Nair, R. D. (2021). On dynamic weighted extropy. *Journal of Computational and Applied Mathematics*, 393: 113507.

Abe, S. (1997). Stability of Tsallis entropy and instabilities of Renyi's and normalized tallis entropies–a basis for q-exponential distributions. *Physics Review E*, 66: 040164.

Abraham, B. and Nair, N. U. (2013). A criterion to distinguish ageing patterns. *Statistics*, 47: 85–92.

Abraham, B. and Sankaran, P. G. (2005). Renyi's entropy for residual lifetime distribution. *Statistical Papers*, 46: 17–30.

Aczel, J. and Daroczy, Z. (1963). Charaklerisierung der entropien positiver ordnung und der shannonschen entropie. *Acta Mathematica Academiae Scientiarum Hungarica*, 14: 95–121.

Aczel, J., Forte, B. and Ng, C. T. (1974). Why the Shannon and Hartley entropies are "natural"? *Advances in Applied Probability*, 6: 131–146.

Ahmad, I. A. and Mugadi, A. R. (2004). Further moment inequalities of life distributions with hypothesis testing applications, the IFRA, NBUC and DMRL classes. *Journal of Statistical Planning and Inference*, 120: 1–12.

Ahmed, A. N., Alzaid, A., Bartoszewicz, J. and Kochar, S. C. (1986). Dispersive and superadditive ordering. *Advances in Applied Probability*, 18: 1019–1022.

Alzaid, A. A. (1994). Ageing concerning of items of unknown age. *Stochastic Models*, 10: 649–659.

Arimoto, A. (1971). Information theoretical considerations on estimation problems. *Information and Control*, 19: 189–184.

Asadi, M. and Ebrahimi, N. (2000). Residual entropy and its characterizations in terms of hazard function and mean residual life function. *Statistics and Probability Letters*, 49: 263–269.

Asadi, M., Ebrahimi, N., Hamedani, G. G. and Soofi, E. S. (2005a). Minimum dynamic discrimination models. *Journal of Applied Probability*, 42: 643–660.

Asadi, M., Ebrahimi, N. and Soofi, E. S. (2005b). Dynamic generalized information measures. *Statistics and Probability Letters*, 71: 85–98.

Asadi, M. and Zohrevand, Y. (2007). On the dynamic cumulative entropy. *Journal of Statistical Planning and Inference*, 137: 1931–1941.

Asha, G. and Rajeesh, C. J. (2015). Characterizations through past entropy measures. *Metron*, 73: 119–134.

Assis, E. M., Borges, S. and Melo, A. B. V. (2013). Generalized q-Weibull model and the bathtub curve. *International Journal of Quality and Reliability Management*, 30: 720–736.

Awad, A. M. and Alawneh, A. J. (1987). Application of entropy to a lifetime model. *IMA Journal of Mathematical Control and Information*, 4: 143–147.

Ayres, R. U. and Martinas, K. (1995). Waste potential entropy: The ultimate ecotoxic. *Economie Appliquee*, 48: 95–120.

Baig, M. A. K. and Dar, J. G. (2008). Generalized residual entropy function and its applications. *European Journal of Pure and Applied Mathematics*, 1: 30–40.

Bain, L. J. (1978). *Statistical Analysis of Reliability and Life-testing Models*. Marcel Dekker, New York.

Baratpour, S. (2010). Characterizations based on cumulative entropy of first order statistic. *Communications in Statistics-Theory and Methods*, 39: 3645–3651.

Baratpour, S. and Khammar, A. H. (2015). Results on Tsallis entropy of order statistics and record values. *iSTATiSTiK: Journal of the Turkish Statistical Association*, 8: 60–73.

Baratpour, S. and Khammar, A. H. (2018). A quantile based generalized dynamic cumulative measure of entropy. *Communications in Statistics-Theory and Methods*, 47: 3104–3117.

Baratpour, S. and Rad, A. H. (2012). Testing of goodness of fit for the exponential distribution based on cumulative residual entropy. *Communications in Statistics-Theory and Methods*, 41: 35–45.

Barlow, R. E. and Campo, R. (1975). Total time on test process and applications to fault tree analysis. In *Reliability and Fault Tree Analysis*, 451–481, Philadelphia. SIAM.

Barlow, R. E. and Hsiung, J. H. (1983). Expected information from a life testing experiment. *Journal of the Royal Statistical Society*, 32: 35–45. Series D.

Baxter, L. A. (1982). Reliability applications of the relevation transform. *Naval Research Logistics Quarterly*, 29: 323–330.

Bebbington, M., Lai, C. D., Murthy, D. N. P. and Zitikis, R. (2009). Modelling N and W shaped hazard rate functions without mixing distributions. *Journal of Risk and Reliability*, 223: 59–69.

Behara, M. and Chawla, J. S. (1974). Generalized gamma entropy. *Statistica Canadiana*, 2: 15–38.

Behara, M. and Nath, P. (1973). Additive and non-additive entropies of finite measurable partitions. In *Probability and Information Theory II*, number 296 in Lecture Notes in Mathematics, Berlin. Springier-Verlag.

Belis, M. and Guiasu, S. (1968). A quantative measure of information in cybernetics. *IEEE Transactions on Information Theory*, 4: 593–594.

Belzunce, F., Guillamon, A., Navarro, J. and Ruiz, J. M. (2001). Kernel estimation of residual entropy. *Communications in Statistics, Theory and Methods*, 30: 1243–1255.

Belzunce, F., Navarro, J., Ruiz, J. M. and Aguila, Y. D. (2004). Some results on residual entropy function. *Metrika*, 59: 147–161.

Belzunce, F., Orlege, E. and Ruiz, J. M. (2005a). A note on replacement policy comparisons from NBUC lifetime of the unit. *Statistical Papers*, 46: 509–522.

Belzunce, F., Riquelme, C. M. and Mulero, J. (2005b). *An Introduction to Stochastic Orders*. Academic Press.

Bercher, J. and Vignat, C. (2008). A new look at the q-exponential distributions via excess statistic. *Physica A: Statistical Mechanics and Applications*, 387: 5422–5432.

Bhat, B. A. and Baig, M. A. K. (2020). Weighted generalized cumulative residual entropy and its dynamic version for lifetime distributions. *Journal of Xian University of Architecture and Technology*, 12: 4231–4243.

Bhattacharjee, A. and Sengupta, D. (1996). On the coefficient of variations of the $\mathcal{L}$ and $\bar{\mathcal{L}}$ classes. *Statistics and Probability Letters*, 27: 177–180.

Bhattacharya, A. (1943). On a measure divergence between two statistical populations defined by their probability distributions. *Bulletin of the Calcutta Mathematical Society*, 35: 99–109.

Bhattacharya, A. (1946). Some analogous to amount of information and their uses in statistical estimation. *Sankhya*, 8: 1–14.

Block, H. W., Borges, W. S. and Savits, T. H. (1985). Age dependent minimal repair. *Journal of Applied Probability*, 22: 370–385.

Block, H. W., Savits, T. H. and Singh, H. (1998). The reversed hazard rate function. *Probability in the Engineering and Information Sciences*, 12: 69–90.

Boekee, D. E. and van der Lubbe, J. C. A. (1980). The r-norm information measure. *Information and Control*, 45: 136–155.

Borges, R. (1967). Zur herleitung der shannonschen information. *Mathematiches Zeitschrift*, 96: 282–287.

Box, G. E. P. and Cox, D. R. (1964). An analysis of transformations. *Journal of the Royal Statistical Society B.*, 26: 211–252.

Burbea, J. and Rao, C. R. (1982). On the convexity of some divergence measures based on entropy functions. *IEEE Transactions on Information Theory*, 28: 489–495.

Cali, C., Longobardi, M. and Psarrakos, G. (2019). A family of weighted distributions based on mean inactivity time and cumulative past entropies. *Ricerche di Mathematica*, DOI: 10.1007/s11587-019-00475-7.

Cali, S., Longobardi, M. and Navarro, J. (2020). Properties for generalized cumulative past measures of information. *Probability in the Engineering and Informational Sciences*, 34: 92–111.

Cao, J. and Wang, Y. (1991). The NBUC and NWUC classes of life distributions. *Journal of Applied Probability*, 28: 473–479.

Chahkandi, M. and Toomaj, A. (2016). Some results on the residual entropy of coherent systems. *Proceedings of the 3rd Seminar on Reliability Theory and Applications*, 81–86.

Chamany, A. and Baratpour, S. (2014). A dynamic discrimination information based on cumulative residual entropy and its properties. *Communications in Statistics: Theory and Methods*, 43: 1041–1049.

Chang, B. H. and Pocock, S. (2000). Analyzing data with dumping at zero. *Journal of Clinical Epidemeology*, 52: 1036–1043.

Chaundy, T. W. and McLeod, J. B. (1960). On a functional equation. *Proceedings of the Edinburg Mathematical Society Notes*, 43: 7–8.

Chernoff, H. (1951). Measure of asymptotic efficiency for tests of a hypothesis based on the sum of observations. *Annals of Mathematical Statistics*, 23: 493–507.

Cox, D. R. and Oakes, D. (1984). *Analysis of Survival Data*. Chapman and Hall, London.

Csiszar, I. (1967). Information type measures of difference of probability distributions and indirect observations. *Studia Scientiarum Mathematicarum Hungarica*, 2: 299–318.

Daneshi, S., Nezakati, A. and Tahmasebi, H. (2019). On weighted cumulative past (residual) inaccuracy for record values. *Journal of Inequalities*, 2019: 134.

Daneshi, S., Nezakati, A. and Tahamasebi, S. (2020). Weighted cumulative residual (past) inaccuracy for min (max) of order statistics. *Statistics, Optimization and Information*, 8: 110–126.

Dar, J. G. and Al-Zahrani, B. (2013). On some characterization results of lifetime distributions using Mathai-Haubold residual entropy. *ISOR–Journal of Mathematics*, 5: 56–60.

Daroczy, Z. (1967). über die Charakterisierung der Shannonschen Entropie. *Statistica*, 27: 199–205.

Daroczy, Z. (1969). On the Shannon's measures of information. *Magyar Tud. Akad. Mat. Fiz. Oszt. Kozl.*, 19: 9–24.

Daroczy, Z. and Katai, I. (1970). Additive zahlen theoretische, funktionen und das mass der information. *Ann. Univ Sci. Budapest. Eotvos Sect. Math.*, 13: 83–88.

Deshpande, J. V. and Kochar, S. C. (1983). Dispersive ordering is the same as tail ordering. *Advances in Applied Probability*, 15: 686–687.

Deshpande, J. V., Kochar, S. C. and Singh, H. (1986). Aspects of positive aging. *Journal of Applied Probability*, 23: 748–758.

Dhillon, B. S. (1981). Life distributions. *IEEE Transactions on Reliability*, 30: 457–460.

Di Crescsenzo, A. and Longobardi, M. (2002). Entropy based measure of uncertainty in past lifetime distributions. *Journal of Applied Probability*, 39: 434–440.

Di Cresenzo, A. and Longobardi, M. (2004). A measure of discrimination between past lifetime distributions. *Statistics and Probability Letters*, 67: 173–182.

Di Cresenzo, A. and Longobardi, M. (2006). On weighted residual and past entropies. *Scientiae Mathematicae Japonicae*, 64(3): 255–266.

Di Cresenzo, A. and Longobardi, M. (2009). On cumulative entropies. *Journal of Statistical Planning and Inference*, 139: 4072–4087.

Di Crescenzo, A. and Longobardi, M. (2015). Some properties and applications of cumulative Kullback-Leibler information. *Applied Stochastic Models in Business and Industry*, 31: 875–891.

Di Crescenzo, A. and Toomaj, A. (2015). Extension of the past lifetime and its connection to the cumulative entropy. *Journal of Applied Probability*, 52: 1156–1174.

Dragomir, S. S., Gluscevic, V. and Pearce, C. E. M. (2001). Inequality, theory and applications. pp. 139–154. In Cho, Y. J., Kim, J. K. and Dragomir, S. S.(eds.). *Approximation for the Csiszar f-divergence via midpoint inequalities*. Nova Science Publications.

Ebrahimi, N. (1996). How to measure uncertainty in residual life distributions. *Sankhya, Series A*, 58: 48–56.

Ebrahimi, N. (1997). Testing whether lifetime distribution is decreasing uncertainty. *Journal of Statistical Planning and Inference*, 64: 9–19.

Ebrahimi, N. (2001). Testing for uniformity of the residual lifetime based on dynamic Kullback-Leibler information. *Annals of the Institute of Statistical Mathematics*, 53: 325–337.

Ebrahimi, N. and Kirmani, S. N. U. A. (1996a). A characterization of proportional hazard models through a measure of discrimination between two residual life distributions. *Biometrika*, 83(1): 233–235.

Ebrahimi, N. and Kirmani, S. N. U. A. (1996b). A measure of discrimination between two residual lifetime distributions and its applications. *Annals of Institute of Mathematical Statistics*, 48: 257–265.

Ebrahimi, N. and Kirmani, S. N. U. A. (1996c). Some results on ordering survival functions through uncertainty. *Statistics and Probability Letters*, 29: 167–176.

Ebrahimi, N. and Pellerey, F. (1995). New partial ordering of survival functions based on notion of uncertainty. *Journal of Applied Probability*, 32: 202–211.

Ebrahimi, N. and Soofi, C. S. (2004). Recent developments in information theory and reliability analysis. *Frontiers in Reliability*, pp. 125–132.

Elabatal, I. (2007). Some ageing classes of life distributions at specific age. *International Mathematical Forum*, 2: 1445–1456.

Erto, P. (1989). Genesis, properties and identification of the inverse Weibull lifetime model. *Statistica Applicato*, 1: 117–128.

Estaban, M. D. and Morales, D. (1995). A summary of entropy statistics. *Kybernetika*, 31: 337–346.

Faddeev, D. K. (1956). On the concept of entropy of a finite probabilistic scheme. *Uspeki Matematicheskikh, Nauk*, 11: 227–231.

Ferreri, C. (1980). Hypoentropy and related heterogeneity and information measures. *Statistica*, 40: 155–168.

Finkelstein, M. (2006). On relative ordering of mean residual lifetime functions. *Statistics and Probability Letters*, 76: 939–944.

Fisher, R. A. (1934). The effect of methods of ascertainment upon the estimation of frequencies. *Annals of Eugenics*, 6: 13–25.

Forte, B. and Daroczy, Z. (1968). A characterization of Shannon entropy. *Bullettino Dell Unione Mathematica, Italiana*, 4: 631–635.

Forte, B. and Ng, C. T. (1973). On characterization of entropies of degree β. *Utilitas Mathematica*, 4: 193–205.

Freimer, M., Mudholkar, G. S., Kollia, G. and Lin, C. T. (1988). A study of the generalised Tukey lambda family. *Communications in Statistics-Theory and Methods*, 17: 3547–3567.

Gilchrist, W. (2000). *Statistical Modelling with Quantile Functions*. Chapman and Hall, CRC, Florida.

Goel, R., Taneja, H. C. and Kumar, V. (2018). Measure of entropy of past lifetime and k-record statistics. *Physica A: Statistical Mechanics and its Applications*, 503: 623–631.

Good, I. J. (1989). Studies in the history of probability and statistics XXXVII, A. M turing's statistical work in world war II. *Biometrika*, 66: 393–396.

Gore, A. P., Paranjpe, S. A., Rajarshi, M. B. and Gadgul, M. (1986). Some methods of summarising survivorship in nonstandard situations. *Biometrical Journal*, 28: 577–586.

Guess, F. and Proschan, F. (1988). Mean residual life: Theory and Applications. In Krishnaiah, P. R. and Rao, C. R. (eds.). *Handbook of Statistics*, 7: 215–224. Elsevier, North Holland.

Guiasu, S. (1986). Grouping data by using weighted entropy. *Journal of Statistical Planning and Inference*, 15: 63–69.

Gupta, R. C. (2007). Role of equilibrium in reliability studies. *Probability in the Engineering and Informational Sciences*, 21: 315–334.

Gupta, R. C. (2009). Some characterization results based on residual entropy function. *Journal of Statistical Theory and Applications*, 8: 45–59.

Gupta, R. C. and Kirmani, S. N. U. A. (1990). The role of weighted distributions in stochastic modelling. *Communications in Statistics-Theory and Methods*, 19(9): 3147–3162.

Gupta, R. C., Taneja, H. C. and Thapliyal, R. (2014). Stochastic comparison of residual entropy of order statistics and some characterization results. *Journal of Statistical Theory and Applications*, 13: 27–37.

Gupta, R. D. and Nanda, A. K. (2002). α and β entropies and relative entropies of distributions. *Journal of Statistical Theory and Applications*, 3: 177–190.

Hankin, R. K. S. and Lee, A. (2006). A new family of non-negative distributions. *Australia and New Zealand Journal of Statistics*, 48: 67–78.

Harkness, W. L. and Shantaram, R. (1969). Convergence of a sequence of transformations of distribution functions. *Pacific Journal of Mathematics*, 31: 403–415.

Hartley, R. V. L. (1928). Transmission of information. *Bell System Technical Journal*, 7: 535–563.

Hashempour, M. and Doostparast, M. (2018). Characterization on the basis of cumulative residual entropy of sequential order statistics. *Journal of Statistical Modelling: Theory and Applications*, 1: 83–92.

Havrda, J. and Charvát, F. (1967). Quantification method of classification processes: Concept of structural α-entropy. *Kybernetika*, 3: 30–35.

Hellinger, E. (1999). Neuebegrundung der theorie det quadratischen formen von un endichen vielin vernaderichen. *J. Rein. Hung. Math.*, 136: 210–271.

Hjorth, U. (1980). A reliability distribution with increasing, decreasing, constant and bathtub shaped failure rates. *Technometrics*, 22: 99–107.

Hollander, M., Park, D. H. and Proschan, F. (1985). Testing whether new is better than used of specific age with randomly censored data. *Canadian Journal of Statistics*, 13: 45–52.

Hooda, D. S. and Saxena, S. (2011). Generalized measure of discrimination between past lifetime distributions. *Pakistan Journal of Statistics and Operations Research*, 7: 265–281.

Hosking, J. R. M. (1990). *L*-moments analysis and estimation of distribution using linear combination of order statistics. *Journal of the Royal Statistical Society*, B52: 105–124.

Hosking, J. R. M. (1996). *Some Theoretical Results Concerning L-moments*. IBM Research Division, Yorktown Heights, New York. Research Report RC 14492, (Revised).

Jain, K. C. and Chhabra, P. (2014). Various relations on new information divergence measures. *International Journal on Information Theory*, 3: 1–18.

Jain, K. C. and Sarawath, R. N. (2012). Series of information divergence measures using f-divergences, convex properties and inequalities. *International Journal of Modern Engineering Research*, 2: 3226–3231.

Jain, K. C. and Srivastava, A. (2007). On symmetric divergence measures of Csiszar f-divergence class. *Journal of Applied Mathematics, Statistics and Informatics*, 3: 55–102.

Jalayeri, S. and Zadeh, K. (2017). An extension of weighted cumulative inaccuracy measure based on doubly truncation. *Pakistan Journal of Statistics and Operations Research*, 13: 103–113.

Jamboori, S. and Yousefzadeh, Y. (2014). On estimating Renyi entropy under progressive censoring. *Communications in Statistics-Theory and Methods*, 43: 2395–2405.

Jeevanand, E. S. and Abdul-Sathar, E. A. (2009). Estimating residual entropy function for exponential distribution from censored samples. *Prob-Stat Forum*, 2: 68–77.

Jeffreys, H. (1946). An invariant form for the prior probability in estimation problems. *Proceedings of the Royal Society of London. Series A. Mathematical and Physical Sciences*, 186: 453–461.

Jiang, R., Ji, P. and Xiao, X. (2003). Ageing property of univariate failure rate models. *Reliability, Engineering and System Safety*, 79: 113–116.

Kang, D. (2015a). Further results on closure properties of LPQE order. *Statistical Methodology*, 25: 23–35.

Kang, D. (2015b). Some remarks on DDRCE class of life distributions. *Sankhya A*, 77: 351–363.

Kaniadakis, G. (2002). Statistical mechanics in the context of special relativity. *Physics Review*, E66(056125).

Kapdistria, S. and Psarrakos, G. (2012). Some extensions of the residual lifetime and its connection to the cumulative entropy. *Probability in the Engineering and Informational Sciences*, 26: 129–146.

Kapur, J. N. (1967). Generalized entropy of order and type. *The Math Seminar*, 4: 78–94.

Kapur, J. N. (1983). Comparative assessment of various measures of entropy. *Journal of Information and Optimization Sciences*, 4: 207–232.

Kapur, J. N. (1988). Some new non-additive measures of entropy. *Bull. UMI*, 7: 253–256.

Kapur, J. N. (1989). *Maximum Entropy Models in Science and Engineering*. Wiley Eastern Limited, New Delhi.

Kayal, S. (2014). Some results on generalized residual entropy based on order statistics. *Statistica*, LXXIV: 383–402.

Kayal, S. (2015). Generalized residual entropy and upper record values. *Journal of Probability and Statistics*, Article ID 640426, DOI: 10.1155/2015/640426.

Kayal, S. (2016). On generalized cumulative entropy. *Probability in the Engineering and Informational Sciences*, 30: 640–662.

Kayal, S. (2018a). On weighted generalized cumulative entropy of order n. *Methodology and Computing in Applied Probability*, 20: 487–503.

Kayal, S. (2018b). Quantile based Chernoff distance for truncated random variables. *Communications in Statistics-Theory and Methods*, 47: 4938–4957.

Kayal, S. (2018c). Quantile based cumulative inaccuracy measure. *Physica A: Statistical Mechanics and its Applications*, 510: 329–344.

Kayal, S. and Moharana, R. (2017). On weighted cumulative residual entropy. *Journal of Statistics and Management Systems*, 20: 153–173.

Kayal, S., Moharana, R. and Sunoj, S. M. (2020). Quantile based study of dynamic inaccuracy measure. *Probability in the Engineering and Information Sciences*, 34: 183–199.

Kayal, S., Sunoj, S. M. and Rajesh, G. (2017). On dynamic generalized measures of inaccuracy. *Statistica*, 77: 133–148.

Kayal, S. and Tripathi, M. R. (2018). A quantile based Tsallis divergence. *Physica A: Statistical Mechanics and its Applications*, 492: 496–505.

Keilson, J. and Sumita, U. (1982). Uniform stochastic ordering and related inequalities. *Canadian Journal of Statistics*, 15: 63–69.

Kerridge, D. F. (1961). Inaccuracy and inference. *Journal of the Royal Statistical Society*, Series B, 23: 184–194.

Khalil, M., Habibirad, A. and Yousefzadeh, F. (2018). Some properties of Lin-Wong divergence on the past lifetime data. *Communications in Statistics-Theory and Methods*, 47: 3464–3476.

Khammar, A. H. and Jahanshahi, S. M. A. (2018a). On weighted cumulative Tsallis entropy and its dynamic version. *Physica A: Statistical Mechanics and its Applications*, 491: 678–692.

Khammar, A. H. and Jahanshahi, S. M. A. (2018b). Quantile-based Tsallis entropy in residual lifetime. *Physica A: Statistical Mechanics and its Applications*, 402: 994–1106.

Khinchin, A. Y. (1953). The concept of entropy in the theory of probability. *Uspekhi Matematicheskikh Nauk*, 8: 3–20.

Khinchin, A. Y. (1957). *Mathematical Foundations of Information Theory*. Courier Corporation.

Kittanch, O. A., Khan, M. A. U., Akbar, M. and Bayoud, H. A. (2016). Average entropy: A new uncertainty measure with application to image segmentation. *The American Statistician*, 70: 18–24.

Klar, B. and Muller, A. (2003). Characterization of classes of life distribution generalizing the NBUE class. *Journal of Applied Probability*, 40: 20–32.

Klefsjö, B. (1982). HNBUE and HNWUE classes of life distributions. *Naval Research Logistics Quarterly*, 29: 615–626.

Klefsjö, B. (1983). A useful ageing property based on Laplace transforms. *Journal of Applied Probability*, 20: 615–626.

Kochar, S. C. (1989). On extensions of DMRL and related partial orderings of life distributions. *Communications in Statistics, Stochastic Models*, 5: 235–245.

Kochar, S. C. and Wiens, D. (1987). Partial orderings of life distributions with respect to their ageing properties. *Naval Research Logistics*, 34: 823–829.

Koski, T. and Persson, L. E. (1992). Some properties of general exponential entropies with application to data compression. *Information Theory*, 62: 103–132.

Kotlarski, I. I. (1972). On a characterization of some probability distributions by conditional expectations. *Sankhya A*, 34: 461–466.

Krakowski, M. (1973). The relevation transform and a generalization of the gamma distribution function. *Revue française d'automatique, informatique, recherche opérationnelle. Recherche opérationnelle*, 7: 107–120.

Krishnan, A. S., Sunoj, S. M. and Nair, N. U. (2020). Some reliability properties of extropy for residual and past lifetime random variables. *Journal of the Korean Statistical Society*, 49: 457–474.

Krishnan, A. S., Sunoj, S. M. and Sankaran, P. G. (2019). Quantile based reliability aspects of cumulative Tsallis entropy in past lifetime. *Metrika*, 82: 17–38.

Krishnan, A. S., Sunoj, S. M. and Sankaran, P. G. (2021). Some reliability properties of extropy and its related measures using quantile function. *Statistica*, 80: 413–437.

Kullback, S. and Leibler, R. A. (1951). On information and sufficiency. *Annals of Mathematical Statistics*, 22: 79–86.

Kumar, P. and Hunter, L. (2004). On an information divergence measure and information inequality. *Carpathian Journal of Mathematics*, 20: 51–66.

Kumar, P. and Johnson, A. (2005). On symmetric divergence measures and information inequality. *Journal of Inequalities in Pure Applied Mathematics*, 6: 1–13.

Kumar, V. (2017). Characterization results based on dynamic Tsallis cumulative residual entropy. *Communications in Statistics, Theory and Methods*, 46: 8343–8354.

Kumar, V. and Rani, R. (2017). Quantile based Tsallis entropy in residual and inactivity time. *Bulletin of the Calcutta Mathematical Society*, 109: 275–294.

Kumar, V. and Rani, R. (2018). Quantile approach of dynamic generalized entropy. *Statistica*, LXXVIII: 105–126.

Kumar, V. and Rekha, R. (2018). A quantile approach of Tsallis entropy of order statistics. *Physica A: Statistical Mechanics and its Applications*, 503: 916–928.

Kumar, V. and Singh, N. (2018). Characterization of life distributions based in generalized interval entropy. *Statistics, Optimization and Information Computing*, 6: 547–549.

Kumar, V., Sreevastava, R. and Taneja, H. C. (2010). Length biased weighted inaccuracy measure. *Metron*, 68: 153–160.

Kumar, V. and Taneja, H. C. (2011a). A generalized entropy based residual life distributions. *International Journal of Biomathematics*, 4: 171–184.

Kumar, V. and Taneja, H. C. (2011b). Some characterization results on generalized cumulative entropy measure. *Statistics and Probability Letters*, 81: 1072–1079.

Kumar, V. and Taneja, H. C. (2012). On length biased dynamic measure of past inaccuracy. *Metrika*, 75: 73–84.

Kumar, V. and Taneja, H. C. (2015). Dynamic cumulative residual and past inaccuracy measure. *Journal of Statistical Theory and Applications*, 14: 399–412.

Kumar, V., Taneja, H. C. and Sreevastava, R. (2011). A dynamic measure of inaccuracy between two past lifetime distributions. *Metrika*, 74: 1–10.

Kundu, A. and Nanda, A. K. (2016). On study of dynamic survival and past entropies. *Communications in Statistics-Theory and Methods*, 45: 104–122.

Kundu, C. (2014). Characterizations based on length biased weighted measure of inaccuracy for truncated random variables. *Applications of Mathematics*, 59: 697–714.

Kundu, C. (2017). On weighted measures of inaccuracy for doubly truncated random variables. *Communications in Statistics-Theory and Methods*, 46: 3135–3147.

Kundu, C., Di Crescenzo, A. and Longobardi, M. (2016). On cumulative residual (past) inaccuracy for truncated random variables. *Metrika*, 79: 335–356.

Kundu, C. and Nanda, A. K. (2010). Some reliability properties of the inactivity time. *Communications in Statistics-Theory and Methods*, 39: 899–911.

Kundu, C. and Nanda, A. K. (2014). Characterizations based on measures of inaccuracy for truncated random variables. *Statistical Papers*, 56: 619–637.

Kundu, C., Nanda, A. K. and Maiti, S. S. (2010). Some distributional results through past entropy. *Journal of Statistical Planning and Inference*, 140: 1280–1291.

Lad, P., Sanfilippo, G. and Agro, G. (2015). Extropy, complementary dual of entropy. *Statistical Science*, 30: 40–58.

Lai, C. D. and Xie, M. (2006). *Stochastic Ageing and Dependence for Reliability*. Springer-Verlag, New York.

Landsberg, P. T. and Vedral, V. (1998). Distributions and channel capacities in generalized statistical mechanics. *Physics Letters A*, (247): 211–217.

Lee, P. M. (1964). On the axioms of information theory. *Annals of Mathematical Statistics*, 35: 415–418.

Li, X. and Zhang, S. (2011). Some new results on Renyi entropy of residual life and inactivity time. *Probability in the Engineering and Informational Sciences*, 25: 237–250.

Li, X. and Zuo, M. J. (2007). Reversed hazard rate order of equilibrium distributions and a related ageing notion. *Statistical Papers*. DOI. 10.1007/S 00362-007-0046-7.

Li, Z. and Li, X. (1998). $\{IFR * t_0\}$ and $\{NBU * t_0\}$ classes of life distributions. *Journal of Statistical Planning and Inference*, 70: 191–200.

Lin, J. and Wong, S. K. M. (1990). A new directed divergence measure and its characterization. *International Journal of General Systems*, 17: 73–81.

Lindley, D. V. and Singpurwalla, N. D. (1986). Multivariate distributions for the life lengths of components of a system sharing common environment. *Journal of Applied Probability*, 23: 418–431.

Mahalanobis, P. C. (1936). On the generalized distance in statistics. *Proceedings of the National Institute of Science*, 2: 49–55.

Mahmoudi, M. and Asadi, M. (2010). On the monotonic behaviour of time dependent entropy of order α. *Journal of the Iranian Statistical Society*, 9: 65–83.

Marshall, A. W. and Olkin, I. (2007). *Life Distributions*. Springer, New York.

Martinas, K. (1997). Entropy and information. *World Features*, 50: 483. Article 287.

Martinas, K. and Frankowicz, M. (2000). Extropy: Reformulation of the entropy principle. *Periodica Polytechnica Ser. Chemical Engineering*, 44: 29–38.

Mathai, A. M. and Haubold, H. J. (2006). Pathway model, Tsallis statistics, super statistics and a generalized measure of entropy. *Physica A: Statistical Mechanics and its Applications*, 375: 110–122.

Mattheou, K., Lee, S. and Karagrigoriou, A. (2009). A model selection criterion based on the BHHJ measure of divergence. *Journal of Statistical Planning and Inference*, 139: 128–135.

Matusita, K. (1967). On the notion of affinity of survival distributions and some of applications. *Annals of the Institute of Statistical Mathematics*, 19: 181–192.

Maya, R., Sathar, E. I. A., Rajesh, G. and Nair, K. R. M. (2014). Estimation of residual entropy of order α with dependent data. *Statistical Papers*, 55: 585–602.

Maya, S. S. and Sunoj, S. M. (2008). Some dynamic generalized information measures in the context of weighted models. *Statistica*, LXVIII: 71–84.

Midhu, N. N., Sankaran, P. G. and Nair, N. U. (2013). A class of distributions with linear mean residual quantile function and its generalizations. *Statistical Methodology*, 15: 1–24.

Midhu, N. N., Sankaran, P. G. and Nair, N. U. (2014). A class of distributions with linear hazard quantile functions. *Communications in Statistics-Theory and Methods*, 43: 3634–3689.

Minimol, S. (2017). On generalized cumulative past entropy measures. *Communications in Statistics-Theory and Methods*, 46: 2816–2822.

Mirali, S., Bharatpour, S. and Fakov, V. (2017). On weighted cumulative residual entropy. *Communications in Statistics-Theory and Methods*, 46: 2857–2809.

Misagh, F. (2012). Some properties of interval entropy function and their applications. *World Applied Sciences Journal*, 20: 1666–1671.

Misagh, F. (2016). On shift dependent cumulative entropy measures. *International Journal of Mathematics and Mathematical Sciences*, Article ID 7213285, DOI: 10.1155/2016/7213285.

Misagh, F. and Yari, G. H. (2011). On weighted interval entropy. *Statistics and Probability Letters*, 81: 188–194.

Misagh, F. and Yari, G. H. (2012). Interval entropy and informative distance. *Entropy*, 14: 480–490.

Misra, N., Francis, J. and Naqvi, S. (2017). Some sufficient conditions for relative ageing of life distributions. *Probability in the Engineering and Informational Sciences*, 31: 83–99.

Moharana, R. and Kayal, S. (2019). On weighted extended cumulative residual entropy of the kth upper record. In *Lecture notes on Data Engineering Communications and Technology*, 21: 223–241.

Muliere, P., Parmigiani, G. and Polson (1993). A note on residual entropy function. *Probability in the Engineering and informational Sciences*, 7: 413–420.

Muth, E. J. (1977). Reliability models with positive memory derived from mean residual life function. pp. 401–435. In Tsokos, C. P. and Shimi, I. N. (eds.). *Theory and Application of Reliability*. Academic Press.

Nair, K. R. M. and Rajesh, G. (1998). Characterizations of probability distributions using the residual entropy function. *Journal of Indian Statistical Association*, 36: 157–166.

Nair, K. R. M. and Rajesh, G. (2000). Geometric vitality function and its application to reliability. *IAPQR Transactions*, 25: 1–8.

Nair, K. R. M., Sankaran, P. G. and Smitha, S. (2011a). Chernoff distance for truncated distributions. *Statistical Papers*, 52: 893–909.

Nair, N. U. and Gupta, R. P. (2007). Characterization of proportional hazards model by properties of information measures. *International Journal of Statistical Science*, 6: 223–231.

Nair, N. U. and Preeth, M. (2008). Multivariate equilibrium distributions of order n. *Statistics and Probability Letters*, 78: 3312–3320.

Nair, N. U. and Preeth, M. (2009). On some properties of equilibrium distributions of order n. *Statistical Methods and Applications*, 18: 453–464.

Nair, N. U. and Sankaran, P. G. (2009). Quantile based reliability analysis. *Communications in Statistics-Theory and Methods*, 38: 222–232.

Nair, N. U. and Sankaran, P. G. (2010). Properties of a mean residual life arising from renewal theory. *Naval Research Logistics*, 57: 373–379.

Nair, N. U. and Sankaran, P. G. (2013a). Characterizations of discrete distributions using reliability concepts in reversed time. *Statistics and Probability Letters*, 83: 1939–1945.

Nair, N. U. and Sankaran, P. G. (2013b). Some new applications of the total time on test transforms. *Statistical Methodology*, 10: 93–102.

Nair, N. U., Sankaran, P. G. and Balakrishnan, N. (2013a). *Quantile-based reliability analysis*. Basel: Birkhauser.

Nair, N. U., Sankaran, P. G. and Preeth, M. (2012a). Reliability aspects of discrete equilibrium distributions. *Communications in Statistics-Theory and Methods*, 41: 500–515.

Nair, N. U., Sankaran, P. G. and Sunoj, S. M. (2013b). Quantile based stop-loss transform and its applications. *Statistical Methods and Applications*, 22: 167–182.

Nair, N. U., Sankaran, P. G. and Vineshkumar, B. (2012b). Modelling life-times by quantile functions using Parzen's score function. *Statistics*, 46: 799–811.

Nair, N. U., Sankaran, P. G. and Vineshkumar, B. (2011b). The Govindara-julu distribution: Some properties and applications. *Communications in Statistics-Theory and Methods*, 41: 4391–4406.

Nair, N. U. and Sudheesh, K. K. (2006). Characterization of continuous distributions by variance bound and its implications to reliability mod-elling and catastrophe theory. *Communications in Statistics-Theory and Methods*, 35: 1189–1199.

Nair, N. U., Sunoj, S. M. and Rajesh, G. (2019). Some new results for resid-ual entropy. *Journal of the Indian Society for Probability and Statistics*, 20: 185–199.

Nair, N. U., Sunoj, S. M. and Rajesh, G. (2021a). Cumulative residual en-tropy of equilibrium distribution of order n. *Communications in Statics-Theory and Methods*, DOI: 10.1080/03610926.2021.1931332.

Nair, N. U., Sunoj, S. M. and Rajesh, G. (2021b). Some aspects of reversed hazard rate and past entropy. *Communications in Statistics-Theory and Methods*, 50: 2106–2116.

Nair, N. U., Sunoj, S. M. and Rajesh, G. (2022). Relation between rela-tive hazard rate residual divergence with some applications to reliability analysis. *Sankhya A*, DOI: 10.1007/s13171-021-00277-w.

Nair, N. U. and Vineshkumar, B. (2010a). L-moments of residual life. *Jour-nal of Statistical Planning and Inference*, 140: 2618–2631.

Nair, N. U. and Vineshkumar, B. (2010b). Reversed percentile residual life and related concepts. *Journal of the Korean Statistical Society*, 40: 85–92.

Nair, R. S., Abdul-Sathar, E. I. and Rajesh, G. (2017). A study on dynamic weighted failure entropy of order α. *American Journal of Mathematical and Management Sciences*, 36: 137–149.

Nanda, A. K. and Das, S. (2006). Study on R-norm residual entropy. *Cal-cutta Statistical Association Bulletin*, 58(231-232): 197–209.

Nanda, A. K. and Das, S. (2011). Dynamic proportional hazard rate and reverse hazard rate models. *Journal of Statistical Planning and Inference*, 41: 2108–2119.

Nanda, A. K., Jain, K. and Singh, H. (1996). Properties of moments of *s*-order equilibrium distributions. *Journal of Applied Probability*, 33: 1108–1111.

Nanda, A. K. and Paul, P. (2006a). Some properties of past entropy and their applications. *Metrika*, 64: 47–61.

Nanda, A. K. and Paul, P. (2006b). Some results on generalized residual entropy. *Information Sciences*, 176: 27–47.

Nanda, A. K., Sankaran, P. G. and Sunoj, S. M. (2014). Renyi's residual entropy: A quantile approach. *Statistics and Probability Letters*, 85: 114–121.

Nanjundan, G. and Pasha, S. (2018). Characterization of zero inflated gamma distribution. *Journal of Computation and Mathematical Sciences*, 9: 1801–1805.

Nath, P. (1968). Inaccuracy and coding theory. *Metrika*, 13: 123–135.

Navarro, J., del Aguila, Y. and Asadi, M. (2010). Some new results on the cumulative residual entropy. *Journal of Statistical Planning and Inference*, 140: 310–322.

Navarro, J. and Psarrakos, G. (2017). Characterizations based on generalized cumulative residual entropy functions. *Communications in Statistics-Theory and Methods*, 46: 1247–1267.

Noughabi, H. A. and Jarrahiferi, J. (2019). On the estimation of entropy. *Communications in Statistics-Theory and Methods*, 31: 88–99.

Nourbaksh, M. and Yari, G. (2017). Weighted Renyi entropy of lifetime distributions. *Communications in Statistics-Theory and Methods*, 46: 7085–7098.

Nourbhakhsh, M., Yari, G. and Mehrali, Y. (2020). Weighted entropy and their estimations. *Communications in Statistics-Simulation and Computation*, 40: 1142–1158.

Oluyede, B. O. (1999). On inequalities and selection of experiments for length-biased distributions. *Probability in Engineering and Informational Sciences*, 13: 169–185.

Ord, J. K., Paul, G. P. and Taillie, C. (1981). The choice of a distribution to describe personal incomes. *Statistical Distributions in Scientific Work*, 6: 193–202.

Ortega, Eva-Maria. (2008). A note on some functional relationships involving the mean inactivity time order. *IEEE Transactions on Reliability*, 1: 172–178.

Osterreicher, F., Vince, L. and Kafke, P. (1999). On powers of f-divergence defining a distance. *Studia Scientiarum Mathematicarum Hungarica*, 26: 415–422.

Pakes, A. G. (1996). Length biasing and laws equivalent to the log-normal. *Journal of Mathematical Analysis and Applications*, 197: 825–854.

Pakes, A. G. and Navarro, J. (2007). Distributional characterizations through scaling relations. *Australia and New Zealand Journal of Statistics*, 49: 115–135.

Park, S., Rao, M. and Shin, W. (2012). On cumulative residual Kullback-Leibler information. *Statistics and Probability Letters*, 82: 2025–2032.

Picard, C. F. (1979). Weighted probabilistic information measures. *Journal of Information and System Science*, 4: 343–356.

Pintacuda, N. (1966). Shannon entropy: A more general derivation. *Statistica*, 26: 509–524.

Psarrakos, G. and Economou, P. (2017). On the generalized cumulative residual entropy of weighted distributions. *Communications in Statistics-Theory and Methods*, 46: 10914–10925.

Psarrakos, G. and Navarro, J. (2013). Generalized cumulative entropy and record values. *Metrika*, 76: 623–640.

Psarrakos, G. and Toomaj, A. (2017). On the generalized cumulative residual entropy with applications in actuarial science. *Journal of Computational and Applied Mathematics*, 309: 186–190.

Qiu, G. (2017). Extropy of order statistics and record values. *Statistics and Probability Letters*, 120: 52–60.

Qiu, G. (2019). Further results on quantile entropy in the past lifetime. *Probability in the Engineering and Informational Sciences*, 33: 146–159.

Qiu, G. and Jia, K. (2018a). Extropy estimators with applications in testing uniformity. *Journal of Nonparametric Statistics*, 30: 182–196.

Qiu, G. and Jia, K. (2018b). Residual extropy of order statistics. *Statistics and Probability Letters*, 133: 15–22.

Qiu, G., Wang, L. and Wang, X. (2019). On extropy of properties of mixed systems. *Probability in the Engineering and Informational Sciences*, 33: 471–486.

Rajesh, G. and Abdul-Sathar, E. I. (2017). Estimation of inaccuracy measure for censored dependent data. *Communications in Statistics-Theory and Methods*, 40: 10058–10070.

Rajesh, G., Abdul-Sathar, E. I., Maya, R. and Nair, K. R. M. (2015). Non parametric estimation of residual entropy function with censored dependent data. *Brazilian Journal of Probability and Statistics*, 29: 866–877.

Rajesh, G., Abdul-Sathar, E. I. and Nair, R. S. (2017). On dynamic weighted residual entropy. *Communications in Statistics-Theory and Methods*, 46: 2139–2150.

Rajesh, G. and Nair, K. R. M. (2016). Residual entropy function in discrete time. *Far East Journal of Theoretical Statistics*, 2: 1–10.

Rajesh, G. and Sunoj, S. M. (2019). Some properties of cumulative Tsallis entropy of order α. *Statistical Papers*, 60: 583–593.

Ramberg, J. S. and Schmeiser, B. W. (1974). An approximate method for generating asymmetric random variables. *Communications of the ACM*, 17: 78–82.

Renjini, K. R., Sathar, E. I. A. and Rajesh, G. (2016a). Bayes estimation of dynamic cumulative residual entropy for Pareto distribution under type II censored data. *Applied Mathematical Modelling*, 40: 8424–8434.

Renjini, K. R., Sathar, E. I. A. and Rajesh, G. (2016b). A study of the effect of loss functions on the Bayes estimates of dynamic cumulative residual entropy for Pareto distribution under upper record values. *Journal of Statistical Computation and Simulation*, 86: 324–329.

Rao, C. R. (1965). On discrete distributions arising out of methods of ascertainments. pp. 320–332. In Patil, G. P. (ed.). *In Classical and Contagious Discrete Distributions*. Pergunon Press and Statistical Publishing Society, Calcutta. Also reprinted in *Sankhya A*, 27: 311–324.

Rao, M. (2005). More on a new concept of entropy and information. *Journal of Theoretical Probability*, 18(14): 967–981.

Rao, M., Chen, Y., Vemuri, B. C. and Wang, F. (2004). Cumulative residual entropy: A new measure of information. *IEEE Transactions on Information Theory*, 50(6): 1220–1228.

Rathie, P. N. (1970). On a generalized entropy and coding theorem. *Journal of Applied Probability*, 7: 124–133.

Renyi, A. (1959). On a theorem of P. Erodos and its applications in information theory. *Mathematica*, 1: 341–344.

Renyi, A. (1961). On measures of entropy and information. In *Proceedings of Fourth Berkeley Symposium on Mathematics, Statistics and Probability, 1960*, volume 1, pages 547–561. University of California Press.

Rezaei, M., Gholizadeh, B. and Zadkhah, I. (2015). On relative reversed hazard rate order. *Communications in Statistics-Theory and Methods*, 44: 300–308.

Roberts, A. W. and Varberg, D. E. (1973). *Convex Functions*. Academic Press.

Rockafeller, R. T. (1970). *Convex Analysis*. Princeton University Press.

Rolski, T. (1975). Mean residual life. *Bulletin of International Statistical Institute*, 46: 266–270.

Salicru, M., Menendez, M. L., Morales, D. and Porodo, D. (1993). Asymptotic distribution of (h, ϕ) entropies. *Communications in Statistics-Theory and Methods*, 22: 2015–2031.

Sankaran, P. G. and Dileep Kumar, M. (2018). A new class of quantile functions useful in reliability analysis. *Journal of Statistical Theory and Practice*, 12: 615–634.

Sankaran, P. G. and Gupta, R. P. (1999). Characterization of lifetime distributions using measure of uncertainty. *Calcutta Statistical Association Bulletin*, 49: 154–156.

Sankaran, P. G. and Sunoj, S. M. (2017). Quantile based cumulative entropies. *Communications in Statistics-Theory and Methods*, 46: 805–814.

Sankaran, P. G., Sunoj, S. M. and Nair, N. U. (2016). Kullback-Leibler divergence: A quantile approach. *Statistics and Probability Letters*, 111: 72–79.

Sankaran, P. G., Thomas, B. and Midhu, N. N. (2015). On bilinear hazard quantile functions. *Metron*, 73: 135–148.

Santanna, A. P. and Taneja, I. J. (1985). Trigonometric entropies, jensen difference divergent measures and error bounds. *Information Sciences*, 35: 145–155.

Sati, M. M. and Gupta, N. (2015). Some characterizations results on dynamic cumulative Tsallis entropy. *Journal of Probability and Statistics*, Article ID 694203, DOI: 10.1155/2015/694203.

Schroeder, M. (2004). An alternative to entropy in the measurement of information entropy. *Entropy*, 6: 388–412.

Sekeh, Y. S. (2015). A short note on the estimation of WCRE and WCE. *arXir 1508.04742*.

Sekeh, Y. S., Borzadaran, G. R. M. and Roknabadi, A. H. R. (2014). Some results on a weighted version of the generalized dynamic entropies. *Communications in Statistics-Theory and Methods*, 43: 2989–3006.

Sengupta, D. and Deshpande, J. V. (1994). Some results on the relative ageing of two life distributions. *Journal of Applied Probability*, 31: 991–1003.

Shaked, M. and Shanthikumar, J. G. (2007). *Stochastic Orders*. Springer, New York.

Shannon, C. E. (1948). A mathematical theory of communication. *Bell System Technical Journal*, 27: 379–423.

Shantikumar, J. and Baxter, L. A. (1985). Closure properties of the relevation transform. *Naval Research Logistics*, 32: 185–189.

Sharma, B. D. and Mittal, D. P. (1975). New non-additive measures of inaccuracy. *Journal of Mathematical Sciences*, 10: 122–133.

Sharma, B. D. and Taneja, I. J. (1975). Entropy of type (α, β) and other generalized additive measures in information theory. *Metrika*, 22: 205–215.

Sharma, B. D. and Taneja, I. J. (1977). Three generalized additive measures of entropy. *Elektronische Informationsverarbeitung und Kybernetik*, 13: 419–433.

Sibson, R. (1969). Information raduis. *Zeitschrift für Wahrscheinlichkeitstheorie und verwandte Gebiete*, 14: 149–160.

Singh, H. and Deshpande, J. V. (1985). On some new ageing properties. *Scandinavian Journal of Statistics*, 12: 213–220.

Singh, S. and Kundu, C. (2018). On weighted Renyi's entropy for doubly truncated distributions. *Communications in Statistics-Theory and Methods*, 48: 2562–2579.

Smith, R. M. and Bain, L. J. (1975). An exponential power life-testing distribution. *Communications in Statistics-Theory and Methods*, 4: 449–481.

Smitha, S. (2010). *A study on the Kerridge's inaccuracy measure and related concepts*. PhD thesis, Cochin University of Science and Technology.

Stein, W. E. and Dattero, R. (1999). Bondesson's functions in reliability theory. *Applied Stochastic Models in Business and Industry*, 15: 103–109.

Suhov, Y. and Sekeh, Y. S. (2015). Weighted cumulative entropies: An extension of CRE and CE. *arXiv:1507.07051*.

Sunoj, S. M., Krishnan, A. S. and Sankaran, P. G. (2017). Quantile based entropy of order statistics. *Journal of the Indian Society of Probability and Statistics*, 18: 1–17.

Sunoj, S. M., Krishnan, A. S. and Sankaran, P. G. (2018a). A quantile based study of cumulative residual Tsallis entropy measure. *Physica A: Statistical Mechanics and its Applications*, 494: 410–421.

Sunoj, S. M. and Linu, M. N. (2012a). Dynamic cumulative residual Renyi's entropy. *Statistics*, 46(1): 41–56.

Sunoj, S. M. and Linu, M. N. (2012b). On bounds of some dynamic information divergence measures. *Statistica*, LXXII: 23–36.

Sunoj, S. M. and Sankaran, P. G. (2012). Quantile based residual entropy function. *Statistics and Probability Letters*, 82: 1049–1053.

Sunoj, S. M., Sankaran, P. G. and Nair, N. U. (2017). Quantile-based reliability aspects of Renyi's information measure. *Journal of the Indian Society of Probability and Statistics*, 18: 267–280.

Sunoj, S. M., Sankaran, P. G. and Nair, N. U. (2018b). Quantile based Kullback-Leibler divergence. *Statistics*, 52: 1–17.

Sunoj, S. M., Sankaran, P. G. and Nanda, A. K. (2013). Quantile based entropy function in past lifetime. *Statistics and Probability Letters*, 83: 366–372.

Tahmasebi, S., Cali, C., Longobardi, M. and Ahmadi, J. (2017). Cumulative residual inaccuracy in upper record values. *Proceedings of the 3rd Seminar on Reliability Theory and its Applications*, Ferdowsi University of Mashhad, Iran, 303–309.

Tahmasebi, S. (2020). Weighted extension of generalized cumulative residual entropy and their applications. *Communications in Statistics-Theory and Methods*, 49: 5196–5219.

Tahmasebi, S. and Eskanderzadeh, M. (2017). An extension of generalized cumulative entropy. *Journal of Statistical Theory and Applications*, 16: 165–177.

Tahmasebi, S., Longobardi, M., Foroghi, F. and Lak, F. (2020). An extension of weighted generalized cumulative past measure of information. *Ricerche di Mathematica*, 69: 53–81.

Taneja, I. J. (2001). General information measures and their applications. Online book: www.mtm.ufsc.br/ Taneja/book/book.html.

Taneja, I. J. (2011a). A sequence of inequalities among difference of symmetric divergence measures. http://arxiv.org/abs/1104.5700 V1.

Taneja, I. J. (2011b). New developments in generalizing information measures. pp. 37–135. In P. W. Hawkes (ed.). *Advances in lmages and Electron Physics*.

Taneja, H. C. and Kumar, V. (2012). On dynamic cumulative residual inaccuracy measures. In Hukins, G. L., Hunter, A. and Korsunsky, A. M. (eds.). *Proceedings of the World Congress on Engineering*, 1: 153–156, WEC, London.

Taneja, H. C., Kumar, V. and Sreevastava, R. (2009). A dynamic measure of inaccuracy between two residual life distributions. *International Mathematical Forum*, 4: 1213–1220.

Thapliyal, R. and Taneja, H. C. (2012). Generalized entropy of order statistics. *Applied Mathematics*, 3: 1977–82.

Thapliyal, R. and Taneja, H. C. (2013a). A measure of inaccuracy in order statistics. *Journal of Statistical Theory and Applications*, 2: 200–207.

Thapliyal, R. and Taneja, H. C. (2013b). Order statistics based on measure of past entropy. *Mathematical Journal of Interdisciplinary studies*, 1: 63–70.

Thapliyal, R. and Taneja, H. C. (2015). On residual inaccuracy of order statistics. *Statistics and Probability Letters*, 97: 125–131.

Toomaj, A., Sunoj, S. M. and Navarro, J. (2017). Some properties of cumulative residual entropy of coherent and mixed systems. *Journal of Applied Probability*, 54: 379–393.

Tsallis, C. (1988). Possible generalization of Boltzmann-Gibbs statistics. *Journal of Statistical Physics*, 52: 479–487.

Tukey, J. W. (1962). The future of data analysis. *Annals of Mathematical Statistics*, 33: 1–67.

van Erven, T. and Harremoes, P. (2014). Renyi divergence and Kullback and Leibler divergence. arXiv: 1206.2459 V2.

van Staden, P. J. and Loots, M. T. (2009). *L-moment Estimation for the Generalized Lambda Distribution*. Third Annual ASEARC Conference, New Casle, Australia.

Varma, R. S. (1966). Generalizations of Renyi entropies order α. *Journal of Mathematical Sciences*, 1: 34–48.

Vasicek, O. (1976). A test for normality based on sample entropy. *Journal of the Royal Statistical Society B.*, 38: 54–59.

Verdu, S. (1998). Fifty years of Shannon's theory. *IEEE Transactions on Information Theory*, 44: 2057–2078.

Verdugo, A. C. G. and Rathie, P. N. (1978). On the entropy of continuous distributions. *IEEE Transaction on Information Technology*, 24: 120–122.

Vinesh Kumar, B., Nair, N. U. and Sankaran, P. G. (2015). Stochastic orders using quantile based reliability functions. *Journal of the Korean Statistical Society*, 44: 221–231.

Vonta, F. and Karagrigoriou, A. (2010). Generalized measures divergence in survival analysis and reliability. *Journal of Applied Probability*, 47: 216–234.

Wang, F., Vemuri, B. C., Rao, M. and Chen, Y. (2003). Cumulative residual entropy: A new measure of information and its application to image alignment. In *Proceedings of the 9th IEEE International Conference on Computer Vision*, volume I, pages 548–553.

Wang, F. and Vemuri, B. C. (2007). Non-rigid multimodel image registration using cross cumulative residual entropy. *International Journal of Computer Vision*, 74: 201–215.

Wei, X. (1992). Relative mean residual life: Theory and related topics. *Micro-electron Reliability*, 32: 1319–1326.

Xiong, A., Shang, P. and Zhang, Y. (2019). Fractional cumulative entropy. *Communications in Nonlinear Science and Numerical Simulation*, 78: 104879, DOI:10.1016/j.cnsns.2019.104879.

Xu, M. and Moura, M. C. (2017). On the q-Weibull distribution for reliability applications. *Reliability Engineering and System Safety*, 158: 93–105.

Yan, L. and Kang, D. (2016). Some new results on Renyi's residual entropy ordering. *Statistical Methodology*, 33: 55–70.

Yue, D. and Cao, J. (2001). The NBUL class of life distributions and replacement policy comparisons. *Naval Research Logistics*, 48: 578–591.

Zang, Y., Shang, P., He, J. and Xiong, H. (2019). Cumulative Tsallis entropy based power spectrum of financial time series. *Chaos*, 29(103118).

Zaradasht, V. (2020). Results on relative mean residual life and relative cumulative residual entropy. *Statistics, Optimization and Information Computing*, 7: 150–159.

Zarezadeh, B. and Asadi, M. (2010). Results on residual entropy of order statistics and record values. *Information Sciences*, 180: 4195–4206.

Zellener, A. (1971). *An Introduction to Bayesian Inference in Econometrics*. Wiley.

Zhang, Y., Shang, P., He, J. and Xiong, H. (2020). Cumulative Tsallis entropy based on multi-scale permuted distribution of financial time series. *Physica A: Statistical Mechanics and its Applications*, 548: 124388.

Zimmer, W., Keats, J. B. and Wang, F. K. (1988). The Burr XII distribution in reliability analysis. *Journal of Quality Technology*, 20(4): 386–394.

Zografos, K. and Nadaraja, S. (2005). Survival exponential entropies. *IEEE Transactions on Information Theory*, 51: 1239–1246.

Index

For Product Safety Concerns and Information please contact our EU
representative GPSR@taylorandfrancis.com
Taylor & Francis Verlag GmbH, Kaufingerstraße 24, 80331 München, Germany

* 9 7 8 1 0 3 2 3 1 4 1 7 4 *